S0-EJV-119

Neoplasia and Cell Differentiation

Neoplasia and Cell Differentiation

Editor
G. V. Sherbet
Department of Biochemical Pathology, University College Hospital Medical School, London

47 figures and 7 tables

S. Karger · Basel · München · Paris · London · New York · Sydney 1974

S. Karger · Basel · München · Paris · London · New York · Sydney
Arnold-Böcklin-Strasse 25, CH-4011 Basel (Switzerland)

Printed in Switzerland by Buchdruckerei City-Druck AG, Glattbrugg
ISBN 3-8055-1581-2

Contents

Preface

There is an increasing awareness, nowadays, of the possibility that epigenetic mechanisms might operate in neoplatic systems. Epigenetic systems share several interesting features with neoplastic systems [1], and the developmental path of tumours is said to correspond closely with that of embryos [2]. There are also important points of difference between the systems, such as for example, in regulation and supercellular organisation which play a prominent part in embryonic development but which are conspicuously absent in neoplastic systems [3]. It has been suggested that neoplastic transformation might be a pathological counterpart of normal differentiation, and that it might arise by a misprogramming of gene activity by epigenetic mechanisms [4].

This book is therefore devoted to a discussion of some epigenetic processes most relevant to the study of neoplasia. The first seven chapters deal with the problems of nuclear differentiation in normal development and in neoplasia, the molecular basis of the operation of epigenetic mechanisms, and the roles played by cell surfaces in embryogenesis, and in neoplastic development and dissemination.

The considerable emphasis placed here on the problems of regeneration is due to the fact that the question of dedifferentiation is central to the problem of neoplasia. It has been observed that many tumours show a progressive anaplasia. The degree of anaplasia is often used as a criterion for assessing the malignancy of tumours. It also forms the basis of histological grading of neoplasms, which is used as a guide in the treatment and prognosis [5, 6]. But are these anaplastic changes analogous to true dedifferentiation which one encounters in regeneration systems? A critical appraisal of the state of differentiation in tumours might therefore be of value to those investigating neoplastic phenomena as well as to students of embryology. Hence, the remaining chapters of the book deal with the problems of differentiation and organisation of tumours *in vitro,* either on their own or in association with embryonic cells. Such experiments have, in fact, provided us with considerable information about the growth characteristics, invasive ability, and the degree of malignancy of tumours. Although not described in detail in this book, tumour and embryonic cell interactions have been used recently to evolve a biological system of grading of tumours which can be used to assess the malignancy of tumours and also as an aid to prognosis [7].

Another aspect of the question of differentiation, which has exciting possibilities, is: If tumours did dedifferentiate, can they be induced to redifferentiate? Can neoplastic growth and differentiation be controlled by epigenetic influences? It was suggested some years ago that the "individuation field", the persistence of which was postulated to be responsible for regeneration, might inhibit growth and promote differentiation of tumours [8, 9]. Although this method of inducing tumour differentiation has not been particularly successful [10–12], other methods such as the combination of tumour tissues with embryonic organisers etc. appear to have achieved some degree of success [13]. These questions have been discussed in some detail. A chapter on teratomas is also included, and this hardly requires justification. WILLIS [14] believes that teratomas are produced from foci of pluripotent embryonic cells. The wide spectrum of differentiation seen in teratomas supports such a view. It has even been suggested that their development is comparable to embryonic development itself with respect to their 'programme of development'.

It is hoped that this book will be helpful to workers engaged in cancer and embryological researches, and will stimulate co-operative efforts between workers in the two fields. For lack of space, it has not been possible to include articles on such topics as embryonic inductions, competence, growth, regulation, morphogenesis etc. It is hoped to produce a second volume dealing with some of these topics if the response of the readership to the present volume is encouraging.

I wish to thank the authors for their valuable contributions, and the publishers, Karger AG, for their cordial co-operation and excellence of production.

References

1 SHERBET, G.V.: Adv. Cancer Res. *13:* 97–167 (1970).

2 FOULDS, L.: In R. W. RAVEN Cancer, vol. 2, pp. 27–44 (Butterworth, London 1958).
3 FOULDS, L. In R.J.C. HARRIS Biological organisation at the cellular and supercellular level, pp. 229–244 (Academic Press, New York 1963).
4 MARKERT, C.L.: Cancer Res. *28:* 1908–1914 (1968).
5 BLOOM, H.J.G.: Brit. J. Cancer *4:* 259–288 (1950); Brit. J. Cancer *4:* 347–367 (1950); Proc. Roy. Soc. Med. *51:* 122 (1958); Brit. J. Cancer *19:* 228–262 (1965); BLOOM, H.J.G. and RICHARDSON, W.W.: Brit. J. Cancer *11:* 359–377 (1965).
6 TOUGH, I.C.K.: CARTER, D.C.: FRASER, J. and BRUCE, J.: Brit. J. Cancer *23:* 294–301 (1969).
7 SHERBET, G.V. and LAKSHMI, M.S.: Unpublished work.
8 WADDINGTON, C.H.: Nature *135:* 606–608 (1935).
9 NEEDHAM, J.: Proc. Roy. Soc. Med. *29:* 1577–1626 (1936).
10 RUBEN, L.N.: J. exp. Zool. *128:* 29–51 (1955); RUBEN, L.N. and BALLS, M.: J. Morph. *115:* 225–237 (1964).
11 ERMIN, R. and GORDON, M.: Zoologica, N.Y. *40:* 53–84 (1955).
12 SCHEREMETIEVA, E.A.: J. exp. Zool. *158:* 101–111 (1965).
13 See pp. 18, 361–362, and 393–395.
14 WILLIS, R.A.: Pathology of tumours (Butterworth, London 1967).

February 1973

G. V. SHERBET
University College Hospital
Medical School
London

Neoplasia and Cell Differentiation, pp. 1–26
(Karger, Basel 1974)

Nuclear Differentiation in the Development of Normal and Neoplastic Tissues

JOHN D. SIMNETT

Department of Pathology, University of Newcastle upon Tyne, Newcastle upon Tyne

Contents

A possible consequence of present-day research into tumor biology is that new therapeutic techniques may be developed based on the concept of reestablishing in the tumorous tissue the control mechanisms present in normal tissue which may lead to more effective and less traumatic methods of treatment than current eradicative procedures. Thus, it is essential to examine what evidence there is for believing that this restoration of normal function is a realistic objective.

Clinical observations on human tumors suggest that the neoplastic state is virtually irreversible and this has contributed toward the idea that tumor cells have suffered some permanent change in their genetic constitution. DULBECCO [1969] argues that cancer cells transmit their neoplastic properties to their daughter cells and that oncogenesis must, therefore, involve changes intrinsic to the cell, probably originating from permanent alterations in the DNA complement of somatic cell nuclei. However, although some types of transplantable neoplasm may be transmitted by a single cell this may represent a late stage in the evolution of the tumor [KLEIN, 1959] and it is by no means proven that neoplasia is generally a function of the individual cell rather than of the tissue. Moreover, similar heritable changes occur during the differentiation of normal cells in the absence of permanent genetic changes. An alternative concept to 'somatic mutation' is that neoplasia may be controlled by epigenetic processes involving abnormal patterns of gene expression. FURTH [1969] points out that pituitary tumors may be caused by damage to pituitary-dependent endocrine organs and concludes that neoplasia results from a derangement of intercellular homeostatic mechanisms, rather than from changes instrinsic to the cell. Homeostatic disturbances may lead to an increased rate of cell proliferation in normal tissues but this is generally self-limiting and some additional mechanism must, therefore, be responsible for the stable state of the proliferative phase in neoplasia.

The controversial question of the relative importance of permanent or epigenetic changes in neoplasia has attracted great attention and a number of reviews of the subject have appeared in recent years. CURTIS [1965] considers the spontaneous appearance of tumors to be part of the ageing process and has shown how this may be related to the gradual accumulation of somatic mutations. The possible role both of germ cell and somatic cell mutation in carcinogenesis has been reviewed by SHAPIRO [1967], while BERGER [1969] has shown that gross chromosomal changes are an almost universal feature of human malignancy. On the other hand, BRAUN [1970] has described a wide range of circumstances in which neoplastic tissues show reversible epigenetic effects, and OLENOV and FEL [1968] have reported epigenetic changes in antigen synthesis by tumor cells which have subsequently been discussed by OLENOV [1970]. Unfortunately, the effect of these reviews has been further to polarize the conflicting theories concerning the genetic mechanisms which control neoplasia. The aim of the present review is to show that neoplastic transformation may result both from stable genomic changes and from labile epigenetic changes and that loss of neoplastic properties can occur both in cells of normal genotype and in cells of altered genetic constitution.

I. Genetic Mechanisms in Normal Development

A. Effects Associated with Permanent Genetic Changes

1. Chromosome Elimination

During the larval development of many insect species, the DNA content of individual cells has been reported to reduplicate without subsequent cell division [RODMAN, 1967] which leads to the production of the giant polytene chromosomes typical of such species as *Drosophila* or *Chironomus.* Further measurements in *Drosophila* larvae [RUDKIN, 1969] showed that chromatin reduplication was incomplete and that in male larvae after five successive replications—by which time the nucleus would be expected to contain 32 times the diploid content of DNA—the observed DNA content was only 22.6 times diploid which amounted to a loss of over 25% of the chromosomal material. The probable explanation is that during successive replications increasing amounts of the centric heterochromatic regions of the chromosome are eliminated [RUDKIN, 1965]. Different regions of the insect chromosome reduplicate in a specific order and the heterochromatic regions replicate late in S phase [BIANCHI and DE BIANCHI, 1969]. Their elimination appears to result from the cell moving out of S phase before reduplication of the heterochromatin is complete, a phenomenon which is particularly obvious in the insect *Planococcus* in which entire chromosomes of male larvae become heterochromatic and, subsequently, fail to replicate [NUR, 1966].

Condensation of chromatin in the heterochromatic regions brings about a stable repression of the corresponding genes, an effect which may be of great importance in the stabilization of cell differentiation [SCHULTZ, 1965; BROWN, 1966], and the significance of chromosome elimination is that it may allow cells to conserve chromatin precursors which would otherwise be wasted in the construction of non-functional genetic material [RUDKIN, 1969].

It would be interesting to know whether a similar process of elimination occurs in vertebrates. FRENSTER *et al.* [1963] found that up to 80% of the DNA of calf lymphocytes was condensed into heterochromatin and that almost all the rapidly labelled RNA was associated with the non-condensed euchromatin, indicating that genes in the heterochromatic regions are not transcribed. In addition, it has been shown that during the S phase of opossum leukocytes some condensed regions of the chromosomes replicate even later than the heterochromatic sex chromosomes [SINHA, 1967a]. The main conditions which promote gene elimination in insects are evidently present in the cells of vertebrates. In mice there is a progressive polyploidization in the nuclei of liver cells, values of up to 32 N being found in larger animals. The actual ratios by which such progression occurred were slightly less than the expected 2:4:8 progression but this could be the result of systematic technical errors [EPSTEIN, 1967] rather than of a partial elimination of DNA. Large discrepancies between the observed nuclear DNA content and the values expected on the basis of complete polyploidization have been reported in a number of mammalian cell types but this is probably due to gene amplification in which functional cells produce extra DNA templates for the synthesis of specific enzymes [PELC, 1968; SOD-MORIAH and SCHMIDT, 1968; PAVAN and BRITO DA CUNHA, 1969]. However, although there is no sound evidence for gross chromosomal loss in normal vertebrate cells, cytological evidence does not exclude the possibility that submicroscopic segments or individual genes may be lost.

2. Nuclear Transplantation Experiments in Amphibia

The transfer of nuclei from differentiating Amphibian cells into unfertilized eggs [BRIGGS and KING, 1952; ELSDALE, GURDON and FISCHBERG, 1960] permits a direct analysis of the concept of stable restriction in genetic expression during development. A large proportion of nuclei from early embryos (blastula or gastrula stages) have the capacity to promote completely normal development on transplantation into the unfertilized egg and the fact that metamorphosed animals from a nuclear transplant clone are histocompatible [SIMNETT, 1964a; VOLPE and MCKINNELL, 1966] indicates that nuclei from a single early embryo are of identical genotype. Normal development can also result from the transplantation of nuclei from later larval stages. Thirty-eight per cent of nuclei from the brain rudiment of closed neural fold stages in *Xenopus* [SIMNETT, 1964b] and 7% of nuclei from the intestinal epithelium of feeding *Xenopus* tadpoles [GURDON, 1962] are capable of promoting normal development of the unfertilized egg. However, nuclear transfer experiments in *Xenopus* [GURDON, 1962; SIMNETT, 1964b]; *Rana pipiens* [BRIGGS and KING, 1957; DI BERARDINO and KING, 1967]; *Bufo* and *Rana temporaria* [NIKITINA, 1946], and the axolotl [SIGNORET, BRIGGS and HUMPHREY, 1962] have shown that the proportion of normal development becomes progressively smaller the later the donor stage in respect of ectodermal [NIKITINA, 1964; SIMNETT, 1964b; DI BERARDINO and KING, 1967], mesodermal [BRIGGS, SIGNORET and HUMPHREY, 1964] and endodermal [BRIGGS and KING, 1957; GURDON, 1962] transplant nuclei. The significance of this restriction in developmental potentiality is at present a matter of controversy. Emphasizing the fact that a proportion of nuclei even from differentiated cells may be totipotent, GURDON and GRAHAM [1967] concluded that cell differentiation during embryogenesis must be initiated by factors which control the rate of gene activity rather than by the stable repression of selected genes. Evidence of such factors comes from experiments in which nuclei are transplanted into oocytes and into unfertilized mature eggs. Nuclei from embryonic ectoderm cells and from adult brain cells, when injected into oocytes, undergo a remarkable enlargement: RNA synthesis is stimulated while DNA synthesis is repressed [GURDON, 1968], though the nuclei apparently retain their ability to divide following maturation of the oocyte [DETLAFF, NIKITINA and STROEVA, 1964] at which time a cytoplasmic factor capable of inducing DNA synthesis is released under the influence of pituitary hormone [GURDON, 1967]. This stimulatory factor persists for some time [GURDON, 1967] and nuclei from adult liver, brain and blood cells, which normally have a very low rate of DNA replication, are thus induced to synthesize DNA when transplanted into mature unfertilized eggs [GRAHAM, ARMS and GURDON, 1966]. In contrast, the cytoplasm of the mature egg, unlike that of the oocyte, possesses the property of suppressing the production of RNA by nuclei of differentiating cells which, prior to transplantation, were synthesizing RNA [GURDON and BROWN, 1965]. It was therefore concluded [GURDON and GRAHAM, 1967] that gene expression during early development was a labile element determined by factors in the cytoplasmic environment and that the increasing proportion of abnormal development obtained by transplanting the nuclei of differentiating cells was due not to stable changes in gene expression but to a failure of nuclei from cells with a relatively long mean generation time [GRAHAM and MORGAN, 1966; FLICKINGER, FREEDMAN and STAMBROOK, 1967] to respond adequately to the synchronizing mechanism in the egg cytoplasm which induces nuclear replication [GRAHAM, 1966].

Results of nuclear transfer experiments in *Rana pipiens* suggest that, contrary to the arguments presented above, stable changes in gene expression may be important in initiating early differentiation [DI BERARDINO and KING, 1967]. Abnormally developing embryos derived from the transplantation of endoderm nuclei have specific defects in ectodermal and mesodermal structures [BRIGGS and KING, 1957] while those derived from ectoderm nuclei have deficiencies in mesodermal and endodermal organs [DI BERARDINO and KING, 1967]. In *Rana* such developmental abnormalities are probably due to gross chromosomal defects arising during the early stages of embryogenesis [HENNEN, 1963; DI BERARDINO and KING, 1967]. By the end of gastrulation the chromosomes of many nuclei contain condensed heterochromatic regions. When such nuclei are transplanted, DNA synthesis in these regions may be delayed, as already explained, and as a result chromosome bridges form during the first cleavage, leading to chromosome breakages which can be detected cytologically at later stages [DI BERARDINO and HOFFNER, 1970]. The existence of heterochromatic regions at early embryonic stages indicates that stable restrictions in gene expression may, contrary to the views of GURDON and GRAHAM [1967], play a part in normal differentiation. While there is no evidence to support the suggestion that actual gene deletions may occur in normally differentiating cells [FISCHBERG, GURDON and ELSDALE, 1958] it is not difficult to imagine how these could occur in abnormal circumstances. For example, during the preneoplastic stage following the application of carcinogenic azo-dyes, the rate of DNA synthesis in liver cells increases [SIMARD, COUSINEAU and DAOUST, 1968] and on the basis of evidence presented above this could conceivably lead to the elimination of late-replicating chromosome regions.

Condensation of chromatin produces a stable repression of genetic information which is evidently important not only in maintaining the differentiated state but also in promoting the process of cell differentiation. However, this repression is not necessarily permanent since many phenomena have been observed which depend on the expression of genes normally existing in a stable repressed form.

B. Phenomena Dependent on Altered Patterns of Gene Expression

Many animals can regenerate lost body parts and in some cases this ability has been ascribed to the presence of a reserve population of undifferentiated cells which, following tissue damage, proliferate and differentiate into the required tissue elements. This idea has now been challenged. As a result of ultrastructural observations on the regeneration of amputated limbs of newts HAY [1959] concluded that the blastema cells from which the regenerate is formed, are derived not from the postulated undifferentiated mesenchyme cells but from highly differentiated elements, principally cartilage cells and striated muscle fibres present at the site of the injury. A similar dedifferentiation of striated muscle fibre has been described during the regeneration of insect limbs [BULLTÈRE, 1968]. It also appears that in *Planaria* regeneration may be due not to undifferentiated neoblasts, as proposed by some authors [STEPHAN-DUBOIS, 1965] but to renewed activity on the part of fully differentiated cells [CHANDEBOIS, 1965]. In the case of both newts and *Planaria* it has been established that the process of regeneration involves differentiated cells which lose many of their specialized functions, for example, blastema cells derived from striated muscle no longer produce myofibrils [HAY, 1959]. No difference can be detected between blastema cells of different tissue origin and it seems possible that a blastema cell derived from one differentiated cell type may redifferentiate into another. More definite examples of metaplastic transformation are to be found in a number of regenerative processes. Severe damage to striated muscle causes the stromal framework to collapse and subsequent repair is by fibrosis, but mild forms of injury such as devascularization or Coxackie virus infection leave the sarcolemmal and endomysial sheaths intact and true muscle regeneration may then occur [GODMAN, 1958]. A few days after injury, mononucleate cells are found in the sarcolemmal tube and these fuse to form normal striated muscle fibres [ALLBROOK, 1962]. At least part of this population of mononucleate cells may be derived from circulating monocytes [BATESON, WOODROW and SLOPER, 1967]. Intramuscular grafts of bladder epithelium or synthetic materials induce the formation of ectopic bone, presumably by causing the local transformation of connective tissue cells into osteocytes [JOHNSON and MCMINN, 1956]. In Amphibia bone fracture gives rise to a hematoma in which erythrocytes dedifferentiate into blastema cells from which new bone differentiates. Similar dedifferentiation of Amphibian erythrocytes can be induced *in vitro* by weak electrical currents and it has been suggested that currents generated by mechanical stress in the fractured bone provide the stimulus for blastema cell production *in vivo* [BECKER and MURRAY, 1967].

Although the molecular basis is not known, it seems probable that in such cases cell transformation involves the unmasking of chromosomal regions previously repressed. This view is supported not only by the experiments of GRAHAM, ARMS and GURDON [1966] in which DNA synthesis in frog erythrocyte nuclei was initiated by transplantation into egg cytoplasm, but also by cell fusion experiments in which transcription of genes in the chick erythrocyte nucleus responsible for specific enzyme synthesis was induced by the influence of irradiated mouse cell cytoplasm [HARRIS and COOK, 1969]. Amphibian and avian erythrocytes are extreme examples of the process of stable restriction in gene expression and it is of particular interest that in these highly differentiated cells genes can be reactivated under the appropriate conditions. The apparent paradox of differentiated cells retaining redundant genetic information is resolved by our increasing knowledge of the importance of dedifferentiation and cell transformation in regeneration.

II. Genetic Mechanisms in Neoplasia

A. Effects Associated with Permanent Genetic Changes

1. Hereditary Predisposition toward Cancer

A number of genetically transmitted forms of neoplasia have been reported both in Man [TURPIN and LEJEUNE, 1969] and in other animals [SHAPIRO, 1969]. Human retinoblastoma appears to be controlled by a dominant mutation [MACKLIN, 1959] and the lack of any consistent gross chromosomal abnormality in this disease [LELE *et al.*, 1963] suggests that only one focus or a small number of foci are involved. Some breeds of dogs are predisposed to malignant melanoma [RAVEN, 1950] and tumors of this type appear in approximately 80% of grey horses on reaching old age [MACFADYAN, 1933]. The melanotic tumors which have been described in hybrid platyfish [GORDON and LANSING, 1943] may be caused by defective competition between two types of pigment cells, the macro- and micromelanophores. Another type of melanoma has been described in a mutant of *Drosophila melanogaster* [ROTTINO and KOPAC, 1966]. In this case the tumor appears at the time of metamorphosis and is evidently derived from a population of circulating blood cells, probably at a critical stage in their differentiation.

The identification of the internal stimuli which induce spontaneous neoplasms is generally an elusive problem on account of the apparent randomness with which such tumors arise. The various types of genetically-determined melanomas described are valuable in this respect since all originate in the course of a definite developmental process and it may, hence, prove possible to identify the specific area of malfunction which leads to neoplastic proliferation.

In addition to melanotic tumors, a number of other heritable neoplastic conditions have been described in *Drosophila* [SHAPIRO, 1967]. One of the most interesting of these is the non-invasive ovarian tumor associated with 'fused'—a mutant gene producing multiple effects which include abnormalities in the structure of wing veins and partial sterility in the females [SMITH and KING, 1966]. During oogenesis a single cystoblast normally proliferates to produce 16 cystocytes, one of which differentiates into the oocyte, but in 'fused' mutants cystoblast proliferation may become uncontrolled, giving rise to as many as 10,000 cells. The induction of tumors is not solely dependent upon genetic factors since the proportion of ovarian chambers which become neoplastic varies from 8 to 60% according to environmental conditions. Such variability is also characteristic of the genetically controlled lung adenoma of strain A mice [BLOOM and FALCONER, 1964] in which the low incidence of apparently spontaneous tumors may be greatly enhanced by exposure to carcinogens [ROGERS, 1951].

2. Hereditary Transmission of Virus-Induced Lymphoma

In a number of strains of mice malignant lymphomas may be induced either by X-rays [LIEBERMANN and KAPLAN, 1959 or by polycyclic hydrocarbons [TOTH, 1963; TOTH and SHUBIK, 1967] and thence transmitted to isogeneic hosts by cell-free tumor extracts. This type of murine tumor is known to be causally associated with the 'C-type' RNA viruses which have been detected in electron micrographs from tumors in a wide variety of species [DMOCHOWSKI, 1966]. The spontaneous appearance of C-type virus has been described in ageing Balb/c mice [HUEBNER and TODARO, 1969] and in virus-free cells of Balb/c mice after several months culture *in vitro* [AARONSON, HARTLEY and TODARO, 1969]. The RNA isolated from C-type virus appears to be homologous with certain sections of the host cell DNA [HAREL *et al.*, 1966] and the virus may be transmitted from mother to offspring either as self-replicating RNA particles or incorporated into the chromosome of the host cell in the form of a DNA provirus which, under the influence of carcinogens or natural ageing processes, may be induced to transcribe for the infective RNA virus [HUEBNER and TODARO, 1969].

3. The Significance of Chromosome Abnormalities in Tumors

There is general agreement about the widespread occurrence of chromosomal abnormalities in human tumors. SPRIGGS [1964] considered that, with the exception of some leukemias, no satisfactory examples of human tumors lacking chromosomal abnormalities had been described. This conclusion was substantiated by BERGER [1969] in a review of cases dating back to 1956 and covering a wide range of human malignancies. However, there are many reasons for questioning the causal role of chromosome defects in initiating neoplasia. Despite the frequency of tumors with such defects the actual proportion of karyotypically abnormal cells within a tumor may be low. In a study of acute leukemia in children [HUNGERFORD, 1961], some patients had no karyotypic abnormalities while in the majority of patients most cells were of normal numerical and morphological karyotype, chromosome abnormalities being observed only in a small proportion of cells. It was concluded that chromosome defects were not a cause of acute leukemia. FRACCARO *et al.* [1968] found that in a cystic adenoma of the human ovary 18% of cells had 1 extra C chromosome while 5% had 2 extra. The fact that the remaining 77% had a presumably normal karyotype is not convincing evidence in support of the causal role of chromosomal defects in initiating neoplasia. Similarly, BENEDICT *el al.* [1969] found 3 aneuploid metaphases, containing 60, 150 and 300 chromosomes in a case of ovarian cystadenoma, while a comparable analysis by CURCIO [1966] showed chromosome numbers up to 458. The total number of metaphases analyzed was not stated but cells of abnormal karyotype probably accounted for only a fraction of the total. It seems improbable that such a small population of variant cells could steer the main mass of karyotypically normal cells toward malignancy. Chromosome defects occur even in non-neoplastic cell populations and their continued existence may depend upon metabolic cooperation with normal cells [HUGHES, 1968].

Karyotype analysis of leukemic cells from experimental animals shows that, although chromosome defects are common, there is no consistent type of abnormality associated with the disease [FORD and MOLE, 1959] such as would be expected if the genetic defect was the primary cause of neoplasia. The lack of such 'common variants' has also been noted in human tumors [BERGER, 1969] but there are some exceptions. In a study of 10 patients with chronic granulocytic leukemia, 9 were found to have the abnormal Ph^1 chromosome. The remaining patient was in a state of remission and the apparent absence of the Ph^1 chromosome may simply have been due to the peripheral blood containing few cells capable of dividing in culture [NOWELL and HUNGERFORD, 1961]. The suggestion that the karyotypic abnormality is the actual cause of leukemia is supported by the fact that cytotoxic drug therapy leading to remission of the disease also results in the disappearance of abnormal cells which subsequently reappear if the patient relapses [DJALDETTI *et al.*, 1966; GARSON, BURGESS and STANLEY, 1969]. A consistent abnormality has also been reported in a radiation-induced mouse leukemia [WALD *et al.*, 1964] and in a transplantable thyroid carcinoma [AL-SAADI, 1968]. CONEN and ERKMAN [1968] believe that many earlier reports of varied karyotype within a single tumor may have been due to poor cytological preparations since in their experience many solid tumors showed a single abnormality while the multiple abnormalities frequently observed in leukemic cells could have arisen from a single abnormal karyotype. The 'apparent chaos of karyotype' found in some tumors is probably due to divergent evolution from an original population of cells possessing a consistent abnormality and some generalities governing the pattern of such aberrations have been suggested [TURPIN and LEJEUNE, 1969]. An additional factor favoring karyotypic heterogeneity may be the occurrence of spontaneous hybridization between cells of different karyotype, a phenomenon which has been described in cell cultures [BARSKI and CORNEFERT, 1962] and which may also occur *in vivo* [SINHA, 1967b].

Until these possibilities have been fully explored it is premature to cite the apparent rarity of common variants as a general argument against the causal role of chromosomal abnormalities in cancer. However, there are many studies which indicate that neoplasia can arise without any cytologically detectable deviation from the normal karyotype. In one strain of mice, analysis of a total of 29 testicular teratomas showed that the tumor cells were indistinguishable from normal testis cells both in morphology and in karyotype [STEVENS and BUNKER, 1964]. Human bladder tumors frequently contain a high proportion of tetraploid and some octaploid nuclei, but the pattern of abnormalities suggests that they are derived from neoplastic cells of normal diploid constitution [LEVI *et al.*, 1969]. Mouse embryo cells cultivated in serum-supplemented medium developed a number of apparently unrelated chromosomal defects. Following implantation *in vivo*, these cell cultures became neoplastic but the resulting tumors retained a similar range of abnormal karyotypes to that observed *in vitro*. From the fact that no selective process occurred during transformation it was concluded that the neoplastic properties could not have been caused by any specific chromosomal abnormality [PARSHAD and SANFORD, 1969]. The same conclusion had been reached by BAYREUTHER [1960] during an extensive survey of karyotypes in tumors from a number of species including Man. Out of 78 spontaneous primary neoplasms, 68 had a normal karyotype, although in 75% of cases this was lost after as little as one passage in a secondary host of the same strain. A preponderance of normal karyotypes was also observed in primary tumors induced by viruses, endocrine imbalance, radiation and chemical carcinogens. In a survey of human patients with acute lymphoblastic or myeloblastic leukemia it was shown [SANDBERG, 1966] that over 50% of the subjects had a normal diploid chromosome constitution in leukemic marrow cells but that the staining properties of the chromosomes differed from those of normal cells, possibly due to reorientation of proteins or RNA associated with the chromosomal DNA.

An important line of evidence concerning the possible causal relationship between genetic abnormalities and neoplasia comes from observations on the concurrent development of karyotypic changes and neoplastic properties in cell cultures. The field has been reviewed by FOLEY [1968] who showed that cultured diploid cells may frequently be neoplastic and conversely that cultured heteroploid cells are not always malignant following transplantation *in vivo.* Virus-transformed cells may acquire neoplastic properties before any detectable karyotypic change and the tumor-forming capacity is not increased by minor shifts toward aneuploidy [YERGANIAN *et al.*, 1965, 1968]. Transplantation studies on Syrian hamsters have shown that many cultured heteroploid cell lines are incapable of producing tumors in experimental hosts [FOLEY *et al.*, 1962].

Although the role of chromosomal abnormalities in initiating neoplasia is a subject for debate it cannot be denied that karyotypic changes do occur in a large proportion of tumors and that they may be important in the further evolution of the disease. In the words of TURPIN and LEJEUNE [1969], 'the chromosomal aberrations are probably neither the cause nor the consequence of cancer; they are the neoplastic process itself'. The karyotype of many tumors suffers progressive changes which may facilitate an evolutionary trend in which neoplastic cells gain greater freedom from control by the normal homeostatic or immunological regulatory mechanisms of the host [HAUSCHKA, 1961] (see also pp. 387–390 in this volume). An example of this effect occurs in mouse tumors where heteroploidy leads to a weakening or loss of specific antigens thus enabling certain populations of tumor cells to survive an immunological attack [HAUSCHKA *et al.*, 1956].

4. Chromosomal Abnormalities and Predisposition to Viral Carcinogenesis

The subject most likely to develop leukemia is a child whose identical twin has already developed the disease [MILLER, 1967]. Other genetically-determined conditions which predispose towards leukemia are Bloom's syndrome which is due to an autosomal recessive gene and in which 1 out of 8 subjects develops leukemia by the age of 30; extra chromosomes as in G, D or F-trisomy, or Klinefelter's syndrome (XXY), and Fanconi's aplastic anemia [MILLER, 1967]. The last-named example is one of a number of inherited non-malignant hematological disorders in which a high proportion of patients possess cell lines of abnormal karyotype [JENSEN, 1968]. In one study it was shown that while 5 out of 7 patients with abnormal cell lines subsequently developed leukemia, this proportion was reduced to 2 out of 13 in subjects who lacked obvious chromosomal abnormalities [NOWELL, 1965]. Consequently, it would appear that people with inherited chromosomal defects, or who acquire such abnormalities as the result of an inherited hematological disorder, are predisposed to leukemia. However, cultured Fanconi cells do not undergo spontaneous neoplastic transformation *in vitro* [TODARO, 1968] and it has been proposed that the chromosomal abnormality is not the direct cause of leukemia but that it renders the cell more susceptible to the action of viral transforming agents [ROWLEY, BLAISDELL and JACOBSEN, 1966]. Certain experimental studies support this hypothesis. Fibroblasts from subjects with G-trisomy (Down's syndrome) or Fanconi's aplastic anemia are more susceptible than cells from normal subjects to *in vitro* neoplastic transformation by SV 40 virus [MILLER and TODARO, 1969]. These authors also observed an increased susceptibility to SV 40 transformation in cells of subjects heterozygous for Fanconi's anemia who have no phenotypic abnormality and they further reported that human adenovirus 12 caused *in vitro* transformation of Fanconi cells but not of normal diploid cells. Viral transformation has also been observed in cultures of human amnion cells, again using SV 40. As uninfected cultures age so the proportion of chromosomal abnormalities increases and this correlates with an increased susceptibility to neoplastic transformation [GAFFNEY, RAMOS and FOGH, 1970].

Normal human leukocytes can be cultured for a maximum of 40 to 45 days but when infected *in vitro* with Epstein-Barr virus they undergo a morphological transformation and develop a long-term growth potential. Sixty-five per cent of such transformed cells were found to have extra C chromosomes and a further 17% had other aberrations [GERBER, WHANG-PENG and MONROE, 1969]. It was concluded that the virus caused the abnormal karyotype but on the basis of the evidence presented it seems equally possible that the abnormalities arose in uninfected cells during the initial period of culture and that such cells were then more susceptible to the subsequent transforming action of the virus.

In addition to the importance of chromosomal changes occurring prior to infection, there is no doubt that the virus itself causes a number of changes in the host cell genome. Hybridization of viral RNA with the DNA from transformed cells permits an estimate to be made of the amount of viral DNA incorporated into the individual host cell. In the case of both SV 40 and polyoma virus it appears that a large number of copies of the viral genome are incorporated into the nucleus: in the case of mouse cells transformed with SV 40 the number varied from 7 up to 44 [WESTPHAL and DULBECCO, 1968] while in transformed hamster cells the average number of SV 40 viral genomes per nucleus was 64 [TAI and O'BRIEN, 1969]. Evidence that the number of viral DNA copies per cell may vary according to the virus and to the species from which the host cell is derived is supported by studies on adenovirus-transformed cells [GREENE, 1970].

It appears that the mechanism of viral carcinogenesis involves more than the simple introduction of new genes into the host cell since on this basis it is difficult to understand why multiple copies of the viral genes are required for successful transformation. Evidently, there is a high degree of mutual regulation between viral and host cell genomes. In cells transformed by adenovirus types 2, less than 10% of the viral genome is transcribed [FUJINAGA and GREEN, 1970] while in the case of mouse embryo cells transcription of a large section of the host cell genome is suppressed following transformation by SV 40 [VALLADARES *et al.*, 1968]. The appearance of gross chromosomal aberrations is an additional factor in viral oncogenesis since, following SV 40 transformation of human diploid cells, there was a preferential loss of the G and D chromosomes [MOORHEAD and SAKSELA, 1965].

These various observations indicate that incorporation of viral DNA does not only serve to introduce new genetic material but also causes a severe disruption of the genetic regulatory circuits of the previously normal host cell. The question of whether changes in the pattern of gene expression can cause neoplasia in the absence of stable genomic alterations will be discussed in the following section.

B. Phenomena Explicable in Terms of Altered Patterns of Gene Expression

1. Field Effects and the Multifocal Origin of Tumors

Histopathological observations on a large number of primary human neoplasms show that tumors arise not from a microscopic focus but from an extensive field of tissue. This effect is well demonstrated by squamous cell carcinomas of the skin and lip and by cutaneous melanomas in which there is frequently an obvious gradation from normal tissue at the periphery, through transitional hyperplastic regions to frank neoplasia at the centre [WILLIS, 1967]. Astrocytomas appear to result from neoplastic proliferation of a considerable part of the brain rather than from discrete foci with subsequent spread [SCHERER, 1940]. In many types of tumor, for example those of the prostate, liver, cervix and testis, neoplastic proliferation develops simultaneously in a number of sites. Multiple foci of mammary carcinoma, often of different histological appearance, may arise in both breasts at the same time, although usually one focus eventually becomes dominant producing a tumor of homogenous appearance [STERNBERG, 1967]. This type of evidence does not accord with the somatic mutation concept of cancer since the probability of causal genetic changes occurring simultaneously in the cells which comprise such large tracts of tissue is infinitesimal. It strongly suggests that the primary change in neoplasia is an altered pattern of gene expression induced simultaneously in a large population of cells by defect in a normal homeostatic mechanism operating at the supracellular level.

2. Anomalous Metabolic Products in the Neoplastic Cell

The concept of changes in gene expression is supported by the fact that many tumors synthesize metabolic products not found in the normal tissue of origin. Examples of osseous, cartilaginous and other types of metaplasia in human and animal tumors have been reviewed by WILLIS [1967] and an interesting case of metaplasia has been described in a mouse lymphoid tumor which was induced to transform into a phagocytizing reticulum cell tumor following serial transplantation [DAWE and POTTER, 1957].

A number of non-pituitary tumors in Man produce adrenocorticotrophic hormone (ACTH) which is normally synthesized only by the pituitary. In some cases, for example pancreatic tumors and oat-cell tumors of the lung, both ACTH and melanocyte-stimulating hormone may be produced [LIDDLE *et al.*, 1965]. Many non-parathyroid tumors in Man are associated with hypercalcemia and there is evidence that this may be caused by increased systemic levels of parathyroid hormone released by the tumor [MUNSON, TASHJIAN and LEVINE, 1965].

Butter yellow, a potent liver carcinogen, induces in liver cells the production of antigens normally found only in the kidney and spleen [WEILER, 1952] while the presence of kidney antigens in liver tumors has been confirmed by DAY [1965]. OLENOV and FEL [1968], comparing the antigenicity of liver and kidney tumors of the rat, showed that tumors contained antigens present in the non-homologous organ but not in the homologous organ.

Two conclusions may be drawn from such data: that normal differentiation does not involve deletion of the gene responsible for the synthesis of the particular end product (i.e., hormone or antigen) and that, in certain cases, neoplastic transformation is accompanied by the unmasking of repressed genetic information. In some cases, however, the reverse process may occur: for example, many melanomas lose the ability to produce melanin. The loss is not necessarily irreversible since both arabinofuranosylcytosine, which inhibits DNA synthesis in the S phase and colcemid, which inhibits mitosis at metaphase, induced the synthesis of melanin in amelanotic mouse melanoma cells [SILAGI, 1969]. A similar effect was obtained at the histological level in the case of an anaplastic transplantable rat kidney tumor which in organ culture regained the ability to produce differentiated kidney tubules [ELLISON, AMBROSE and EASTY, 1969]. In both cases, part of the genetic information required for the functioning of the differentiated cell ceased to be expressed following neoplastic transformation but this was evidently due to the masking of gene expression rather than to gene deletion.

In plasma cell tumors there is a number of phenomena which correspond closely to changes occurring during the differentiation of normal plasma cells [POTTER, 1968]. Immunoglobulins of normal plasma cells are composed of two types of polypeptide chains—the so-called light and heavy chains. Functional immunoglobulins generally contain 2 identical light and 2 identical heavy chains and differentiation of the normal plasma cell requires both selection of the genes which produce the polypeptides typical of that cell, and also stabilization of the program for immunoglobulin synthesis in such a way that it is passed on unchanged to the daughter cells. Animals which are heterozygous for a locus which controls immunoglobulin synthesis may produce chains from both alleles, but if the female parent is immunized against the paternal allotype the F1 offspring, even though heterozygous, make immunoglobulin derived only from maternal genes. The phenomena of allelic suppression, gene selection and stabilization may all be observed in plasma cell tumors and in some cases these neoplastic cells may produce functional antibodies which bind to antigens.

3. Evidence for the Loss of Neoplastic Properties

There is a large number of well authenticated cases in which human tumors have undergone an unexpected and spontaneous remission [BOYD, 1966]. In one review, over 180 cases of human cancer of different origin were described which were considered to have sufficient documentation, including histopathological confirmation, to be accepted as authentic examples of spontaneous regression [EVERSON and COLE, 1966]. The pattern of antigens in a tumor is frequently different from that found in normal tissues of the host [SOUTHAM, 1967] and there is evidence that the host can mount a spontaneous or induced immunological reaction to its own tumor [SOUTHAM, 1965]. For example, in the virus-induced rabbit papilloma rejection appears to be mediated by an immune reaction to antigens of autologous papilloma cells [KREIDER, 1963]. It therefore seems probable that many cases of spontaneous remission have an immunological basis. FREI [1964] examined this interpretation in spontaneously regressing epidermal tumors of mice, induced by local application of dimethylbenzanthracene (DMBA). Since the degree of host sensitization caused by a tumor antigen was expected to increase with time it was assumed that mice in which tumors were induced after a short period of DMBA treatment would have more remissions than those which developed tumors later. In fact the reverse was found and FREI concluded that tumor regression could not have been caused by immunological destruction.

There are many other examples which show that reversal of malignancy can occur without the destruction of cells which is characteristic of immunological rejection. Single cells from a malignant embryonal carcinoma were implanted into host mice of the same strain. These cloned cells multiplied and developed into a variety of apparently normal somatic tissues [KLEINSMITH and PIERCE, 1964] and it was shown that these grafts had no adverse effect on the host after a 6-month-period [PIERCE, 1970]. This complete reversal of malignancy has also been described in the crown gall teratoma of the tobacco plant, cells of which can be induced to differentiate into normal plants by serial grafting into growing shoot tips [BRAUN, 1959].

A number of examples of reversal have been described in neuroblastoma of human infants. These regressions in all cases followed treatment in the form of immuno- [CUSHING and WOLBACH, 1927], chemo- [WILKERSON, VAN DE WATER and GOEPFERT, 1967] or radiotherapy [DYKE and MULKEY, 1967] which, however, did not eradicate the tumor tissue but caused it to differentiate into a mass of non-malignant and apparently normal elements which included mature ganglion cells. Explantation of human neuroblastoma and culture *in vitro* caused rapid differentiation into morphologically normal ganglion cells and it was concluded that this form of neoplasia must be due to a biochemical defect which blocks the differentiation of immature neuroblasts [GOLDSTEIN, BURDMAN and JOURNEY, 1964]. This idea is supported by cybernetic studies in which it was shown that cancer may arise from the blockage of genetic information connected with the functioning of the mature cell [TSANEV and SENDOV, 1969].

In the examples so far discussed the malignant tissue was derived from embryonic or multipotential cells, but regression has also been observed in tumors derived from mature tissues [BRAUN, 1970]. In the newt, *Triturus cristatus,* potentially lethal invasive tumors, arising in the germinative layer of the dermal mucus glands, may be induced by polycyclic hydrocarbons but a significant proportion of primary tumors and of metastases show spontaneous regression during which neoplastic cells differentiate by the formation of pigment and connective tissue cells, cornified layers and mucus glands [SEILERN-ASPANG and KRATOCHWIL, 1962]. These authors also describe a chemically-induced tumor of the planarian, *Dendrocoelum lacteum,* which regresses and differentiates when the surrounding tissues are induced to regenerate following amputation [SEILERN-ASPANG and KRATOCHWIL, 1963]. A very similar phenomenon may, it is claimed, occur in fragments of the Lucké renal adenocarcinoma of the frog *Rana pipiens,* implanted into the limb. Following limb transection, tumor cells were reported as differentiating into normal muscle, cartilage and connective tissue [ROSE and WALLINGFORD, 1948] and when tumor fragments were labelled with tritium before implantation, normal labelled cells were subsequently observed [MIZELL, 1965]. A related phenomenon, also described in Amphibia, is the induction of accessory structures, composed of normal tissues, in response to a variety of carcinogenic agents [BREEDIS, 1952; BALLS and RUBEN, 1968]. One interpretation of this result is that transformed precancerous cells are reverting to the normal state without becoming frankly neoplastic.

III. Biological Mechanisms Exploitable in Tumor Therapy

A. Reversal of Neoplasia in Relation to Karyotype

Since qualitative changes in gene activity may occur in normal differentiated cells it is not difficult to imagine how similar changes in a tumor cell of normal genetic constitution might cause that cell to lose its neoplastic properties. It is important to define whether neoplastic cells of abnormal genotype can also undergo reversal since, on account of the prevalence of karyotypic changes in human tumors, any form of therapy which aimed at reversal rather than eradication would otherwise appear to be of limited application.

The serial passage of a tissue-culture line of ascites tumor cells *in vivo* caused a decrease in its malignant properties and in some cases the hitherto fatal tumor regressed completely [GUERIN and KITCHEN, 1960]. A number of similar cases have been quoted by DAWE [1968]. Transplantable tumor cell lines are generally of highly abnormal karyotype and these examples thus show that partial or complete reversal of neoplasia can occur in genetically abnormal cells. This phenomenon can also be observed in cells transformed by the incorporation of viral genomes. Polyoma-virus transformed cells maintained in culture were cloned onto layers of fixed normal or transformed cells. Cells cloned on normal layers produced over 80% of colonies composed of variant cells which had regained the property of contact inhibition, while cells cloned onto transformed layers produced a maximum of 20% variant clones. Variant cells were found to have lost the ability to produce tumors after inoculation *in vivo,* although they still contained viral DNA as indicated by the continued production of viral RNA and antigen [RABINOWITZ and SACHS, 1968].

Cells of a highly malignant line were fused with a line of low malignancy and the resulting hybrids were of high malignancy [BARSKI and CORNEFERT, 1962]. The hybrids of neoplastic and non-neoplastic cells were similarly shown to be malignant [SCALETTA and EPHRUSSI, 1965]. On the basis of these results it was suggested that malignancy is a dominant characteristic, though the suppression of melanin production (normally a dominant feature) in melanoma cell hybrids [SILAGI, 1967] seems to argue against such a simple interpretation. Moreover, the malignant quality is not always 'dominant' since, in some combinations, cell hybridization may suppress malignancy. Fusion with a non-malignant fibroblast line suppressed the neoplastic properties of three different types of mouse ascites tumor cell [HARRIS *et al.,* 1969; HARRIS and KLEIN, 1969] although the resulting hybrids must have had highly abnormal karyotypes.

Evidence from these three sources, the *in vivo* passage of cultured neoplastic cells, variant types in virus-transformed cells and the loss of malignancy following hybridization, thus supports the view that neoplastic reversal is not confined to genetically normal cells.

B. Some Factors of Potential Importance in Therapy

Many of the experimental methods used to induce neoplastic reversal, for example serial passage or cell hybridization, are obviously not directly applicable as therapeutic methods. One promising line of research is to identify the extracellular diffusible factors which regulate the rate of cell division in normal tissues since such substances, when injected into experimental animals, often inhibit the growth of homologous neoplastic tissues [MOHR *et al.*, 1968; RYTOMAA and KIVINIEMI, 1970]. The fact that many such factors are not species-specific [BULLOUGH *et al.*, 1967; CHOPRA and SIMNETT, 1970; SIMNETT, FISHER and HEPPLESTON, 1969] would facilitate the large-scale purification required if they are to be of practical therapeutic value. However, tumor tissues differ in one radically important way from normal tissues and it is by no means certain that, as a general rule, they would respond to regulatory substances prepared from normal cells.

As an extension of the concept of competition between individual organisms as a determining force in evolution it was similarly proposed that the normal functioning of the organism was maintained by continuous competition between its component parts [see HERTWIG, 1918]. Approaching evolutionary theory from the point of view of embryonic differentiation, HERTWIG [1918] argued that the theory of internal competition could not adequately explain the perfect physiological integration required for the normal development of the organism, and DRIESCH [1908] reached the same conclusion from a consideration of the mechanisms responsible for limb regeneration in Amphibia or for the regulation of development in isolated blastomeres of sea-urchin embryos. The concept of integration rather than competition as an internal factor in evolution has been reviewed by WHYTE [1965] and support for this idea comes from a number of sources. The persistence of polymorphic characters, including blood groups in Man [NIJENHUIS, 1965] and shell patterns in snails [CLARKE, 1968] is difficult to explain on the basis of environmental selection and it may be due rather to the stability of particular gene complexes which are favored by the overall genetic environment within the individual. The existence of such interactive complexes is suggested by the fact that individuals of different genotype may, nevertheless, have the same phenotype which is altered only if new mutant genes are introduced [DUN and FRASER, 1959; RENDEL, 1968] or if the organism is subjected to extreme environmental pressures [WADDINGTON, 1952; DAVITASHVILI, 1969]. The changes in phenotype which frequently accompany the re-arrangement of genes along the chromosome, and the regular pattern in the timing of DNA synthesis in different chromosomal regions is further evidence of the importance of gene cooperation [LIMA DE FARIA, 1962] while the observation that changes in the sequence of amino-acids following gene mutation are not random suggests that the process of mutation itself may be controlled by adjacent genetic material [ECK, 1962].

It is evident that neoplastic cells, even those showing gross karyotypic abnormalities, must retain certain essential gene complexes if they are to continue functioning [TURPIN and LEJEUNE, 1969] and in this respect there is no sharp distinction between normal and malignant cells. It is at higher levels of organization that the fundamental differences are manifested. Such phenomena as the uncontrolled proliferation, invasiveness and metastasis of neoplastic cells are examples of the type internal competition once erroneously ascribed to the behavior of normal cells [HERTWIG, 1918]. The emergence of a dominant histological type in multifocal mammary carcinoma [STERNBERG, 1967] shows that tumors can compete successfully not only with normal tissues but with other neoplastic tissues of the same origin. Another aspect of this competitive ability is that neoplastic cells are responsive to the selective pressures which result in tumor progression [HAUSCHKA, 1961]. The force of selection is such that malignancy may re-emerge in cells whose neoplastic qualities have been suppressed by hybridization with non-malignant cells [HARRIS and KLEIN, 1969].

The responsiveness of neoplastic cells to selective pressure may be of value in tumor therapy and it is conceivable that karyotypically abnormal cells, which have a more flexible evolutionary potential [HAUSCHKA and LEVAN, 1953] may prove to be more amenable than genetically normal cells to forms of treatment designed to induce neoplastic reversal. An interesting selection experiment was performed by POLLACK, GREEN and TODARO [1968]. Neoplastically transformed mouse embryo fibroblasts were treated in culture with FUdR which killed the dividing cells but left the nonproliferating cells unaffected. The progeny of these treated cells showed an increased contact-inhibition *in vitro* and these variant cells also had a reduced tumor-producing ability following transplantation *in vivo*. Both irradiation and chemotherapy place the tumor cells at a selective disadvantage and usually produce their beneficial effects not by eradicating the tumor but by inducing it to differentiate [SPEAR, 1946; DYKE and MULKEY, 1967; WILKERSON, VAN DE WATER and GOEPFERT, 1967]. Many types of tumor, in particular the anaplastic or undifferentiated forms [SPEAR, 1946], are unresponsive to these types of treatment and the reason may be that the single negative selective force provided by the therapeutic agent may present too stark a choice: if the tumor is not completely eradicated it will tend to evolve into a more resistant form [KOLLER, 1959]. According to GLÜCKSMANN [1956], 'in responsive tumors cells prevented from proliferation are able to do something constructive (i.e. differentiate) whereas in incurable tumors the cells are left idle'. Thus, the most logical tactic might be to apply negative selection (either by chemical antimetabolites or irradiation), simultaneously with a positive stimulus to normal differentiation which would provide the tumor with a third evolutionary possibility in addition to eradication or progression. That such a form of combined therapy might be valuable is suggested by recent successes in the treatment of choriocarcinoma and Burkitt's lymphoma [Brit. med. J., 1968] which may be caused by synergistic action between the extrinsic therapeutic agent and an unidentified physiological control mechanism present in the patient.

References

AARONSON, S.A.; HARTLEY, J.W., and TODARO, G.J.: Proc. nat. Acad. Sci. U.S. *64:* 87 (1969).
ALLBROOK, D.: J. Anat., Lond. *96:* 137 (1962).
AL-SAADI, A.: Cancer Res. *28:* 739 (1968).
BALLS, M. and RUBEN, L.N.: Progr. expt. Tumor Res. *10:* 238 (1968).
BARSKI, G. and CORNEFERT, F.: J. nat. Cancer Inst. *28:* 801 (1962).
BATESON, R.G.; WOODROW, D.F., and SLOPER, J.C.: Nature *213:* 1035 (1967).
BAYREUTHER, K.: Nature *186:* 6 (1960).
BECKER, R.O. and MURRAY, D.G.: Trans. N.Y. Acad. Sci. *29:* 606 (1967).
BENEDICT, W.F.; ROSEN, W.C.; BROWN, C.D., and PORTER, I.H.: Lancet *ii;* 640 (1969).
BERGER, R.: Path. Biol. *17:* 1133 (1969).
BIANCHI, N.O. and DE BIANCHI, M.S.A.: Genetics *61* Suppl.: 275 (1969).
BLOOM, J.L. and FALCONER, D.S.: J. nat. Cancer Inst. *33:* 607 (1964).
BOYD, W.: The spontaneous regression of cancer. (Thomas, Springfield 1966).
BRAUN, A.C.: Proc. nat. Acad. Sci. U.S. *45:* 932 (1959).
BRAUN, A.C.: Amer. Scientist *58:* 307 (1970).
BREEDIS, C.: Cancer Res. *12:* 861 (1952).
BRIGGS, R. and KING, T.J.: Proc. nat. Acad. Sci. U.S. *38:* 455 (1952).
BRIGGS, R. and KING, T.J.: J. Morph. *100:* 269 (1957).
BRIGGS, R.; SIGNORET, J., and HUMPHREY, R.R.: Develop. Biol. *10:* 233 (1964).
Brit. med. J. *4:* 438 (1968).
BROWN, S.W.: Science *151:* 417 (1966).
BULLIÈRE, D.: J. Microscopie *7:* 647 (1968).
BULLOUGH, W.S.; LAURENCE, E.B.; IVERSEN, O.H., and ELGJO, K.: Nature *214:* 578 (1967).

CHANDEBOIS, R.: in V. KIORTSIS and H.A.L. TRAMPUSCH Regeneration in animals and related problems, pp. 131–142 (North Holland, Amsterdam 1965).
CHOPRA, D.P. and SIMNETT, J.D.: Nature *225:* 657 (1970).
CLARKE, B.C.: in E.T. DRAKE Evolution and environment, pp. 351–368 (Yale Univ. Press, New Haven 1968).
CONEN, P.E. and ERKMAN, B.: Canad. med. Ass. J. *99:* 348 (1968).
CURCIO, S.: Arch. Ostet. Ginec. *71:* 436 (1966).
CURTIS, H.J.: Cancer Res. *25:* 1305 (1965).
CUSHING, H. and WOLBACH, S.B.: Amer. J. Path. *3:* 203 (1927).
DAVITASHVILI, L.S.: Evolution *23:* 513 (1969).
DAWE, C.J.: Nat. Cancer Inst. Monogr. *29:* 229 (1968).
DAWE, C.J. and POTTER, M.: Amer. J. Path. *33:* 603 (1957).
DAY, E.D.: Proc. Amer. Ass. Cancer Res. *6:* 13 (1965).
DETTLAFF, T.A.; NIKITINA, L.A.,and STROEVA, O.G.: J. Embryol. exp. Morph. *12:* 851 (1964).
DI BERARDINO, M.A. and HOFFNER, N.: Develop. Biol. *23:* 185 (1970).
DI BERARDINO, M.A. and KING, T.J.: Develop. Biol. *15:* 102 (1967).
DJALDETTI, M.: PADEH, B.; PINKHAS, J., and de VRIES, A.: Blood *27:* 103 (1966).
DMOCHOWSKI, L.: in Y. ITO Subviral carcinogenesis, pp. 362–407 (Nagoya, Editorial Committee, First International Committee on Tumor Viruses 1966).
DRIESCH, H.: The science and philosophy of the organism (Black, London 1908).
DULBECCO, R.: Harvey Lect. *63:* 33 (1969).
DUN, R.B. and FRASER, A.S.: Austr. J. biol. Sci. *12:* 506 (1959).
DYKE, P.C. and MULKEY, D.A.: Cancer *20:* 1343 (1967).
ECK, R.V.: J. theoret. Biol. *2:* 139 (1962).
ELLISON, M.L.; AMBROSE, E.J., and EASTY, G.C.: Exp. Cell Res. *55:* 198 (1969).
ELSDALE, T.R.; GURDON, J.B., and FISCHBERG, M.: J. Embryol. exp. Morph. *8:* 437 (1960).
EPSTEIN, C.J.: Proc. nat. Acad. Sci. U.S. *57:* 327 (1967).
EVERSON, T.C. and COLE, W.H.: Spontaneous regression of cancer (Saunders, Philadelphia/London 1966).
FISCHBERG, M.; GURDON, J.B., and ELSDALE, T.R.: Exp. Cell Res., Supl. *6:* 161 (1958).
FLICKINGER, R.A.; FREEDMAN, M.L., and STAMBROOK, P.J.: Develop. Biol. *16:* 457 (1967).
FOLEY, G.E.: Nat. Cancer Inst. Monogr. *29:* 217 (1968).
FOLEY, G.E.; HANDLER, A.H.; ADAMS, R.A., and CRAIG, J.M.: Nat. Cancer Inst. Monogr. *7:* 173 (1962).
FORD, C.E. and MOLE, R.H.: Progr. nucl. Energy, Ser. VI *2:* 11 (1959).
FRACCARO, M.; MANNINI, A.; TIEPOLO, L.; GERLI, M. and ZARA, C.: Lancet *i:* 613 (1968).
FREI, J.V.: Cancer Res. *24:* 1083 (1964).
FRENSTER, J.H.; ALLFREY, V.G., and MIRSKY, A.E.: Proc. nat. Acad. Sci. U.S. *50:* 1026 (1963).
FUJINAGA, K. and GREENE, M.: Proc. nat. Acad. Sci. U.S. *65:* 375 (1970).
FURTH, J.: Harvey Lect. *63:* 47 (1969).
GAFFNEY, E.V.; ROMOS, L., and FOGH, J.: Cancer Res. *30:* 871 (1970).
GARSON, O.M.; BURGESS, M.A., and STANLEY, L.G.: Brit. med. J. *2:* 556 (1969).
GERBER, P.; WHANG-PENG, J., and MONROE, J.H.: Proc. nat. Acad. Sci. U.S. *63:* 740 (1969).
GLÜCKSMANN, A.: Brit. J. Radiol. *29:* 483 (1956).
GODMAN, G.C.: in S.L. PALAY Frontiers in cytology (Yale Univ. Press, New Haven 1958).
GOLDSTEIN, M.N.; BURDMAN, J.A. and JOURNEY, L.J.: J. nat. Cancer Inst. *32:* 165 (1964).

GORDON, M. and LANSING, W.: J. Morph. *73:* 231 (1943).
GRAHAM, C.F.: J. Cell Sci. *1:* 363 (1966).
GRAHAM, C.F.; ARMS, K., and GURDON, J.B.: Develop. Biol. *14:* 349 (1966).
GRAHAM, C.F. and MORGAN, R.W.: Develop. Biol. *14:* 439 (1966).
GREENE, M.: Fed. Proc. *29:* 1265 (1970).
GUERIN, L.F. and KITCHEN, S.F.: Cancer Res. *20:* 344 (1960).
GURDON, J.B.: J. Embryol. exp. Morph. *10:* 622 (1962).
GURDON, J.B.: Proc. nat. Acad. Sci. U.S. *58:* 545 (1967).
GURDON, J.B.: J. Embryol. exp. Morph. *20:* 401 (1968).
GURDON, J.B. and BROWN, D.D.: J. molec. Biol. *12:* 27 (1965).
GURDON, J.B. and GRAHAM, C.F.: Sci. Progr., Oxford *55:* 259 (1967).
HAREL, J.; HAREL, L.; GOLDÉ, A., and VIGIER, P.: Acad. Sci. C.R., Paris *263:* 745 (1966).
HARRIS, H. and COOK, P.R.: J. Cell. Sci. *5:* 121 (1969).
HARRIS, H. and KLEIN, G.: Nature *224:* 1315 (1969).
HARRIS, H.; MILLER, O.J.; KLEIN, G.; WORST, P., and TACHIBANA, T.: Nature *223:* 363 (1969).
HAUSCHKA, T.: Cancer Res. *21:* 957 (1961).
HAUSCHKA, T.S.; KVEDAR, B.J.; GRINNELL, S.T., and AMOS, D.B.: Ann. N.Y. Acad. Sci. *63:* (Art. 5) 683 (1956).
HAUSCHKA, T.S. and LEVAN, A.: Exp. Cell Res. *4:* 457 (1953).
HAY, E.D.: Develop. Biol. *1:* 555 (1959).
HENNEN, S.: Develop. Biol. *6:* 133 (1963).
HERTWIG, O.: Das Werden der Organismen (Fischer, Jena 1918).
HUEBNER, R.J. and TODARO, G.J.: Proc. nat. Acad. Sci. U.S. *64:* 1087 (1969).
HUGHES, D.T.: Nature *217:* 518 (1968).
HUNGERFORD, D.A.: J. nat. Cancer Inst. *27:* 983 (1961).
JENSEN, M.K.: Acta med. scand. *183:* 535 (1968).
JOHNSON, F.R. and MCMINN, R.M.H.: J. Anat., Lond. *90:* 106 (1956).
KLEIN, G.: Cancer Res. *19:* 343 (1959).
KLEINSMITH, L.J. and PIERCE, G.B.: Cancer Res. *24:* 1544 (1964).
KOLLER, P.C.: in R.W. RAVEN Cancer, vol. V, pp. 28–53 (Butterworth, London 1959).
KREIDER, J.W.: Cancer Res. *23:* 1593 (1963).
LELE, K.P.; PENROSE, L.S., and STALLARD, H.B.: Ann. hum. Genet. *27:* 171 (1963).
LEVI, P.E.; COOPER, E.H.; ANDERSON, C.K., and WILLIAMS, R.E.: Cancer *23:* 1074 (1969).
LIDDLE, G.W.; GIVENS, J.R.; NICHOLSON, W.E., and ISLAND, D.P.: Cancer Res. *25:* 1057 (1965).
LIEBERMAN, M. and KAPLAN, H.S.: Science *130:* 387 (1959).
LIMA de FARIA, A.: J. theoret. Biol. *2:* 7 (1962).
MACFADYEAN, J.: J. comp. Path. *46:* 186 (1933).
MACKLIN, M.T.: Arch. Ophthal., Chicago *62:* 842 (1959).
MILLER, R.W.: Cancer Res. *27:* 2420 (1967).
MILLER, R.W. and TODARO, G.J.: Lancet *i;* 81 (1969).
MIZELL, M.: Amer. Zool. *5:* 215 (Abstract) (1965).
MOHR, U.; ALTHOFF, J.; KINZEL, V.; SÜSS, R., and VOLM, M.: Nature *220:* 134 (1968).
MOORHEAD, P. and SAKSELA, E.: Hereditas *52:* 271 (1965).
MUNSON, P.L.; TASHJIAN, A.H., and LEVINE, L.: Cancer Res. *25:* 1062 (1965).
NIJENHUIS, L.E.: Genetica *36:* 208 (1965).
NIKITINA, L.A.: Dokl. Akad. Nauk. SSSR. *156:* 1468 (1964).

NOWELL, P.C.: Arch. Path. *80:* 205 (1965).
NOWELL, P.C. and HUNGERFORD, D.A.: J. nat. Cancer Inst. *27:* 1013 (1961).
NUR, U.: Chromosoma *19:* 439 (1966).
OLENOV, J.M.: Tsitologiya *12:* 3 (1970).
OLENOV, J.M. and FEL, V.J.: J Embryol. exp. Morph. *19:* 299 (1968).
PARSHAD, R. and SANFORD, K.K.: J. nat. Cancer Inst. *43:* 71 (1969).
PAVAN, C. and BRITO da CUNHA, A.: Genetics, Suppl. *61:* 289 (1969).
PELC, S.R.: Nature *219:* 162 (1968).
PIERCE, G.B.: Fed. Proc. *29:* 1248 (1970).
POTTER, M.: Cancer Res. *28:* 1891 (1968).
RABINOWITZ, Z. and SACHS, L.: Nature *220:* 1203 (1968).
RAVEN, R.: Ann. Roy. Coll. Surg. Engl. *6:* 28 (1950).
RENDEL, J.M.: in E.T. DRAKE Evolution and environment, pp. 341–349 (Yale Univ. Press, New Haven 1968).
RODMAN, T.C.: Genetics *55:* 375 (1967).
ROGERS, S.: J. exp. Med. *93:* 427 (1951).
ROSE, S.M. and WALLINGFORD, H.M.: Science *107:* 457 (1948).
ROTTINO, A. and KOPAC, M.I.: Progr. exp. Tumor Res. *8:* 66 (1966).
ROWLEY, J.D.; BLAISDELL, R.K., and JACOBSON, L.O.: Blood *27:* 782 (1966).
RUDKIN, G.: Genetics *52:* 470 (1965).
RUDKIN, G.: Genetics, Suppl. *61:* 227 (1969).
RYTOMAA, T. and KIVINIEMI, K.: Europ. J. Cancer *6:* 401 (1970).
SANDBERG, A.A.: Cancer Res. *26:* 2064 (1966).
SCALETTA, L.J. and EPHRUSSI, B.: Nature *205:* 1169 (1965).
SCHERER, H.J.: Amer. J. Cancer *40:* 159 (1940).
SCHULTZ, J.: Brookhaven Symp. Biol. *18:* 116 (1965).
SEILERN-ASPANG, F. and KRATOCHWIL, K.: J. Embryol. exp. Morph. *10:* 337 (1962).
SEILERN-ASPANG, F. and KRATOCHWIL, K.: Wien. klin. Wschr. *19:* 337 (1963).
SHAPIRO, N.I.: Usp. Sovrem. Biol. *63:* 163 (1967).
SIGNORET, J.; BRIGGS, R., and HUMPHREY, R.R.: Develop. Biol. *4:* 134 (1962).
SILAGI, S.: Cancer Res. *27:* 1953 (1967).
SILAGI, S.: J. Cell. Biol. *43:* 263 (1969).
SIMARD, A.; COUSINEAU, G., and DAOUST, R.: J. nat. Cancer Inst. *41:* 1257 (1968).
SIMNETT, J.D.: Exp. Cell Res. *33:* 232 (1964a).
SIMNETT, J.D.: Develop. Biol. *10:* 467 (1964b).
SIMNETT, J.D.; FISHER, J.M., and HEPPLESTON, A.G.: Nature *223:* 944 (1969).
SINHA, A.K.: Experientia *23:* 889 (1967a).
SINHA, A.K.: Exp. Cell. Res. *47:* 443 (1967b).
SMITH, P.A. and KING, R.S.: J. nat. Cancer Inst. *36:* 445 (1966).
SPEAR, F.G.: Brit. med. Bull. *4:* 2 (1946).
SOUTHAM, C.M.: Europ. J. Cancer *1:* 173 (1965).
SOUTHAM, C.M.: Progr. exp. Tumor Res. *9:* 1 (1967).
SPRIGGS, A.I.: Brit. J. Radiol. *37:* 210 (1964).
SOD-MORIAH, U.A. and SCHMIDT, G.H.: Exp. Cell. Res. *49:* 584 (1968).
STÉPHAN-DUBOIS, F.: in KIORTSIS, V. and TRAMPUSCH, H.A.L. Regeneration in animals and related problems, pp. 112–130 (North Holland, Amsterdam 1965).

STERNBERG, W.H.: in A. SEGALOFF, K.K. MEYER, and S. DEBAKEY Current concepts in breast cancer, pp. 19–30 (Williams & Wilkins, Baltimore 1967).
STEVENS, L.C. and BUNKER, M.C.: J. nat. Cancer Inst. *33:* 65 (1964).
TAI, H.T. and O'BRIEN, R.L.: Virology *38:* 698 (1969).
TODARO, G.J.: Nat. Cancer Inst. Monogr. *29:* 271 (1968).
TOTH, B.: Proc. Soc. exp. Biol. Med. *112:* 873 (1963).
TOTH, B. and SHUBIK, P.: Cancer Res. *27:* 43 (1967).
TSANEV, R. and SENDOV, B.: J. theoret. Biol. *23:* 124 (1969).
TURPIN, R. and LEJEUNE, J.: Human afflications and chromosomal aberrations (Pergamon, Oxford 1969).
VALLADARES, Y.; ALVAREZ, Y.; TABARES, E., and PINTADO, T.: Med. exp. *18:* 283 (1968).
VOLPE, E.P. and MCKINNELL, R.G.: J. Hered. *57:* 167 (1966).
WADDINGTON, C.H.: Nature *169:* 278 (1952).
WALD, N; UPTON, A.C.; JENKINS, A.K., and BORGES, W.H.: Science *143:* 810 (1964).
WEILER, E.: Z. Naturforsch. *7:* 324 (1952).
WESTPHAL, H. and DULBECCO, R.: Proc. nat. Acad. Sci. U.S. *59:* 1158 (1968).
WHYTE, L.L.: Internal factors in evolution (Tavistock Publications, London 1965).
WILKERSON, J.A.; WATER, J.M. van de, and GOEPFERT, H.: Cancer *20:* 1335 (1967).
WILLIS, R.A.: Pathology of tumors, 4th ed. pp. 105–124 (Butterworths, London 1967).
YERGANIAN, G.; CHO, S.S.; HO, T., and NELL, M.A.: Proc. Symp. Mutational Process, pp. 349–360, Prague (1965).
YERGANIAN, G.; NELL, M.A.; CHO, S.S.; HAYFORD, A.H., and HO, T.: Nat. Cancer Inst. Monogr. *29:* 241 (1968).

Author's address: Dr. JOHN D. SIMNETT, The University of Newcastle upon Tyne, Department of Pathology, The Royal Victoria Infirmary, *Newcastle upon Tyne, NE1 4LP* (England)

Neoplasia and Cell Differentiation, pp. 27–59
(Karger, Basel 1974)

Molecular Aspects of Nucleo-Cytoplasmic Relationship in Embryonic Development

A.A. NEYFAKH

Institute of Developmental Biology of the U.S.S.R. Academy of Sciences, Moscow

Contents

I. Introduction

The relationship between the nucleus and the cytoplasm in the cell depends not only on the fact that the former is a generative and the latter, a somatic part of the cell, but also on their being a manifestation of compartmentalization of the cell. The cell is divided by the nuclear membrane into two areas largely differing in concentration of many kinds of ions, in the set of enzymes, in the rate of exchange of many metabolites [SIEBERT, 1971]. It has not been demonstrated that these differences are created by the active transport mechanisms, of the type of the pumps of the cell membrane. It is quite probable, however, that the nuclear membrane does help to maintain the differences existing between its inner and outer space. And, again, the membrane is permeable in both directions for some macromolecules—some proteins and RNA—and is an insurmountable boundary for other macromolecules—many proteins, DNA and giant RNA molecules. As to the biochemical aspect of the problem of the nucleus–cytoplasm relationship, it would have been solved had we known what substances penetrate the nuclear membrane, how they form and their function at the place of their destination. Unfortunately, today we know more about the nature and role of the compounds released from the nucleus than about those transported inside it.

Considering the role of the nucleo–cytoplasmic relationship in development we should not underestimate such an important parameter as *time*. In differentiated cells the role of nuclei is essentially that of maintaining the existing state and the changes in time reflect only inertia of regulation or functional periodicity. Therefore, in such cells, the nucleo—cytoplasmic relationship should resemble a closed circuit, i.e., the cytoplasmic factors determine differential activity of genes; RNA transcribed from these genes go to the cytoplasm to be translated there and to determine specific properties of the cells, including that of maintenance of nuclear differentiation. Thus, at all steps of realization, one and the same message is circulating, now in the form of active genes, now as mRNA or protein. In the embryonic cells the period of realization and the period of different steps of differentiation are of the same order and last for hours. A situation may arise thereby that in the cell possessing a set of specific proteins, different proteins are being synthesized, while in the nucleus the third set of genes have just been activated which are responsible for the traits to be realized in the future. On the other hand, in the cytoplasm of the cells of early embryos, translation may simultaneously proceed on mRNA which have just been synthesized, on those transcribed at the preceding step of development and even on the templates stored in the egg as early as in oogenesis. Regulation of so complex a system is rather difficult to visualize as a simple switching on and off of the genes. Therefore, the existence of control at every successive step, i.e., transcription, processing and release into the cytoplasm, translation etc., whose elements may be revealed in the embryo, is in the long run not surprising [NEYFAKH, 1971].

Finally, the very system of differential gene activation in higher animals in embryonic development is bound to be distinguished by some specificity and to include such embryological phenomena as competence and induction.

Many aspects pertaining to the problem of nucleo—cytoplasmic relationship in development and, at the same time, meagerness of unambiguous evidence, at least in comparison with that for *E. coli*, does not allow one even a bird's eye view of the problem as a whole. However, a scrutiny of at least some issues may be helpful.

II. Quantity of DNA

Little is known about how DNA is quantitatively related to real genes in the eukaryotic cells because it is undetermined whether each gene is one nucleotide sequence or whether this sequence is repeated several times in the chromosome. The extent of such repetitions detected by the rate of renaturation of DNA [BRITTEN and KOHNE, 1967], DNA-RNA renaturation [ANANIEVA *et al.*, 1968] or by formation of a ring in DNA fragments [THOMAS *et al.*, 1970] varied from 20% to 80% in different species and is about 40% in mammals. It is a small value in the genetic sense: assuming that the quantity of copies in each repeating gene is -10^3-10^4, and their share is 40%, only 0.004–0.04% of genes are repeated. Some of these repetitions are centromeric and terminal sites of chromosomes [JONES, 1970; PARDUE and GALL, 1970] which seem to carry no genetic information and have only a structural role. Some of the repeated sequences should belong to the genes of ribosomal and transport RNA (see below); some others may be associated with the regulatory sites of operons which may amount to a considerable portion of the genome and which are believed to possess, at least partially, structural similarity [GEORGIEV, 1969; BRITTEN and DAVIDSON, 1969]. Finally, repeated nucleotide sequences may also indicate only slightly different genes determining, for example, varying parts of light and heavy chains of immunoglobulins [BRITTEN and KOHNE, 1968; NEZLIN and ROCHLIN, 1970], evolutionally related enzymes of similar primary structure, etc.

All this should be taken to mean that the repetition of true, i.e., transcribed and translated, genes *along* the chromosome is not great or, as has been proved by genetic analysis, practically negligible. But it cannot be ruled out that the whole genome may be repeated *across* the chromosome in a multiple way, i.e., it is contained not in one but in several identical DNA chains. It has been shown by relevant calculations that in *Drosophila* the quantity of DNA per haploid gene (one disk of chromosomes of salivary glands corresponding to one gene) is many times higher than is necessary to provide information for one protein [BERENDES, 1969].

There is no need to describe the difficulties encountered by those who attempted to reconcile genetic and radiobiological evidence (one DNA chain along the chromosome) with optical and electron microscopy data (many chains). All the hypotheses put forward so far seem too complicated to be convincing.

In the cells of different types, different sets of genes carrying information about specific tissue proteins are transcribed. It is natural that these differences between the cells cannot be predetermined in the structure of the genome and are created by the mechanism of differential gene activation which is external in relation to the genome. However, in almost all types of cells, intensive transcription on 'non-translated' genes occurs, on which rRNA and tRNA are synthesized. Synthesis of these RNA's should take place in every cell at constant ratios and with much higher intensity than the synthesis of each of kinds of mRNA.

In such case, regulation may be partially abolished due to peculiarities of the structure of the genome – the genes which are intensively transcribed in all cells may be repeatedly presented in the genome at the ratios providing functionally required ratio of their products. The genes for the 28S and 18S rRNA form the nucleolar organizer in which they are repeated hundreds of times: in frog, 450 times [BROWN and WEBER, 1968a, b]; in rat, 330 times [QUINCEY and WILSON, 1969]; in man, 300 times [JEANTEUR and ATTARDI, 1969]. The genes of the 5S rRNA and tRNA are scattered in the genome, singly or in small clusters, [BROWN and WEBER, 1968a] and are presented: in frog for the 5S RNA by 26,000 sequences and for tRNA by1,000 sequences; in rat by 1,660 and 13,000 sequences, respectively [QUINCEY and WILSON, 1969]. The fact that in frog the number of repeated sequences for 5S genes is much higher than for 28S and 18S is likely to be due to a much higher rate of transcription in the nucleolus, the more so since it is determined by specific nucleolar RNA-polymerase [MAUL and HAMILTON, 1967; ROEDER and RUTTER, 1969]. We do not know whether only these non-translated genes are presented in the cell by repeated sequences. Generally speaking, such repetitions should also be expected for many 'house-keeping' enzymes (energy metabolism, synthesis of macromolecules, reparatory enzymes, etc.). However, some of them have been proved to possess Mendelian heredity and, consequently, uniqueness.

Repetition of intensively functioning genes seems to take much load off the regulatory mechanisms. For example, in the cells of rat liver, every minute 650 molecules of the 28S, 18S and 5S RNA each, and 11,000 molecules of tRNA should be synthesized to keep the quantities of ribosomes and tRNA at a constant level. But if repeated sequences are taken into account, the cell is to synthesize only 2 molecules of the 28S and 18S rRNA, 0.4 molecules of 5S rRNA and 1 molecule of 4S tRNA per gene per minute [QUINCEY and WILSON, 1969].

But if we are speaking about developmental processes and, particularly, about the earliest stage–oogenesis–, the role of the oocyte nucleus becomes much more complicated. In the species in which RNA synthesis in oogenesis occurs in their own germinal vesicle and the egg is of considerable size (e.g., fish and amphibians), the quantity of rRNA and tRNA per single cell should be enormous. Gene amplification, recently discovered in amphibians and other animals, is an example of a device to intensify synthesis of these kinds of RNA. In the nucleolus region selective replication of DNA occurs which, being repeated hundreds and hundreds of times, creates in the germinal vesicle thousands (in amphibians) of additional nucleoli each of which participates in the synthesis of 28S and 18S RNA [BROWN and DAWID, 1968].

It was believed that amplification of only the nucleolar organizer takes place, whereas intensification of RNA synthesis on scattered genes of the 5S rRNA and 4S tRNA is achieved only by intensification of their transcription. DENIS [1971] has shown that these genes, too, are capable of amplification. Hybridization of the 5S RNA and 4S tRNA with total DNA isolated from ovaries was much greater than hybridization of the same RNA with DNA from erythrocytes. This means that the relative content of DNA sequences complementary to the 5S and 4S RNA in the ovary tissues is on an average higher – and in the early oocytes, where these genes are likely to be amplified, by hundreds of times higher – than in the unchanged genome of erythrocytes.

Selective replication of DNA of the nucleolar organizer is in itself a phenomenon of exclusive importance, showing, in principle, the possibility of such a mode of regulation of gene activity. If the data of DENIS on amplification of single genes of the 5S and 4S RNA are confirmed, it would be logical to expect that such replication is possible for the translated genes on which mRNA is synthesized. This suggestion is confirmed in a way by measurement, not very accurate though, of DNA in germinal vesicles which exceed the sum made up of the tetraploid quantity of the chromosomal DNA and that of the amplified rRNA genes [BROWN and DAWID, 1968]. This mode of regulation has been, quite naturally, discussed with respect to differentiated cells as well. The change in the quantity of DNA which should occur thereby may not exceed the resolution of the methods employed, as even a 100-fold amplification of 100 single genes will result in not more than 1% increase in the quantity of DNA.

This idea was carried to its extreme by BELL [1969, 1971] who suggested that in the differentiating cells, the genes specific for the given tissue are amplified and this DNA is released as DNP complexes into the cytoplasm where it serves as templates for RNA synthesis. In chicken fibroblasts he found DNP particles with a 16S sedimentation constant containing 7S DNA after prolonged incubation with high concentrations of labelled thymidine. But there are many arguments against this concept: RNA synthesis occurs chiefly in the nucleus and not in the cytoplasm, the double-stranded DNA of the above size may only code polypeptides of a molecular weight of 4,000–6,000. Finally, it was shown experimentally that labelled DNA in the cytoplasm fraction is most probably associated with partial destruction of nuclei [WILLIAMSON, 1970]. The labelled DNA was not found in the cytoplasm of fish embryo cells in this laboratory [KOTOMIN and IVANCHIK, 1973].

The chromosomal DNA of the oocytes themselves undergoes serious structural changes, forming 'lamp brushes' which seem to be capable of active RNA transcription. It is unknown whether the quantity of DNA changes in 'lamp brush' chromosomes.

Finally, when the oocyte is growing the number of mitochondria and, consequently, mitochondrial DNA in its cytoplasm increases rapidly. And if in the usual cells the quantity of mitochondrial DNA is not more than 1—2% of the nuclear DNA, in the eggs it is roughly proportional to their volume. In the small eggs of sea urchin it corresponds to 6 haploid sets of chromosomal DNA (i.e., by hundreds or, maybe, thousands of times more than in the differentiated cell). In the eggs of amphibians and fish the quantity of mitochondrial DNA exceeds that of chromosomal by thousands of times and in the eggs of birds by millions of times. It is believed that some portion of DNA is localized not in mitochondria but in the yolk platelets [PIKO *et al.*, 1967; BALTUS *et al.*, 1968]. Mitochondrial DNA is circular and differs from chromosomal in size and nucleotide sequence, as was shown by hybridization experiments [DAWID and WOLSTENHOLME, 1968].

In the course of early development the quantity of mitochondria in fish and amphibia practically does not increase [ABRAMOVA *et al.*, 1966], but the number of nuclei and nuclear DNA increases progressively. The size of cells decreases and by the stage of development when the cells achieve the normal size (blastula in sea urchin – 1,000 cells; gastrula in amphibia – 40,000 cells), the volume of the cytoplasm and the quantity of mitochondria in it decreases so drastically that the quantity of mitochondrial DNA per cell is again only 1—2%.

Thus, in the course of oogenesis and early development, the number of DNA templates and their qualitative composition undergoes some, as yet insufficiently studied, changes which seriously affect dependent syntheses, and most of all, that of RNA.

III. Synthesis of RNA

1. RNA in Oogenesis

Amplification of the ribosomal RNA genes in early oocytes makes possible an enhancement of synthesis of this kind of RNA. As was shown by DENIS [MAIRY and DENIS, 1970], in Amphibia at the beginning of oogenesis 5S RNA are preferentially synthesized which, however, do not enter the ribosomes but form RNP particles sedimenting at 40S. This is the way to preserve the 5S rRNA for many months up to the moment when the formation of yolk begins. At this step of oogenesis higher quantities of 28S and 18S rRNA are produced and also, probably, ribosomal proteins. The 40S RNP particles distintegrate and the 5S rRNA which are released form, together with two other kinds of rRNA, new ribosomes which constitute the bulk (95%) of the RNA of the egg. The formation of ribosomes at these steps of oogenesis has also been confirmed by electron microscopy [THOMAS, 1970].

tRNA, which according to DENIS [1971], are also made on amplified templates are kept during the first half of oogenesis as 40S RNP complexes. Thus, in the course of oogenesis a protein-synthesizing machinery is created which functions in oogenesis and at the early stages of embryonic development.

In the 'lamp brush' chromosomes considerable quantities of mRNA are also synthesized which exist in a masked form (mmRNA according to TYLOR [1967]: DENIS [MAIRY and DENIS, 1970] believes that they also form RNP complexes sedimenting at 55S.

It is no easy task to estimate the number of genes functioning in oogenesis, principally because we do not yet know how to determine the number of genes in a genome. The first measurements of the number of active genes made by hybridization of RNA with DNA give the number of repeated genes as about 2% of the genome [CRIPPA *et al.,* 1967]. But DAVIDSON and HOUGH, [1969], using special methods (preliminary removal of repeated sequences), made an attempt to determine the number of unique genes transcribed in oogenesis. His figure for the number of active genes (10^4) is 1% of the total number of genes (10^6), if the latter is determined by dividing the total content of chromosomal DNA into the average size of a gene.

Specific behavior of RNA in oogenesis which is expressed in very high intensity of transcription, long-term storage of synthesized RNA and its relatively quick utilization in early embryogenesis, should be ascribed chiefly to the extremely low nucleo-cytoplasmic ratio. Indeed, during the rather short, early stages of development the cells of the embryo are large and each nucleus is surrounded by such quantity of cytoplasm which it is incapable of providing with all the necessary ribosomal and tRNA and with the templates for the synthesis of non-tissue-specific proteins. Organo- and tissue-specific proteins cannot, naturally, be prepared in the egg in advance and have to be synthesized on the templates transcribed in the differentiating cells. Therefore, all the components which can be pre-synthesized, are formed in oogenesis. As an oocyte possesses just one nucleus, formation of a large egg should take months and even years. The possibility of accumulation of RNA should be ensured by special mechanisms. Judging by the fact that mRNA formed in oogenesis are utilized in early development, they should not be very long-lived, hence the necessity arises of special modes of its storage in the egg.

In many species, for example, in insects, RNA in oogenesis is synthesized not by their own nuclei but by those of nurse (but not follicular) cells. It is of interest that in this case these surrounding cells are not somatic cells of the ovary (like follicular cells), but those of germ line formed together with the oocyte from one parent cell by 3–4 divisions. These cells are not very numerous (7–31 in different species) [see DAVIDSON, 1968, p. 193]. MASUI [1967] and DETTLAFF and SKOBLINA [1969] showed that the action of gonadotropin on maturation of the oocyte is at the beginning actinomycin-sensitive, i.e., dependent on synthesis of RNA. This synthesis occurs not in the germinal vesicle (the egg can mature without it) but in the follicular cells which synthesize, under the action of the gonadotropic hormone, another, progesterone-like hormone inducing the process of maturation of the egg.

But the germinal vesicle is far from being inert – when maturation occurs it breaks up and its content is distributed all over the cytoplasm. The role of these substances for subsequent development was directly proved only experimentally, when the development of semilethal 0/0 mutants may be improved to a large degree if they are injected with the content of the vesicles of non-mutant oocytes [BRIGGS and CASSENS, 1966]. It is not clear, however, whether it is RNA or proteins formed in oogenesis that are active substances in this case. The breakage of the germinal vesicle is associated in time with some other processes: appearance of mosaicity in worm egg [YATSU, 1903]; the ability of the egg cytoplasm to induce DNA synthesis in the transplanted nuclei [GRAHAM *et al.,* 1966].

Finally, the specific role of RNA in oogenesis was demonstrated by SMITH [1966] who showed that on the vegetative pole of the axolotl egg there is a factor which determines the appearance of germ cells and which is sensitive to UV irradiation at 257 nm. The granules of the germ plasm in the eggs of many animals which are responsible for the maintenance of the specific properties of germ cells (their determination, absence of diminution of chromatin or elimination of chromosomes) also contain RNA.

2. Synthesis of RNA in Early Development

During the cleavage of the egg, in the majority of animals mitoses follow one another so promptly that the G_1 and G_2 phases are practically absent and that of DNA synthesis (the S phase) is of minimum duration. This does not exclude RNA synthesis but largely limits its period. In such common embryological objects as amphibians and fish, no RNA synthesis occurs at the time of cleavage. This synthesis may be revealed by biochemical [KAFIANI and TIMOFEEVA, 1964; KAFIANI *et al.,* 1969] and autoradiographic [BACHVAROVA and DAVIDSON, 1966; KOSTOMAROVA and ROTT, 1970] methods only at the late mid-blastula stage and almost immediately achieves a high level per nucleus.

As to the sea urchin egg, the question about the onset of RNA synthesis is more difficult. Many authors have reported incorporation of RNA precursors in sea urchin eggs at the earliest stages of development (2–4 blastomeres and even before the beginning of cleavage) [see NEMER, 1967; SLATER and SPIEGELMAN, 1970]. However, some data, i.e., on cleavage of enucleated eggs [HARVEY, 1936]; behavior of lethal hybrids [see CHEN, 1967]; effect of heavy doses of irradiation [NEYFAKH, 1964] and other nuclear inhibitors [GROSS and COUSINO, 1964; NEYFAKH, 1965], provide unambiguous proof that the early stages of development (till the mid-blastula stage) do not require direct genetic control, i.e., function of nuclei. If this is the case, RNA synthesis in nuclei at the beginning of cleavage should ensure the synthesis of proteins which are not directly associated with morphogenesis (e.g., nuclear proteins, according to KEDES and GROSS) or used at later stages of development [KEDES *et al.,* 1969].

Autoradiographic studies reveal RNA synthesis in sea urchin eggs only at the 16 blastomeres stage [CZIHAK, 1965]. This could hardly be due to the low sensitivity of the autoradiographic method, as even some slight RNA synthesis which has been established to occur in the whole embryo biochemically, should be very intensive in each nucleus which are few (2–8) at these stages. Some discrepancy between the autoradiographic and biological data and the biochemical evidence is explained in many papers that have recently appeared, by the fact that early RNA synthesis in sea urchin eggs is cytoplasmic, i.e., mitochondrial [HARTMAN and COMB, 1969; CRAIG, 1970; CHAMBERLAIN, 1970]. Finally, WILT [1970] has just reported that in the species of sea urchin he studied, nuclear synthesis begins at the 16 blastomeres stage. However HOGAN and GROSS [1972] found autoradiographically the incorporation of a precursor in RNA in both nuclei and cytoplasm at early cleavage stages.

In some groups of animals, for example, *Nematoda* and mammals, early cleavage proceeds very slowly in spite of the high temperature. This is in agreement with the fact that in these species RNA synthesis begins as early as after the first or second cleavage. It is quite possible that in small eggs there cannot be stored a great quantity of RNA for the protein-synthesizing machinery and the templates and their synthesis starts from the very beginning of development.

As to the type of RNA synthesized in the early development, until recently it was believed, much under the influence of the evidence of BROWN and GURDON [1964] for Amphibians, that beginning from the late blastula mRNA is synthesized, then tRNA and, finally, at late gastrula, rRNA. At present, the validity of this scheme is being verified for other animals. For example, in sea urchin according to EMERSON and HUMPHREYS [1970] the rRNA synthesis is initiated very early and at the same intensity as at later stages. However, as there are few nuclei in the embryo at early stages, it is difficult to estimate the absolute value of rRNA synthesis against the background of the intensive mRNA synthesis. Later, when the rate of mRNA synthesis per nucleus decreases and the number of nuclei increases, one may reveal rRNA synthesis which was described above as the beginning of its synthesis. In mammals, rRNA synthesis is believed to take place at the 4 blastomeres stage [MINTZ, 1964]. Early rRNA synthesis was also reported in the case of trout [MELNIKOVA *et al.,* 1971]. There is some evidence that in loach, too, the synthesis of tRNA starts before that of mRNA at the early blastula stage [TIMOFEEVA *et al.,* 1971].

Intensity of RNA synthesis in early development may be calculated per nucleus or per embryo and both estimations will be sensible. The former value describes the function of genome. It may be derived from the biochemical data, knowing the number of cells at a given stage and also the share of synthesizing cells. For example, in fish RNA synthesis starts only in the basal cells of blastoderm adhering to the yolk and then, gradually, spreads to the outer layers [KOSTOMAROVA and ROTT, 1970]. In sea urchin RNA synthesis begins only in the micromeres but remains very uneven at the blastula stage [MARKMAN, 1961]. (However, unevenness of RNA synthesis at early stages was not confirmed [HYNES, GROSS, 1970]).

Relevant calculations have shown that after having begun, RNA synthesis per nucleus does not increase more than twofold. The same figure is furnished by the autoradiographic investigations. Subsequently, in sea urchin, the synthesis of RNA decreases markedly [EMERSON and HUMPHREYS, 1970].

RNA synthesis per embryo increases several tens of times. This increase is not just nominal, it reflects a real event – a greater number of templates per unit of cytoplasm mass. Indeed, the mass of the embryo does not increase at all in early development (Amphibians, sea urchin) or increases very slowly (fish). For example, in loach embryos protein of blastoderm increases by 1.5 times [ABRAMOVA, VASILIEVA, 1973] from mid-blastula (6 h of development) till early gastrula (12 h), and the number of cells increases within this time 23-fold [ROTT and SHEVELEVA, 1968], i.e., each cell decreases in size 16 times and the quantity of mRNA per unit of cytoplasm volume increases respectively.

The composition of the mRNA synthesized, i.e., the changes in the composition of active genes in early development, may be judged by the molecular weight distribution of mRNA (centrifugation in the sucrose gradient), or by the adsorption properties (elution from the MAC column), or by its ability to hybridize with DNA. In the course of development of sea urchin the average size of the molecules synthesized increases and they become more heterogeneous [SLATER and SPIEGELMAN, 1970]. This should be taken to mean that the number of active genes increases. The method of competitive hybridization has certain limitations, i.e., it allows determination only of repeated sequences of nucleotides. Nevertheless, the data obtained by this method show that in loach embryos the share of mRNA transcribed from the repeated genes decreases [RACHKUS *et al.,* 1969a]. In sea urchin [WHITELLY *et al.*, 1967] and loach [RACHKUS *et al.*, 1969b] mRNA synthesized at the early stages (blastula—gastrula) competes well with that accumulated in the egg in oogenesis, i.e., at these stages the same genes are active that have been active in oogenesis. But the results for Amphibians have proved to be different [DENIS, 1966, 1968]: mRNA syntheized in the embryos at the late blastula stage has no common sequences with the RNA of the egg; that means that in embryos a new set of genes begins to function.

As the development proceeds, for example, in loach, more and more new genes switch on during gastrulation and organogenesis [RACHKUS *et al.*, 1969b]. The method of competitive hybridization allows one to reveal at every subsequent stage new populations of mRNA which do not compete with RNA synthesized at the previous stage. But the intensity of mRNA synthesis in each nucleus does not increase – actually it decreases. Evidently, the rate of transcription of each of the active genes markedly decreases, although such decrease may remain unaccounted for when calculation is made per unit of cytoplasm mass.

3. Utilization of RNA in Early Development

RNA synthesized in oogenesis is gradually spent as the development proceeds and is replaced by that synthesized by the nuclei of the embryo with its own, and not mother, genotype. This holds for all kinds of RNA but the replacement occurs at different rates and at different stages.

The quantity of ribosomal RNA, which makes up the major portion of RNA, per egg does not change in the course of development of loach and only at the late stages of organogenesis it begins to rise slowly [AJTKHOGHIN *et al.*, 1964; TIMOFEEVA and KAFIANI, 1964]. However, in the embryo itself, i.e., in blastoderm isolated from the yolk, the content of RNA increases rapidly at the stages of early and mid-blastula, which means that the ribosomes stored in the yolk move to the cells of the embryo, or, to be more exact, that the upper layer of the yolk becomes a part of the blastoderm [AJTCHOGHIN *et al.*, 1964]. At this time no new rRNA is synthesized, hence, until the end of gastrulation (the onset of rRNA synthesis), no measurable degradation of the existing ribosomes occurs. At the later stages, if such destruction occurs, its rate is lower than that of formation of new ribosomes.

In amphibian embryos the situation seems to be similar. rDNA-deficient anucleolate mutants of *X. laevis* proceed to develop normally and, hence, with the normal number of functioning ribosomes till the tail bud stage [BROWN and GURDON, 1964], i.e., much later than the synthesis of rRNA begins in normal embryos [BROWN and LITTNA, 1964b]. (Suppression of protein synthesis affects development immediately.) This means that the ribosomes formed in Amphibia in oogenesis may function for a long time without undergoing degradation. It is possible to suggest that in anucleolate mutants the absence of rRNA synthesis prolongs the life-time of ribosomes formed in oogenesis. Evidently, in the normal development they are naturally replaced somewhere between the onset of rRNA synthesis (late gastrula) and the tail bud stage, i.e., during neurulation. That the ribosomes are stored for some months during oogenesis and are then replaced by the new ones within several hours or one day prompts the idea that their life-time is determined not by the factor of time as such but by their function, i.e., by the quantity of protein molecules synthesized on them.

All we know about tRNA is that synthesis in Amphibians begins when the embryo contains about 15—17,000 cells and that tRNA amounts to 40% of the DNA content. The initiated tRNA synthesis is parallel to that of DNA, i.e., to the increasing number of cells, the ratio being maintained at the 0.4: 1 level inherent in adult cells [BROWN and LITTNA, 1966; WOODLAND and GURDON, 1968]. This means that the concentration of tRNA per unit of cytoplasm volume remains constant till the late blastula stage (the onset of synthesis) and after that rises until the size of the cells decreases to that of cells of a tadpole capable of feeding itself. GURDON believes that the maintenance of the 0.4:1 ratio may be a mechanism of regulation of tRNA synthesis [GURDON and WOODLAND, 1968].

The data on utilization of mRNA obtained in experiments on competitive hybridization yield, unfortunately, little information concerning the mRNA transcribed from the unique genes. And it seems that it is the unique genes that determine the major morphogenetic processes.

In loach mRNA synthesized in oogenesis begins to be spent soon after fertilization and its quantity gradually decreases, reaching the minimum at the late blastula stage, i.e., soon after the synthesis of new mRNA has started, including participation of the genes which were active in oogenesis. This results in the level of competition, i.e., the number of types of mRNA being higher [RACHKUS *et al.,* 1969b].

Similar experiments were performed with Amphibians and it was shown that mRNA stored in oogenesis is spent during early development. DAVIDSON showed that in the egg there is approximately 47 mμ g of mRNA DAVIDSON *et al.,* 1966]. Out of this quantity about 60% is spent during early development and a considerable amount of it remains in the embryo during and even after gastrulation [CRIPPA *et al.*, 1967]. It was shown autoradiographically, using tropical frog with synchronous oogenesis, that RNA synthesized at the 'lamp brush' stage is found several months later, after the beginning of development, not only around the nuclei but also inside blastomere nuclei [DAVIDSON and HOUGH, 1969].

These data are yet difficult to interpret. However, some indirect evidence of utilization of mRNA in development may be obtained when studying the dependence of protein synthesis upon nuclear control.

IV. Nuclear and Cytoplasmic Control of Protein Synthesis

Dependence of protein synthesis on genetic control may be investigated in the experiments where the function of nuclei is suppressed or the nuclei are inactivated. This can be achieved using actinomycin D or heavy doses of ionizing irradiation. The mechanisms of action of the two agents are different (actinomycin D blocks RNA synthesis and X-rays damage DNA itself), they produce similar effects on protein synthesis [NEYFAKH and KRIGSGABER, 1968]. By blocking the formation of new templates for translation, one may find out to what degree the cells were provided with mRNA at the moment when the nuclei were inactivated, i.e., determine the time of participation of the synthesized mRNA in the protein synthesis.

1. Dependence of Protein Synthesis on the Function of Nuclei

It was shown in the first paper of GROSS and COUSINO [1964] that in the eggs of sea urchin treated with actinomycin before fertilization, protein synthesis in the first hours of development does not depend on the function of nuclei. In these experiments the incorporation of the labelled precursors at first increased even more intensively than in the control. This means that the mature egg contains an mRNA store providing for the protein synthesis at the stages when the nuclei of the embryo itself do not yet synthesize mRNA. The decrease in the protein synthesis in the actinomycin-treated eggs at the blastula stage showed that at these stages protein synthesis is controlled by the activity of the embryos own nuclei. However, in recent years it was demonstrated that actinomycin affects the spectrum of the proteins synthesized at the earliest stages of development, i.e., the synthesis of nuclear proteins which is, consequently, determined by the nuclei of the embryo, is blocked and the total synthesis is maintained at the level of control or even higher, at the expense of the templates stored in oogenesis [KEDES *et al.*, 1969]. These data and also the experiments demonstrating that normal early development is possible for the embryos with inactivated nuclei, testify to the fact that protein synthesis at the early stages is largely controlled by the mother genotype functioning in oogenesis.

Experiments on actinomycin- or X-ray-induced nuclei inactivation in sea urchin or fish embryos at different stages of development have shown that the genotype begins to participate in the protein synthesis at the early-mid blastula stage [KRIGSGABER and NEYFAKH, 1972]. From this moment, both the templates stored in oogenesis and synthesized by the nuclei of the embryo are translated. Soon after actinomycin or X-ray treatment protein synthesis decreases but is still maintained for a long time at a high level [NEYFAKH and KRIGSGABER, 1968; NEYFAKH *et al.*, 1968]. It means that in normal development some portion of the templates is promptly spent and the remaining templates are long-lived. But we are not yet in a position to say whether these short- and long-lived mRNA code one and the same set of proteins, or if mRNA synthesized on different genes differ also in the time of functioning in the cytoplasm.

Studying the intensity of incorporation of labelled amino acids into proteins in loach embryos irradiated at different stages, one may establish at what stage nuclei begin to control protein synthesis [NEYFAKH *et al.*, 1968]. X-ray inactivation of nuclei at different stages of early development (up to the mid-blastula stage) causes the same effect: the synthesis of protein rises slowly, like in the control, to achieve, finally, only the level of late blastula. Morphological development of the embryos irradiated before the mid-blastula stage is also arrested at late blastula. It should be taken to mean that from fertilization till mid-late blastula protein synthesis is not controlled by the nuclei of the embryo and, hence, is supported by the templates stored in oogenesis.

In the next period, on the contrary, the effect of nuclei inactivation is strictly dependent on the moment of irradiation: the later the embryos were irradiated the higher is the protein synthesis intensity and the closer it is to that of the control. Consequently, at these stages protein synthesis is controlled by the nuclei of the embryo itself, although the templates stored in oogenesis continue to be translated for some time [NEYFAKH *et al.*, 1968].

Thus, nuclear control is not permanent both for morphogenesis and for protein synthesis. In fish embryos it is absent during cleavage and starts at the mid-blastula stage. This is in good agreement with the onset of mRNA synthesis at the same stage [KAFIANI and TIMOFEVA, 1964; KOSTOMAROVA and ROTT, 1970]. There is some evidence showing that at the late blastula – early gastrula stages the morphogenetic function of the nuclei is interrupted, although mRNA synthesis and its participation in translation at this time is continuous [NEYFAKH, 1964].

Because the mRNA synthesized in the nuclei is transferred to the cytoplasm gradually and is translated for a long time, many hours may elapse between gene activity and realization of its function in protein synthesis. Within this time the embryo undergoes certain stages of development. Thus, the nuclei are operative much earlier than the stages which they will control. This is evident in oogenesis when there is a gap of months between the synthesis of mRNA and its translation, but it is no less evident in embryogenesis itself as the hours elapsing between transcription and translation may be filled up with such important events as, for example, gastrulation.

There is the other approach to the problem of genetical control of protein syntheis which consists in the study of the appearance of paternal proteins in the hybrid embryos. The specific for the father species bands of some dehydrogenases were shown electrophoretically in amphibia and bird hybrid embryos. The bands appear at certain, not too early, stages of development [WRIGHT and SUBTELNY, 1971; OHNO *et al.,* 1969]. The determination of expression of paternal genes can also be done by using the thermostability of proteins as a genetical marker on the crosses of fishes which are different with respect to this trait. Thus, the increase of thermostability of aldolase in hybird embryos of loach × zebra fish o meaning a switching on of father genes is expressed at the somite stage. It was shown by this method that egg aldolase is replaced by new tissue-specific types of this enzyme (muscle, brain, etc.) during organogenesis [NEYFAKH *et al.,* 1973]. These examples show that genetical (nuclear) control determines not only certain organ-specific proteins essential for morphogenesis, but also the replacement of house-keeping enzymes making up the main part of cell proteins.

2. Cytoplasmic Control over Protein Synthesis

It is natural that cytoplasmic control should mean only regulation of the protein synthesis rate and also – which is less obvious – selection for translation of mRNA molecules among those released into the cytoplasm.

We may judge about the translation control only by the case when a change in the protein synthesis rate (and especially, its increase) does not involve nuclear activity. A classical example of such a process is the sharp increase in the protein synthesis rate after sea urchin eggs have been fertilized. The mechanism of this process has been studied in detail and although no final conclusions have been drawn, it has become clear that the rate of synthesis is controlled by activation of ribosomes [MAGGIO *et al.*, 1968] and demasking of mRNA stored in oogenesis [HULTIN, 1964].

In sea urchin and loach embryos, after inactivation of nuclei the protein synthesis tends to have the same pattern as that of the control embryos [NEYFAKH *et al.*, 1968; NEYFAKH and KRIGSGABER, 1968]. And if the decrease in protein synthesis rate may be ascribed to the deficiency of the mRNA templates (after irradiation or inactivation), the following increase in the protein synthesis unambiguously testifies to the existence of control at the level of translation. This control may be complex or simple, but it is determined by the stage of embryonic development which, in turn, is determined by the genetic machinery of the preceding period. But when these regulatory mechanisms are at work they do not directly depend upon the nuclei [KRIGSGABER and NEYFAKH, 1972].

The artificial changes of the nucleo–cytoplasmic relationship which can be created in fish embryos, allows one to demonstrate the dual nature of control over the protein synthesis: genetic control, determined by the synthesis of specific mRNA, and translation control which is effected in the cytoplasm and which, ultimately, is also governed in a genetic fashion but at earlier stages.

Thus, haploid andro- and gynogenetic loach embryos may be easily obtained by irradiating eggs or sperm with rather heavy doses of X-rays (10–40 krad) [NEYFAKH, 1964]. The similar pattern of morphological development of haploids is the same as that of diploid embryos, and it is only of haploids (with intact and irradiated cytoplasm), indicates that it is the genetic machinery that is put out of operation by such heavy doses, the cytoplasmic structures, including the protein synthesizing apparatus, remaining unaffected [NEYFAKH *et al.*, 1968]. At the early stages, the development of haploids is the same as that of diploid embryos, and it is only by the end of gastrulation that the rate of development slows down and morphological abnormalities appear which result in the haploid syndrome and death of late embryos or early larvae [NEYFAKH and RADZIEVSKAJA, 1967].

The number of cells of haploid embryos in the period of synchronous divisions is the same as in the control, then gradually increases and becomes twice as high as in diploids, the normal nucleoplasmic ratio being thereby restored [ROTT and SHEVELEVA, 1968]. RNA synthesis in the nuclei of haploid embryos is twice as low as in diploids [KAFIANI *et al.,* 1969]. Therefore, at the beginning of RNA synthesis in haploid embryos it is twice as low as that in diploid. These differences disappear in a compensatory way as the number of nuclei increases. This should be taken to mean that in early loach embryos there is no feedback mechanism in transcription regulation. However, protein synthesis in the haploid embryos is the same as in diploids, i.e., twofold fewer newly formed templates are involved in the translation, running at the same total rate as in diploids. Consequently, it is not the quantity of mRNA (up to 50% of the normal quantity) that is the bottle-neck of the protein synthesis rate; or else, in the cytoplasm of the embryo cells there are ways and means to intensify translation of each template. Therefore protein synthesis rate is determined at the cytoplasmic level and only total inhibition of mRNA synthesis (by actinomycin) or its impairment (by X-rays) is required to affect protein synthesis.

The second model allowing to study the role of nuclear and cytoplasmic control in protein synthesis is loach × goldfish hybrid embryos *(Misgurnus fossilis ♀ × Carassius auratus ♂).* Loach eggs fertilized by sperm of goldfish or other *Cyprinidae* develop into hybrids possessing distinct traits of both species [NEYFAKH and RADZIEVSKAJA, 1967]. They live up to the larval stage and die at the beginning of feeding. A karyological study of these hybrids revealed elimination in them of some chromosomes, evidently of the father species. It was demonstrated by some experiments that father genes in these hybrids function not only at the end of embryonal development, determining goldfish traits, but at the very beginning of the gene function, at mid-blastula.

If goldfish sperm are used to fertilize loach eggs pre-treated with heavy doses of X-rays, the result will be a nucleocytoplasmic, androgenetic haploid hybrid possessing no loach chromosomes, and goldfish nuclei will be surrounded with loach cytoplasm. One could expect that these hybrid embryos would develop as usual haploids or somewhat worse as a part of the chromosomes is lost. In such hybrids, however, practically no nuclear function is observed: these embryos show seemingly normal cleavage, forming blastula just like enucleated embryos (when both gametes have been irradiated), but like enucleated embryos, fail to reach the gastrula stage [NEYFAKH and RADZIEVSKAJA, 1967]. As there is here genetic incompatibility of the nucleus and cytoplasm, one could expect that the function of the nuclei, i.e., the process of transcription, will be suppressed. However, incorporation of labelled uridine in mRNA of the hybrid haploids begins at the same stage and increases at the same rate as in haploid loach embryos whereas it is practically absent in enucleated embryos [NEYFAKH *et al.*, 1968]. Nevertheless, the protein synthesis in these hybrids proceeds just as in enucleated embryos, i.e., lasts no longer than the late blastula stage at which their morphological development is arrested [NEYFAKH *et al.*, 1968]. Further analysis shows that in these embryos the newly synthesized mRNA forms almost no polyribosomes and protein synthesis proceeds only on the templates stored in oogenesis [TIMOFEEVA *et al.*, 1971].

It is hardly possible that in haploid hybrids all mRNA synthesized on the goldfish chromosomes is defective, or that it cannot be translated at foreign ribosomes, for if this were the case, the diploid hybrids would have possessed no traits of the father species. It is possible that elimination of some part of chromosomes and incompatibility of some goldfish genes with loach cytoplasm results in an impairment in the protein-synthesizing machinery which controls, however, only translation of new mRNA. This model may be taken as example of genetic control over a cytoplasmic mechanism of regulation at the translation level.

V. Transport of RNA into Cytoplasm

It has become clear in recent years that our knowledge of an important step in the nucleo-cytoplasmic relationship, i.e., the transport of RNA from the nucleus to the cytoplasm and the events preceding and following this process—is scanty. Even more scanty is our knowledge of the special aspects of this problem associated with development. RNA—both mRNA and rRNA, and as it seems, tRNA—are synthesized not as molecules ready for translation or assembly of ribosomes, but as giant molecules which, prior to leaving the nucleus, break into fragments. One or several of these fragments go into the cytoplasm and others undergo degradation. One of the best known examples of such maturation of precursors is the processing of rRNA which, in warmblooded animals, is synthesized as a giant molecule: 45S (M.W. = 4.1×10^6 daltons), and then, gradually, loses degrading fragments of a molecular weight of 0.3×10^6–1×10^6, breaking into 2 molecules: 28S and 18S [WIENBERG and PENMAN, 1970]. In cold-blooded eukaryotes the precursor is much smaller – 38–40S – and only 10–20% of the precursors undergo degradation [PERRY *et al.*, 1970].

This phenomenon is also known in the case of mRNA. For example, in reticulocytes the mRNA precursor coding hemoglobin is a giant molecule – 60S – in which the mRNA of hemoglobin proper amounts to not more than 3–5%. As a consequence, only part of the RNA synthesized is transferred to the cytoplasm. In reticulocytes it is not more than 20% [SCHERRER, 1970], but in the nucleated erythrocytes of birds this figure is, evidently, much less [ATTARDI *et al.,* 1966; GASARYAN *et al.,* 1971]. In brain cells, about 60% of RNA is transported to the cytoplasm [STEVENIN *et al.,* 1969]. It is obvious that these figures are the mean values of ratios of degrading and transportation to the cytoplasm of parts of the RNA molecules synthesized in the nucleus.

That the processing precedes transport is indicative of the fact that the two processes are interrelated and that only the RNA molecules which have undergone processing leave the nucleus to go to the cytoplasm. But in the life of the cell there is an event – mitosis – when all the above regularities are violated and the entire content of the nucleus is mixed up with the cytoplasm. We have shown with fish embryonic cells and hamster fibroblast cultures that at the beginning of mitosis the newly synthesized RNA leaves the nucleus to go to the cytoplasm where it is evenly distributed. Part of the RNA seems to remain bound to chromosomes. This transport has been described by many authors [see RAO and PRESCOTT, 1970]. It is essential that all, or almost all, labelled RNA has been shown to return to the nuclei of daughter cells [NEYFAKH, 1970; NEYFAKH and KOSTOMAROVA, 1971; NEYFAKH *et al.,* 1971].

This phenomenon is unequivocally demonstrated when between RNA synthesis and mitosis there elapses a minimal time (tens of minutes) and the RNA, which is released into the cytoplasm in the interphase, does not disguise the effect. In amebae, however, wherein synthesis of labelled RNA lasts for several days, transport of RNA to the cytoplasm has been proved to occur during mitosis and accumulation of some part of labelled RNA in daughter nuclei has also been reported [RAO and PRESCOTT, 1970]. This experiment did not answer the question which RNA returned to the nucleus: those that left it at the beginning of mitosis or those that had been in the cytoplasm prior to mitosis. In our experiments pre-mitotic synthesis of labelled RNA in amebae lasted only several hours, and in this case, almost all of labelled RNA was detected in just divided daughter nuclei [YUDIN and NEYFAKH, 1973].

Results of our investigation [ABRAMOVA and NEYFAKH, 1973] are interpreted to mean that RNA which is transferred to the cytoplasm at the beginning of mitosis, preserves its 'nuclear' character, i.e., contains giant molecules which are absent in the interphase cytoplasm. There are grounds to believe that the newly synthesized giant RNA molecules remain bound with chromatin until they have undergone some specific changes, i.e., processing, breaking this linkage and liberating monocistronic fragments; after that they go to the cytoplasm. Similar binding of newly synthesized RNA with chromatin was found experimentally in pigeon reticulocytes [GASARYAN *et al.,* 1971]. During mitosis normal structure of the cell and, first of all, that of chromatin, is violated; the latter becomes spiralized and stops binding nuclear RNA. Therefore, non-processed RNA molecules may go to the cytoplasm. After division, when in the daughter nuclei chromatin aquires its interphase structure again, the RNA molecules of 'nuclear' nature go back to the nuclei (probably just by diffusion). Then the RNA-chromatin bonds, broken during mitosis, are restored. A certain role in this process may belong to the nuclear membrane, although it was demonstrated in the experiments with amebae that the membrane is no obstacle for RNA moving in both directions [GOLDSTEIN *et al.,* 1969].

The data on RNA transport from the nucleus to the cytoplasm are not abundant. It is important that the rate of RNA transport changes in the course of development. In sea urchin this rate at early stages is not high, about 6% per hour [ARONSON and WILT, 1969]; labelled RNA remains in the nuclei for a long time without undergoing degradation. At later stages after gastrulation, labelled RNA is transported from the nuclei at a much higher rate [KIJIMA and WILT, 1969]. The relative rate of RNA transport was proved to change 10–15-fold from the earlier to the late stages [SINGH, 1968].

An important feature of the RNA transport in embryonic development is that it lasts for a long time. We have demonstrated with loach embryos that mRNA synthesized in the nuclei at the mid-late blastula stage is transferred to the cytoplasm for several hours, during which the embryo is undergoing gastrulation [KOSTOMAROVA *et al.,* 1972]. Thus, the rate of transport may affect morphogenesis, determining the time of translation of the templates synthesized earlier.

The second aspect of the problem of nucleocytoplasmic RNA transport is the question of RNP complexes formed both in the nucleus and cytoplasm. In eukaryotic cells, except tRNA, there seems to exist no free RNA which are not associated with protein. The role of these proteins has not been elucidated. In the nuclei, mRNA are bound with the protein molecules – informofers [SAMARINA *et al.,* 1968; GEORGIEV and SAMARINA, 1969].

SPIRIN also found in the cytoplasm of some objects, including those of loach, RNP complexes of different size but with a constant RNP-protein ratio of 1 : 4, which was responsible for their characteristic buoyant density: 1.3. The same ratio has been shown to exist for nuclear RNP. SPIRIN called cytoplasmic RNP particles, informosomes [SPIRIN, 1966], a name determining their function of information transporting particles. But their real function is still unknown. It is an established fact now that informosomes are not an artifact arising during homogenation, although RNA of different nature readily forms complexes with the specific group of cytoplasmic proteins. The linkage of mRNA with protein in informosomes is somewhat stronger than the non-specific linkage of, for example, bacterial rRNA with the same proteins. But neither is it clear whether GEORGIEV's RNP-complexes in nuclei and SPIRIN's informosomes in cytoplasm are different or identical bodies, nor whether protein plays in such complexes a specific or any other role, for example, that of protecting them from the action of ribonucleases. Does all RNA, having left the nuclei, go through the informosome stage, thus, playing a regulatory role in the process of translation, or (an idea supported by many authors) having left the nuclei, does RNA immediately become a part of polyribosomes? Elucidation of these problems is of great importance for molecular biology, but it is even more necessary for an understanding of the events occurring between transcription and translation in embryonic cells.

VI. Regulation

This is undoubtedly one of the most difficult and urgent ones, and not merely for unravelling the question of the nucleo-cytoplasmic relationship in development. A special aspect of this problem associated with development is that 2 close, but not identical, notions are understood by regulation: regulation as maintenance of some stable level (for example, rate of protein synthesis or respiration intensity) and as effectuation of some succession of changes resulting in differentiation (for example, the change of active genes or alterations of protein synthesis rate in development). For the latter notion, which is of greater importance for the developmental problem, the term 'program' is more appropriate. Effectuation of this 'program' is controlled not only by on/off switching genes but also at the successive stages of realization of genetic information.

1. Levels of Regulation in Development

a) Transcription

Regulation of transcription, or differential activity of genes, is the first and most important point, since it is at this level, that specificity of differentiation is determined via the choice of the genes to be transcribed. The general embryological background of this problem will be dealt with below. At this point, it seems appropriate to consider an aspect of this problem which is important for understanding the earliest stages of development: how is the switching-on of the genes controlled at the mid-late blastula stage, which is remarkable in fish and Amphibia? It appears, that in the cytoplasm of the egg and blastomeres during cleavage there is an inhibitor or, there is no activator of transcription. This was demonstrated by GURDON when RNA synthesis was promptly switched over to that of DNA in the nuclei of brain cells transplanted into the frog egg [GURDON, 1967]. This mechanism of gene activation may be partially associated with RNA-polymerase activity, as it was demonstrated [AKHALKATZI *et al.*, 1970; 1972]. Nuclei isolated from early and mid-blastula, as well as *in vivo,* have largely different rates of RNA synthesis if it proceeds at the expence of activity of endogenous RNA-polymerase. But in the presence of an excess of RNA-polymerase from *E. coli,* RNA synthesis greatly enhances and becomes equal in the nuclei of both developmental stages.

Nevertheless, these are likely to be the mechanisms responsible for switching on of genes but not for the time of the onset of gene activation. The true mechanism determining the onset of transcription should be sought in a certain 'clock': having consulted this 'clock' the nuclei may then 'judge' whether the respective stage has begun. This clockwork mechanism, or the value regularly changing in the course of development, may be the ratio of volumes of the nuclei or the quantity of DNA it contains to the volume of the cytoplasm. During cleavage this ratio grows in a regular way and as soon as a certain threshold is achieved, transcription begins [NEYFAKH and ROTT, 1968]. This suggestion is favored by some experiments made in this laboratory in which the onset of RNA synthesis was compared in haploid and diploid embryos. In haploid (obtained by irradiating sperm or eggs with heavy doses), the quantity of DNA per nucleus is twice less than in diploids and, therefore, the nucleus-cytoplasm volume ratio equal to that of diploids is achieved by one division later. Accordingly, mRNA synthesis in haploids also begins one division later [KOSTOMAROVA and ROTT, 1970].

b) Transport to Cytoplasm

The second step of realization of genetic information is RNA transport from the nucleus to the cytoplasm. There is, as yet, no direct proof that this process is really a point of control. But this may well be the case since transport is a relatively slow process and is, evidently, co-ordinated with processing accompanied by selection—at least, concerning RNA fragments going and not going to the cytoplasm. In the early stages of development, however, even the rate of mRNA release into the cytoplasm may be of importance for the control of development to be realized. This rate as well as, probably, the relative number of molecules released from the nucleus changes considerably in the course of development. In sea urchin [ARONSON and WILT, 1969] and in loach, at the beginning of synthesis, the release is extremely slow and it is possible that the major portion of RNA remains in the nucleus. At later stages the rate of release increases manifold.

In loach, genetic control over gastrulation is achieved at the mid-late blastula stage [NEYFAKH, 1964], and mRNA making this process possible, is being synthesized at the same time [KAFIANI and TIMOFEEVA, 1964]. The release of this RNA lasts for several hours, i.e., it proceeds as gastrulation goes on [KOSTOMAROVA *et al.,* 1972]. Thus, between the rate of processing and the rate of morphological development there may exist a causal relationship.

c) Translation

The existence of regulation at the translation level is self-evident and has been demonstrated by many authors to occur in bacteria and in differentiated cells of higher organisms. It also seems to be the case in embryonic development. It is possible that this regulation is at work only with the total protein synthesis, as follows from the compensatory increase in protein synthesis on old templates observed in sea urchin after the synthesis of the new ones has been inhibited [KEDES *et al.,* 1969]. But it cannot be excluded that different life-times of mRNA, possible heterogeneity of ribosomes, and other factors may also regulate synthesis of proteins in a qualitative way, the proteins being, naturally, those whose templates have been translated and passed to the cytoplasm. We know at least of one example of such regulation. During spermatogenesis, RNA synthesis stops before meiosis or right after it [MONESI, 1965; HENNING, 1968]. Consequently, the whole, extremely complex process of *spermiogenesis* occurs using the pre-existing information, i.e., by way of synthesizing protein on templates transcribed some days earlier [HESS, 1967]. One may suggest that as differentiation of spermatid proceeds, different proteins are being synthesized in it. Thus, it was demonstrated with *Drosophila* that when the sperms formed are maturing, the lysine-rich histones are replaced, in a puromycin-sensitive way, by arginine-rich histones which are functional analogues of protamines [DAS *et al.*, 1964]. Thus, in this case, translation control is responsible for storage of the unused templates during almost the entire spermiogenesis, as well as for their coming into operation at the last step of this process. Some events of this kind are likely to occur in enucleate erythrocytes of mammals whose morphological maturation, also requiring specific proteins, finishes in the cells deprived of nuclei. In these cases, similarly to nuclei that have been inactivated in sea urchin or loach embryos, the existence of a translation control over the program of qualitative and quantitative changes in protein synthesis, is obvious. There are, however, no grounds to believe that such control is absent when the nuclei are functioning.

d) Post-Translation Processes

We know very little about what is going on in the cell after protein has been synthesized. It is an established fact that the enzymatic function of proteins is not unequivocal but also depends upon intracellular localization of the protein. Localization of proteins, in turn, depends not only on the properties of the enzyme itself, but also on the properties of membrane proteins and, possibly, the proteins which are responsible for the enzyme-membrane binding. There are cases when localization of enzymes is controlled genetically. For example, in studying the lethal mutation 'g' in axolotl the electrophoretic distribution of protein as well as the activity of esterases *in vitro* turned out to be normal, whereas *in vivo,* the reactions controlled by these enzymes are greatly impaired. Moreover, distribution of esterases between the membrane and hyaloplasm and some properties of membranes themselves also change. It is believed that the 'g' gene controls one of the membrane proteins responsible for the binding of the enzyme and thereby affecting its function in the cell [TOMPKINS, 1970]. In 2 mice lines, genetic differences have been demonstrated which were expressed in differences of intracellular localization of glucuronidase—in hyaloplasm, or bound with ergastoplasm [GENSHOW and PAIGEN, 1967]. In this case too, there is a special genetic control of enzyme localization.

There are cases, however, when enzymatic activity in development regularly changes without direct nuclear control. For example, an increase in the rate of glycolysis during development of loach occurs due to a decrease in the activity of enzymes involved in gluconeogenesis which act in the direction opposite to that of glycolysis. The decrease in their activity is believed to occur at the cost of their natural degradation not compensated by synthesis [MILMAN and YUROVITSKY, 1967].

The character of activity of aspartateaminotransferase, which changes irregularly in the earliest development of sea urchin, depends neither on nuclear function, nor on protein synthesis and is exactly similar to the control. These changes take place in eggs whose cleavage is completely inhibited by puromycin [BOTVINNIK and NEYFAKH, 1969; ABRAMOVA and NEYFAKH, 1971].

Finally, the processes of organization of submolecular structures – problems that are beyond the scope of this paper – are also predetermined genetically by the properties of the proteins they contain; but these processes are realized and, possibly, controlled in the development outside the nuclei and to a large degree autonomously.

Thus, realization of genetic information proceeds through a number of successive stages which, sometimes, are not directly regulated by the genetic apparatus. In embryonic development, and in the process of differentiation in general, this regulation involves a program implying regular successive changes governing development.

2. Embryological Approach to the Problem of Differential Gene Activation

It is evident that the scheme accounting for differential gene activity should be applicable to embryonic development, since it is part of ontogenesis, during which process the function of most genes manifests itself. But among numerous schemes of genetic regulation encountered in higher animals it is difficult to find one that meets the requirements of embryonic development.

JACOB and MONOD's scheme of regulation of gene activity in bacteria cannot be applied unmodified to higher organisms. Nevertheless, the very principle of intergene interactions, the mode of participation of external factors and the set of elements of the regulated system – regulator gene, operator, structural gene, regulator protein, inductor – should be applicable in one or the other way to the scheme of gene activation of higher animals.

In the schemes proposed by BONNER [1965] and MEDVEDEV [1968], the switching-on of genes in development is described as a series of operations meant to specify the address and, thereby, limit the possibilities of development. These schemes, and also the cascade regulation scheme of PONTECORVO [1963] for bacteria, have an advantage in that they reflect the real process of step-wise specialization in the course of differentiation at the genetic level.

The hypothesis of the structure of the operon in higher animals [GEORGIEV, 1969] has the attraction of being detailed and based on experimental material. According to this scheme, the operon consists of structural gene(s) and the foregoing regulatory sites (operators). During transcription, the regulatory sites are first to be read, then follows the structural gene. Subsequently, a giant RNA molecule is formed which, during processing, degrades into mRNA proper, carrying information about protein and regulatory fragments not transported to the cytoplasm, and degrading in the nucleus. The large size of the regulatory portion of the operon may be interpreted to mean either the existence of many modes of operon activation (via several operators) or, else, that many factors are required for one operon to be read. Such a scheme means that the transcription of structural genes depends upon and is associated not with one regulator, as in bacteria, but with many.

BRITTEN and DAVIDSON's concept [1969] is based on the idea that each structural gene is associated with many regulatory ones, and vice versa. Such a system of integration allows one to explain the switching-on of one and the same gene in the differently specialized cells, i.e., the possibility of the functioning of each gene in different sets of active genes.

Each of the above concepts seems to reflect some real features of the regulation system in higher organisms, although no claim is laid to building up an overall theory which could be applicable to embryonic development in particular. Embryonic development has some peculiarities which should be accounted for by a system of regulation of gene function in development of higher organisms.:

1. Embryonic development occurs via progressive differentiation of the branching tree type. This, evidently, should be the scheme for the switching-on genes determining differentiation. For example, operation of the genes specific for the early embryonic anlage (for example, mesoderm) should pave the way to switching of the genes responsible for further specialization of a muscle or a cartilage cell.

2. Competence, i.e., ability to differentiate in few directions determined by differentiation factors (for example, induction of neural tube in ectoderm) should be reflected in the phenomenon of gene competence, i.e., the ability of a limited number of genes out of all the possible ones to be

switched on. For example, target tissue is competent to the action of the hormone which is capable of inducing in the cells, activation of some genes specific for the given tissue. In other cells the same hormone may activate other genes or be totally inactive. However, competence is the property not of genes themselves, not even of the nucleus, but of the whole cell. This follows from the experiments on transplantation of nuclei into the egg [see GURDON and WOODLAND, 1968] and on hybridization of somatic cells [HARRIS, 1968]. These experiments proved the nuclear function to be almost limitlessly capable of changing under the action of different cytoplasmic environments.

3. Not very high informational capacity of differentiation factors which, nevertheless, results in gene activation of extreme specificity. It is known, for example, that embryonic induction of nerve or mesoderm tissue may be caused in the competent ectoderm by proteins of widely different origin [TIEDEMANN, 1968]. In some cases an induction may be caused by much simpler substances than proteins [LASH, 1968]. Specific genes may be activated by hormones of very simple structure (thyroxine, steroids). Yet, it goes without saying, that, to pick out one gene among tens and hundreds of thousands of others, requires an agent of extremely high specificity to recognize the gene and put it into operation. This could be done only by a complex and specific protein which can recognize a certain nucleotide sequence, or, as was postulated by BONNER [1965], and BRITTEN and DAVIDSON [1969], by RNA which is complementary to a given sequence.

These requirements, actually, pre-determine the scheme which could be suggested at the present-day level of knowledge.

Direct selection of a gene and putting it into operation is realized by a special regulatory protein capable of recognizing the nucleotide sequences of the operator of the gene to be activated. Such regulatory proteins have been proved to exist and were isolated from bacteria and complex phages. Integration of the scheme may be accomplished by one regulatory protein being able to switch one of several genes (due to the identical nucleotide sequences in their operators) or one gene being switched on by several regulatory proteins (due to similarity of the recognition sites). But unlike the case of bacteria, in higher organisms such protein is activated not by the enzyme's substrate but by a substance which is either a differentiation factor or a hormone. That proteins which bind hormones and even transport them to chromatin exist, has been reported by a number of authors [JENSEN *et al.*, 1968; MATTYSSE and PHILLIPS, 1969; O'MALLEY *et al.*, 1970].

In the cells of higher organisms, not all kinds of regulatory proteins need be present, as is the case in bacteria, but only a small part of them. This means that only a small part of the genes will be put into operation if one or several of these proteins are activated by a respective inductor. Consequently, at the molecular level, competence of the cell means only that several regulator genes are active and the products of their activity – regulatory proteins – are present in the cytoplasm. A change in competence, then, will mean an altered set of active regulator genes and replacement of regulatory proteins in the cytoplasm. Such a concept does not contradict the facts that a) inductors do not affect isolated nuclei (where there are no regulatory proteins) [JOHR and VARNER, 1968]; b) whatever the origin of the nuclei, they can change their function in a new cytoplasmic environment as they are subjected to the action of new regulatory proteins; c) the cell is competent only toward a limited number of external agents which can make it develop in only one of few directions.

Hence, differentiation begins when the inductor activates some regulatory proteins out of those present in the cytoplasm, the latter switching on a definite set of structural genes determining specific features of the differentiating cells. Simultaneously, new regulator genes come into play, creating new competence for the next stage of development.

Thus, the cell itself possesses a major part of the information required for certain genes to be chosen out of many others; its competence to develop along a certain line is realized as a set of regulatory proteins possessing sufficient specificity to recognize a gene. The choice of a development pathway is done by the inductor, the information capacity of which may not be very high. In the course of development the cell receives from outside, a considerable amount of information, portion by portion, in the form of differentiation factors. It is like dialing a lengthy phone number, a signal after signal, each being just one of ten figures.

The above scheme seems to be in agreement with the main peculiarities of embryonic development: progressive differentiation is attained by the fact that different cells possessing equal competence are acted upon by different differentiation factors to direct them to different pathways of development. Competence can be explained at the genetic level, using the elements of JACOB and MONOD. Finally, the problem of the switching-on of specific genes by the action of factors of low specificity is clarified. It may not seem worthwhile at this point, to elaborate this scheme; however, there are grounds to believe that any detailed mechanisms of gene activation, should they be established, will fit it.

References

ABRAMOVA, N.B.; LICHTMAN, T.V. and NEYFAKH, A.A.: Fed. Proc. Suppl. *25:* 489 (1966).
ARAMOVA, N.B. and NEYFAKH, A.A.: Ontogenesis (Russian) *2:* 71 (1971).
ABRAMOVA, N.B. and NEYFAKH, A.A.: Exp. Cell Res. *76* (in press) (1973).
ABRAMOVA, H.B. and VASILYEVA, M.: Ontogenesis (Russian) (in press) (1973).
AKHALKATSI, R.G.; TIMOFEEVA, M.J. and KAFIANI, C.A.: Biokhimia *35:* 1193 (1970).
AKHALKATSI, R.G.; KAFIANI, C.A. and CASARIAN, K.G.: in SBARSKY and GEORGIEV Cell nucleus, p. 212 (Nauka, Moscow 1972).
AJTKHOGHIN, M.A., BELITSINA, N.V. and SPIRIN, A.S.: Biokhimiya (Russian) *29:* 169 (1964).
ANANIEVA, L.N.; KOZLOV, Yu.V.; RISKOVA, A.P. and GEORGIEV, G.P.: Molek. Biol. (Russian) *2:* 736 (1968).
BELL, E.: in Control mechanism of growth and differentiation, p. 127 (University Press Cambridge 1971).
BERENDES, H.D.: Ann. Embryol. Morphogenes., Suppl. *1:* 153 (1969).
BONNER, I.: in The molecular biology of development, pp. 133–143 (Clarendon Press, Oxford 1965).
BOTVINNIK, N.M. and NEYFAKH, A.A.: Exp. Cell Res. *54:* 287 (1969).
BRIGGS, R. and CASSENS, G.: Proc. nat. Acad. Sci. U.S. *55:* 1103 (1966).
BRITTEN, R.I. and DAVIDSON, E.H.: Science *165:* 349 (1969).
BRITTEN, R.I. and KOHNE, D.E.: Science *161:* 529 (1968).
BROWN, D.D. and DAWID, J.B.: Science *160:* 272 (1968).
BROWN, D.D. and GURDON, J.B.: Proc. nat. Acad. Sci. U.S. *51:* 139 (1964).
BROWN, D.D. and LITTNA, E.: J. molec. Biol. *8:* 669 (1964a).
BROWN, D.D. and LITTNA, E.: J. molec. Biol. *8:* 687 (1964b).
BROWN, D.D. and LITTNA, E.: J. molec. Biol. *20:* 95 (1966).
BROWN, D.D. and WEBER, C.S.: J. molec. Biol. *34:* 661 (1968a).
BROWN, D.D. and WEBER, C.S.: J. molec. Biol. *34:* 681 (1968b).
CHAMBERLAIN, J.P.: Biochim. biophys. Acta *213:* 183 (1970).
CHEN, P.S.: in Biochemistry of animal development, vol. 2, p. 115 (Academic Press, New York 1967).
CRAIG, S.P.: J. molec. Biol. *47:* 615 (1970).
CRIPPA, M.; DAVIDSON, E.H. and MIRSKY, A.E.: Proc. nat. Acad. Sci. U.S. *57:* 885 (1967).
CZIHAK, G.: Naturwissenschaften *52:* 141 (1965).
DAVIDSON, E.H.: Gene activity in early development (Academic Press, New York 1968).
DAVIDSON, E.H.; CRIPPA, M.; KRAMER, F.R. and MIRSKY, A.E.: Proc. nat. Acad. Sci. U.S. *56:* 856 (1966).
DAVIDSON, E.H. and HOUGH, B.R.: Proc. nat. Acad. Sci. U.S. *63:* 342 (1969).
DAVIDSON, E.H. and HOUGH, B.R.: J. exp. Zool. *172:* 25 (1969).
DAS, C.C.; KAUFMAN, B.R. and GAY, H.: Nature *204:* 128 (1964).
DAWID, I.B.: J. molec. Biol. *12:* 581 (1965).
DAWID, I.B. and WOLSTENHOLME, D.R.: Biophys. J. *8:* 65 (1968).
DENNIS, H.: J. molec. Biol. *22:* 285 (1966).
DENNIS, H.: Adv. Morphogenes. *7:* 115 (1968).
DENNIS, H.: Personal communication (1971).
DETTLAFF, T.A. and SCOBLINA, M.N.: Ann. Embryol. Morphogenes. Suppl. *1:* 133 (1969).

EMERSON, C.P. and HUMPHREYS, T.: Develop. Biol. *23:* 81 (1970).
GANSCHOW, R. and PAIGEN, K.: Proc. nat. Acad. Sci. U.S. *58:* 938 (1967).
GASARYAN, K.G.; LIPASOVA, V.A.; KIRIANOV, C.J.; ANANJANZ, T.G. and ERMAKOVA, N.G.: Molec. Biol. (Russian) *5:* 680 (1971).
GEORGIEV, G.P.: J. theoret. Biol. *25:* 473 (1969).
GEORGIEV, G. P. and SAMARINA, O. P.: Embryol. Morphologenes. Suppl. *1:* 81 (1969).
GOLDSTEIN, L.; RAO, M.V.N. and PRESCOTT, D.M.: Ann. Embryol. Morphogenes. Suppl. *1:* 189 (1969).
GRAHAM, C.F.; ARMS, K. and GURDON, J.B.: Develop. Biol. *14:* 349 (1966).
GROSS, P.R. and COUSINEAU, G.H.: Exp. Cell Res. *33:* 368 (1964).
GURDON, J.B.: Proc. nat. Acad. Sci. U.S. *58:* 545 (1967).
GURDON, J.B. and BROWN, D.D.: J. molec. Biol. *12:* 27 (1965).
GURDON, J.B. and WOODLAND, H.R.: Biol. Rev. *43:* 233 (1968).
HARRIS, H.: Nucleus and cytoplasm (Oxford Univ. Press, 1970).
HARTMAN, I.F. and COMB, D.A.: J. molec. Biol. *41:* 155 (1969).
HARVEY, E.B.: Biol. Bull. *71:* 101 (1936).
HENNING, W.: Proc. nat. Acad. Sci. U.S. *38:* 227 (1968).
HESS, O.: Exp. Biol. Med. *1:* 90 (1967).
HOGAN, B. and GROSS, P.B.: Exp. Cell Res. *72:* 101 (1972).
HYNES, R.O. and GROSS, P.R.: Develop. Biol. *21:* 383 (1970).
HULTIN, T.: Develop. Biol. *10:* 305 (1964).
JATSU, N.: Biol. Bull. *6:* 123 (1903).
JEANTEUR, P. and ATTARDI, G.: J. molec. Biol. *45:* 305 (1969).
JENSEN, E.V.; SUZUKI, T.; KAWASHIMA, T.; STUMPF, W.E.; JUNGBLUT, P.W. and DESOMBRE, E.R.: Proc. nat. Acad. Sci. U.S. *59:* 632 (1968).
JONES, K.W.: Nature *225:* 912 (1970).
JOHRI, M.M. and VARNER, J.E.: Proc. nat. Acad. Sci. U.S. *59:* 269 (1968).
JUDIN, A.L. and NEYFAKH, A.A.: Exp. Cell Res. (in press) (1973).
KAFIANI, C.A. and TIMOFEEVA, M.J.: Dokl. Akad. Nauk SSSR (Russian) *164:* 721 (1964).
KAFIANI, C.A.; TIMOFEEVA, M.J.; MELNIKOVA, N.L. and NEYFAKH, A.A.: Biochim. biophys. Acta *169:* 274 (1969).
KAFIANI, C.A.; TIMOFEEVA, M.J.; NEYFAKH, A.A.; MELNIKOVA, N.L. and RACHKUS, J.A.: J. Embryol. exp. Morph. *21:* 295 (1969).
KEDES, L.H.; GROSS, P.R.; COGNETT, R. and HUNTER, A.L.: J. molec. Biol. *45:* 337 (1969).
KIJIMA, S. and WILT, F.H.: J. molec. Biol. *40:* 235 (1969).
KOSTOMAROVA, A.A. and ROTT, N.N.: in ZBARSKY, J.B. The cell nucleus and its ultrastructure, pp. 258–262 (Nauka, Moscow 1970).
KOTOMIN, A.V. and IVANCHIK, T.A.: Molek. Biol. (Russian) (in press) (1973).
KRIGSGRABER, M.R. and NEYFAKH, A.A.: J. Embryol. exp. Morph. *28:* 491 (1972).
KRIGSGRABER, M.R.; IVANCHIK, T.A. and NEYFAKH, A.A.: Biokhimiya *33:* 1214 (1968).
LASH, J.V.: J. Cell Physiol., Suppl. *1:* 35 (1968).
MAGGIO, R.; VITTORELLI, M.L.; CAFFARELLI-MORMINO, J. and MONROY, A.: J. molec. Biol. *31:* 621 (1968).
MAIRY, M. and DENIS, H.: Arch. in Physiol. Biochim. *78:* 599 (1970).
O'MALLEY, B.W.; SHERMAN, M.R. and TOFT, D.O.: Proc. nat. Acad. Sci. U.S. *67:* 501 (1970).
MARKMAN, B.: Exp. Cell Res. *23:* 118 (1961).

MASUI, J.: J. exp. Zool. *166:* 365 (1967).
MATTYSSE, A.G. and PHILLIPS, C.: Proc. nat. Acad. Sci. U.S. *63:* 897 (1969).
MAUL, G.G. and HAMILTON, T.H.: Proc. nat. Acad. Sci. U.S. *57:* 137 (1967).
MEDVEDEV, J.A.: in Molecular Genetical mechanisms of development (Russian) (Medenica, Moscow 1968).
MELNIKOVA, N.L.; TIMOFEEVA, M.I.; ROTT, N.N. and IGNATIEVA, G.M.: Ontogenesis *3:* 85 (1971).
MILMAN, L.S. and YUROVITZKY, Y.G.: Biochim, biophys. Acta *148:* 362 (1967).
MINTZ, B.: J. exp. Zool. *157:* 85 (1964).
MONESI, V.: Exp. Cell Res. *39:* 197 (1965).
NEYFAKH, A.A.: Nature *201:* 880 (1964).
NEYFAKH, A.A.: Curr. Topics dev. Biol. *6:* 45 (1971).
NEYFAKH, A.A.: in LOPASHOV, G.V., NEYFAKH, A.A. and STROEVA, O.G. Cell differentiation and induction mechanisms, pp. 38 (Nauka, Moscow 1965).
NEYFAKH, A.A.; ABRAMOVA, N.B. and BAGROVA, A.M.: Exp. Cell Res. *65:* 345 (1971).
NEYFAKH, A.A. and KOSTOMAROVA, A.A.: Exp. Cell Res. *65:* 340 (1971).
NEYFAKH, A.A.; KRIGSGRABER, M.R. and ILJIN, M.I.: Dokl. Akad. Nauk SSSR (Russian) *181:* 253 (1968).
NEYFAKH, A.A. and KRIGSGRABER, M.R.: Dokl. Akad. Nauk SSSR (Russian) *183:* 493 (1968).
NEYFAKH, A.A.; KOSTOMAROVA, A.A. and BURAKOVA, T.A.: Exp. Cell Res. *72:* 223 (1972).
NEYFAKH, A.A. and RADZIEVSKAYA, V.V.: Genetics (Russian) *3:* 80 (1967).
NEYFAKH, A.A. and ROTT, N.N.: J. Embryol. exp. Morph. *20:* 129 (1968).
NEYFAKH, A.A.; TIMOFEEVA, M.I.; KRIGSGRABER, M.R. and SVETAYLO, N.A.: Genetika (Russian) *4:* 90 (1968).
NEYFAKH, A.A.; GLUSHANKOVA, M.A.; KOROBZOVA, N.S. and KUSAKINA, A.A.: Develop. Biol. (in press) (1973).
NEMER, M.: in Progress in nucleic acids research and molecular biology, vol. 7. p. 243 (Academic Press, New York 1967).
NEZLIN, R.S. and ROKHLIN, O.V.: J. All Union Mendeleev's Chem. Soc. *15:* 666 (1970).
OHNO, S.; CCHRISTIAN, L.; STENIUS, C.; CASTRO-SIERRA, E. and MURAMOTO, J.: Biochem. Genet. *2:* 361 (1969).
PARDUE, M.L. and GALL, I.G.: Science *168:* 1356 (1970).
PERRY, R.P.; TSAI-YING CHENG; FREED, I.I.; GREENBERG, J.R.; KELLEY, D.E. and TARTOF, K.D.: Proc. nat. Acad. Sci. U.S. *65:* 609 (1970).
PIKO, L.; TYLER, A. and VINOGRAD, J.: Biol. Bull. *132:* 68 (1967).
PONTECORVO, G.: Proc. roy. Soc. B. *158:* 1 (1963).
QUINCEY, R. V. and WILSON, S. H.: Proc. nat. Acad. Sci. U.S. *64:* 981 (1969).
RACHKUS, J.A.; TIMOFEEVA, M.I. and KAFIANI, C.A.: Molek. Biol. (Russian) *3:* 438 (1969).
RACHKUS, J.A.; KUPRIYANOVA, N.S.; TIMOFEEVA, M.I. and KAFIANI, C.A.: Molek. Biol. (Russian) *3:* 617 (1969).
RAO, M.V.N. and PRESCOTT, D.M.: Exp. Cell Res. *62:* 286–292 (1970).
ROEDER, R.C. and RUTTER, W.J.: Nature *224:* 237 (1969).
ROTT, N.N. and SHEVELEVA, G.A.: J. Embryol. exp. Morph. *20:* 141 (1968).
SAMARINA, O.P.; LUKAMIDIN, E.M.; MOLNAR, J. and GEORGIEV, G.P.: J. molec. Biol. *33:* 251 (1968).
SCHERRER, K.: Exp. Biol. Med. *1:* 244 (1967).

SCHERRER, K.; MARCAUD, L.; ZAIDELA, F.; LONDON, J. and GROSS, F.: Proc. nat. Acad. Sci. U.S. *56:* 1571 (1966).
SHMERLING, J.G.: Biokhimiya *30:* 113 (1965).
SIEBERT, G.: in SBARSKY and GEORGIEV Cell nucleus, p. 219 (Nauka, Moscow 1972).
SINGH, U.N.: Exp. Cell Res. *53:* 537 (1968).
SLATER, D.V. and SPIEGELMAN, S.: Biochim. biophys. Acta *213:* 194 (1970).
SMITH, L.D.: Develop. Biol. *14:* 330 (1966).
SOLOMON, J.: Biochim. biophys. Acta *24:* 584 (1957).
SPIRIN, A.S.: Curr. Topics dev. Biol. *1:* 1 (1966).
STEVENIN, J.; MANDEL, P. and JACOB, M.: Proc. nat. Acad. Sci. U.S. *62:* 490 (1969).
THOMAS, Ch.: Biochim. biophys. Acta *224:* 99 (1970).
THOMAS, C.A., jr.; HAMKALO, B.A.; MISRA, D.N. and LEE, C.S.: J. molec. Biol. *51:* 621 (1970).
TIEDEMANN, H.: Exp. Biol. Med. *1:* 8 (1967).
TIEDEMANN, H.: J. Cell Physiol., Suppl. *1:* 129 (1968).
TIMOFEEVA, M.J.; NEYFAKH, A.A. and STROKOV, A.A.: Genetika (Russian) *7:* 93 (1971).
TIMOFEEVA, M.J.; and KAFIANI, C.A.: Biokhimiya *29:* 110 (1964).
TIMOFEEVA, M.J.; SOLOVJEVA, I.A. and SOSINSKAYA, I.: Ontogenesis (Russian) (in press) (1973).
TYLER, A.: Develop. Biol. Suppl. *1:* 170 (1967).
WHITELEY, A.H.; MCCARTHY, B.J. and WHITELEY, H.R.: Proc. nat. Acad. Sci. U.S. *55:* 519 (1966).
WHITELEY, A.H.; MCCARTHY, B.J. and WHITELEY, H.R.: Develop. Biol. *21:* 216 (1970).
WEINBERG, R.A. and PENHAM, S.: J. molec. Biol. *47:* 169 (1970).
WILLIAMSON, R.: J. molec. Biol. *51:* 157 (1970).
WILT, F.H.: Develop. Biol. *23:* 444 (1970).
WOODLAND, H.R. and GURDON, J.B.: J. Embryol. exp. Morph. *19:* 363 (1968).
WRIGHT, D.A. and SUBTELNY, S.: Develop. Biol. *29:* 119 (1971).

Author's address: Dr. A.A. NEYFAKH, Institute of Developmental Biology of the USSR Academy of Sciences, Vavilov St. 26, *Moscow* (USSR)

Neoplasia and Cell Differentiation, pp. 60–105
(Karger, Basel 1974)

Factors Controlling the Initiation and Cessation of Early Events in the Regenerative Process[1]

BRUCE M. CARLSON

Department of Anatomy, University of Michigan, Ann Arbor, Mich.

Contents

I. Introduction

A regenerated vertebrate appendage represents the culmination of a long and complex series of developmental events. Following the amputation of a limb, the animal must first protect the remaining tissues of the stump from the external environment by means of a wound healing process. Yet, if the healing process is too effective in terms of rapid healing of all damaged tissues, regeneration is inhibited. As in any damaged structure, non-viable debris resulting from the traumatic event must be cleaned up. Next there occurs a set of reactions almost unique to a regenerating appendage. Following the initial 'clean-up' phase, the remaining tissues at

1 Original work described in this review was supported by grants from the University of Michigan Cancer Institute and by a grant from the Muscular Dystrophy Associations of America.

the distal end of the stump undergo a period of prolonged loss of normal adult structure, and appearing in their place is a population of cells whose structure is quite reminiscent of embryonic cells. These cells migrate distally and aggregate beneath the wound epidermis into a homogeneous mass called the regeneration blastema. Many of the steps leading to the formation of a blastema reveal another characteristic of regenerating systems. They are either wholly dependent upon or are strongly influenced by the relationship of the regenerating structure to the body as a whole. Thus, nervous connections and hormonal environment, for example, represent intrinsic ingredients of the total regenerating system. Once established, the regeneration blastema continues to increase rapidly in size by means of cell proliferation.

Differentiative processes next assume ascendency, but it must be kept in mind that this phase includes not only actual cytodifferentiation, but also the laying down of a definitive morphogenetic pattern of tissues, the rudiments of which may be apparent before morphologically evident cytodifferentiation has occurred. Here again the relationship of the regenerate to the body must be stressed. A regenerating structure can only become an arm within a certain territory of the body. It must also receive information concerning symmetry and axiation from the body to which it is attached. Although the duration and mechanism of this morphogenetic signal-calling continues to be debated, it is an inescapable fact that at some time the cells of the regenerating limb are given information which enables them to reproduce exactly the structure which they are bound to replace.

Although it is impossible to categorize briefly the individual problems and facets of the entire regenerative process, many of them represent implicit components of two general questions. 1. What is required to initiate and to sustain the regenerative process? 2. What determines the form of the regenerate? In this review I am going to confine myself almost entirely to the first question. The fascinating problem of morphogenesis is of such magnitude that justice could not be done to it in the same writing. The established blastema will constitute the end point of the present review, and major emphasis here will be placed upon both the stimulus for the regenerative process and the means by which the stimulus is translated into events leading to the formation of the blastema. SCHMIDT [1968] has reviewed most thoroughly our knowledge of the morphological and chemical features of the regenerating limb. Rather than reiterating much of that material, I intend, instead, to concentrate upon factors which have been shown to control certain phases of the early regenerative process.

II. Descriptive Review of Early Limb Regeneration

In order to provide a morphological basis for this discussion, a brief histological description of limb regeneration in the adult newt up to the stage of blastema formation will be given below.

Amputation is followed by a rapid cessation of bleeding and by a partial retraction of the muscle and overlying skin away from the level of amputation. Epidermal healing is normally completed within 24—48 hours unless retraction of the soft tissues has been so extreme that a significant length of bone protrudes. In this case, several more days may be required for complete epidermal continuity to be established. The early wound epidermis is not underlain by a basement membrane and maintains direct contact with the underlying mesodermal tissues throughout much of the early regenerative process. Beneath the wound epithelium, the major morphological activity of the first few days after amputation is the degeneration of tissue fragments injured and/or devascularized by the initial trauma (fig. 1a). Degeneration of extravasated erythrocytes and phagocytosis of cellular debris by leukocytes are prominent features of this stage. As was illustrated by SINGER and SALPETER [1961], the wound epidermis itself may participate actively in phagocytosis by sending tongues of epithelium into the underlying damaged tissues, surrounding pockets of debris and then extruding this material to the outside of the limb. Toward the end of the first week, nerve fibres have penetrated the wound epidermis.

Early in the second week the period of phagocytosis and demolition is gradually replaced by a stage characterized by the appearance of

---

Fig. 1 a–d. Regeneration of the forearm in adult newts. *a* The amputation surface in this 5-day regenerate is covered by a wound epithelium which is sending some tongues toward a small area of underlying inflammation. Except for the distal ends of some of the cut muscle fibres, there is little loss of structure in the limb stump. *b* 12-day regenerate. Soft tissue destruction is well underway, and there is considerable osteoclastic erosion of bone. The wound epithelium is considerably thicker than normal. Occasional dedifferentiated cells can be seen in areas of intense tissue destruction. *c* Typical view of the distal musculature during the period of dedifferentiation (18 days). Most of the muscle fibers, with the exception of the few in the center of the photomicrograph, have lost large amounts of cytoplasm. The nuclei associated with these muscle fibers are increasing in diameter and show some characteristic changes in the chromatin pattern. The distal end is at the bottom of the photograph. *d* A homogenous appearing blastema is well established by 25 days. Note the characteristic regular columnar shape of the basal epidermal cells along the lateral edges of the regenerate. Cartilage, not derived from the blastema, has already differentiated alongside the distal portions of the radius and ulna.

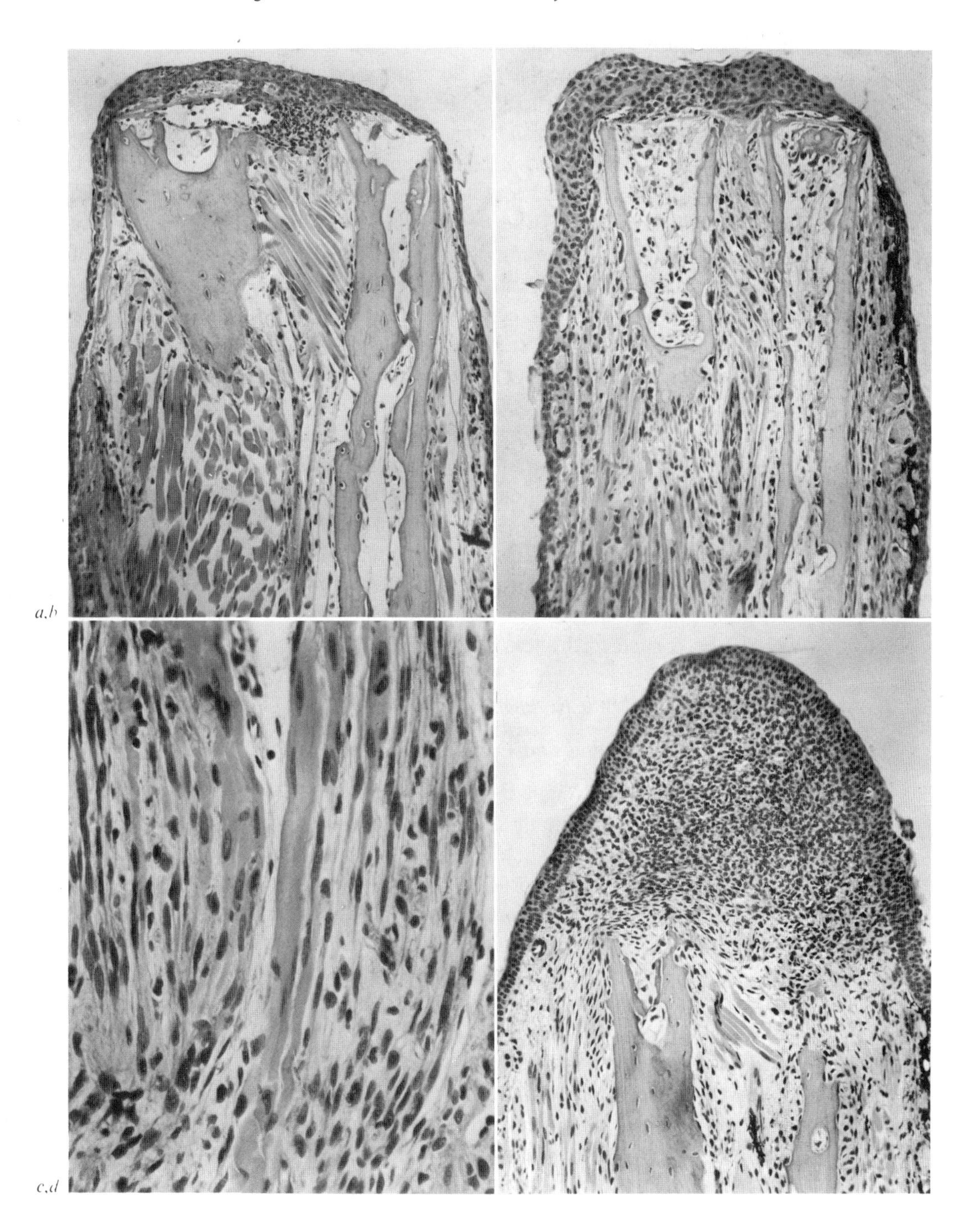
a,b
c,d

morphologically dedifferentiated cells. Early during this period the last characteristic event of the phase of demolition occurs. This is the osteoclastic destruction of skeletal structures. In my laboratory osteoclasts regularly appear first on the 10th day after amputation. At this time the earliest dedifferentiated cells are seen in the subdermal connective tissue and in the distal areas of damaged muscle (fig. 1b). In a very striking fashion, the wave of dedifferentiation sweeps proximally, leaving in its wake the empty supporting elements of the muscular tissue (fig. 1c). The bony skeleton of newts is greatly eroded, but not completely removed. During the dedifferentiative phase, the wound epidermis undergoes a substantial thickening from the usual 2—3 layers to almost five times that figure. The thickened area of epidermis is usually called the apical epidermal cap.

By the end of the 3rd week the proximal course of the dedifferentiative process has stopped, and the mesenchymatous cells left in its wake begin to accumulate beneath the apical epidermal cap. As a result of further aggregation and multiplication of these cells, a homogenous cellular mass, known as the regeneration blastema, is established (fig. 1d). This structure provides the raw materials from which the regenerate is finally formed. Since factors related to the differentiation and morphogenesis of the regenerate will not be covered in this review, this morphological description will not be carried to the termination of the regenerative process.

III. Normal Limb Regeneration

A. Wound Healing and Demolition

It is extremely difficult to relate causally any of the morphological events occurring immediately after amputation with subsequent regeneration, and few experimental studies have been devoted to analysis of this phase of the regenerative process. One of the major difficulties is that, barring some obvious experimental manipulations, there are no morphological differences between regenerating limbs and those in which regeneration has been inhibited. Thus, there is a great need for biochemical, autoradiographic or ultrastructural comparison between normally regenerating and inhibited limbs during this period. The fundamental question which remains to be answered is whether amputation sets off an immediate chain of events which lead directly to a full regenerative process or whether the newly amputated limb responds to the trauma by creating a metabolic climate which is compatible with the initiation of the actual regenerative phase by a later series of stimulatory events.

The concept of a specific wound hormone which acts as an all or nothing trigger for regeneration has been in the literature for years. Wound hormones have generally been considered to have been liberated by the cells damaged by the trauma of amputation. In 1941 NEEDHAM found that immediate immersion of freshly amputated tadpole tails into a solution of beryllium nitrate completely inhibited subsequent regeneration. However, if treatment was delayed by an hour, no inhibition occurred. NEEDHAM interpreted these results to favor the concept of the inhibition of a transient wound factor which operates only a short time after the initial trauma. THORNTON [1949, 1950, 1951] conducted further experiments with this inhibitor and concluded that rapid covering of the amputation surface by a wound epithelium prevented the beryllium from contacting the internal tissues, thus obviating the need for inactivation of a wound hormone as an explanation for the inhibitory action. KARCZMAR [1946] proposed that an "amputation surface factor" is released at the time of amputation and that its concentration increases early in the regenerative process. This assertion, however, was not backed up by any direct evidence.

An early post-amputational process which has been shown to have a direct relationship to the presence or absence of regeneration is epidermal wound healing. Normally, the amputation surface is covered by a wound epithelium within several hours to several days depending upon the size of the limb. During the initial migratory phase of healing, the epidermis appears to follow the general rules of wound healing [reviewed by WEISS, 1961; Lash, 1955, 1956], and it will not be further discussed here. The wound epithelium is underlain by neither a dermal layer nor a basement membrane. There is little question that the covering of the amputation surface by a wound epidermis is an indispensable condition for regeneration. Limb regeneration in *Ambystoma* larvae is completely inhibited by frequent removal of the wound epidermis [THORNTON, 1957]. It has been known for years [TORNIER, 1906; TAUBE, 1921; GODLEWSKI, 1928; EFIMOV, 1931] that full thickness skin flaps sewn over the amputation surface do not permit regeneration. As of yet, there have been no critical studies which allow one to determine exactly when the wound epidermis becomes important. As will be discussed later, the presence of a wound epidermis seems to be essential for the initiation of dedifferentiation, at least in adults, but whether it begins to exert a specific action relating to regeneration as soon as wound closure is completed or whether there is a delay before the epidermis becomes an effective participant is presently not known.

One rather well documented role of the early wound epithelium is that of phagocytosis. TABAN [1955], BODEMER [1958] and SINGER and SALPETER [1961] have described epidermal tongues extending into the underlying areas of damaged mesodermal tissues and have experimentally demonstrated the passage of cellular debris through the epidermis to the outside. This phagocytic phenomenon has not been demonstrated to have any direct relationship to the presence or absence of regeneration.

The relationship between the wound epidermis and destructive processes in the underlying tissues was the subject of considerable research by Russian investigators during the early 1930's. Almost all of this work was performed on axolotls. EFIMOV [1931] confirmed earlier work which showed that a skin flap covering the amputation surface inhibits regeneration and that destructive processes are rather quickly brought to a halt. Even intensive supplementary trauma inflicted upon the mesodermal portions of skin-covered amputation stumps did not lead to any progression of the early regenerative phases. ORECHOWITSCH and BROMLEY [1934] investigated the histolytic properties of tissues during early phases of regeneration and also noted that if an amputation surface is covered by skin, destructive processes in the underlying tissues decrease within a few days. POLEZHAEV [POLEZHAEV, 1936a; POLEŽAJEW and FAWORINA, 1935] also supported the concept of epidermal participation in histolysis. From his experiments of allowing skin from different regions of the body to cover the limb stump, he concluded that epithelia from the limbs and tail have certain properties not shared by epithelium from the head or back (which do not support regeneration). POLEZHAEV [1936a] proposed an interacting system whereby mesodermal trauma acts upon the wound epidermis to elicit a histolytic property of the epidermis itself. The activated epithelium then acts back upon the mesodermal tissues causing a progression of the histolytic reaction. As evidence for the need for mesodermal trauma to stimulate the histolytic properties of the epidermis, POLEZHAEV cited earlier experiments [POLEZHAEV, 1933b] in which he covered limb stumps with skin flaps and inhibited regeneration for 1—2 months. Following removal of the full thickness skin, histolysis and subsequent regeneration did not occur without additional trauma to the underlying tissues. POLEZHAEV [1936a, p. 282] summarized this activity by attributing two properties to the early wound epidermis: 1) the intensification of processes of destruction of mesodermal tissues, and 2) the attraction to the wound epidermis of cells of mesodermal origin. This latter statement presaged the definitive work by THORNTON (see below) on the role of the apical epidermal cap in the early stages of blastema formation.

It is generally recognized that in a regenerating limb, the phase of demolition is characterized by a lytic reaction far in excess of that which would occur in a similarly traumatized non-regenerating system. The biochemical and histochemical evidence has been summarized by SCHMIDT [1968, chap. 8] and will not be repeated here. In most cases, increases in lytic (especially proteolytic) activity have not been detected in the days immediately following amputation, but rather, enzyme activity is usually greatest during overt dedifferentiation or in the early blastemal phase. The recent studies on collagenase activity in regenerating limbs of the newt [GRILLO *et al.,* 1968; DRESDEN and GROSS, 1970] have also borne out this general pattern of activity. Little is known about tissue interactions or other signals leading to the phase of heightened destructive activity except for ORECHOWITSCH and BROMLEY's [1934] demonstration of the need for a wound epithelium for the maintenance of full activity. The relationship between innervation and events during the period of demolition has scarcely been considered, although THORNTON's [1954] demonstration of excessive resorption following minor trauma in denervated larval limbs certainly deserves more attention.

One must summarize this section with the unsatisfactory conclusion that almost nothing is known about the stimulus for the excessive destruction of formed tissues during the period of demolition and that no direct causative factors (with the possible exception of the wound epidermis) have been adequately demonstrated from the time of amputation to the time of obvious demolition of formed tissues. As tissue destruction progresses, cells of a rather characteristic morphology appear in the areas of greatest activity. These cells are usually called dedifferentiated cells by workers in this field, and their presence heralds the next phase of regeneration–the period of dedifferentiation. For several days destructive processes are even heightened, and at the same time the number of dedifferentiated cells increases. Thus, it is impossible to separate these processes.

B. Dedifferentiation

The earliest stage at which one can say with any reasonable degree of assurance that regeneration will occur is the period of dedifferentiation. The term dedifferentiation itself has been the source of considerable controversy because it has been applied to a number of normal developmental and neoplastic processes without adequate definition and without necessarily connoting the same thing. As originally used in the literature of limb regeneration [BUTLER, 1933], dedifferentiation refers to a morphological process by which the tissues comprising the apical part of an amputated limb lose their adult structure, as well as the concomitant appearance in the region of histolysis of cells with a characteristic mesenchymatous structure. The appearance of these cells is the same as those which are found in the regeneration blastema. Their fine structure has been described in *Ambystoma* larvae by HAY [1959, 1962, 1966] and in the adult newt by SALPETER and SINGER [1960], HAY [1966] and NORMAN and SCHMIDT [1967]. The description of dedifferentiation given above is quite general, but nevertheless, it represents a characteristic morphological stage in a regenerating limb (fig. 1c) which is easily recognized and is agreed upon by virtually all students of regeneration, even though there are differing opinions concerning the designation of this phase.

Exactly what occurs at the cytological level during the overall phase of dedifferentiation has not been fully determined. A long standing, and perhaps a majority viewpoint, holds that dedifferentiation is a process by which fully differentiated nuclei in mature tissues divest themselves of their specialized cytoplasmic or extracellular trappings and revert to a morphologically less specialized state. These 'dedifferentiated' cells then accumulate apically to form the regeneration blastema. There is considerable support for the occurrence of such a process in skeletal tissues [BUTLER, 1933; THORNTON, 1953; TRAMPUSCH and HARREBOMÉE, 1965; EGGERT, 1966, and others], and this has even been confirmed by direct observation of living animals [HAY, 1962]. Whether or not skeletal muscle undergoes a like process has been a matter of considerable debate, especially since the discovery of the satellite cell [MAURO, 1961]. Despite considerable support for this concept over the years [TOWLE, 1901; THORNTON, 1938; CHALKLEY, 1954, 1959; HAY, 1959, 1962; LENTZ, 1969], there has remained an opposition viewpoint [WEISS, 1939; MANNER, 1953; NICHOLAS, 1955; TOTO and ANNONI, 1965; SCHMIDT, 1968] which has doubted either the identification of the nuclei considered to be derived from muscle fibers or the claims for viability of the nucleated fragments which break off from damaged skeletal muscle fibers. The latter group has generally considered 'dedifferentiation' to represent the activation of various connective tissue cells or reserve elements rather than the process described above. It is not within the scope of this review to become embroiled in the controversy of the cytology of the dedifferentiative process or the origin and subsequent potencies of 'dedifferentiated' cells. The importance of the resolution of this problem in terms of cytogenetic implications and in determining the exact site of action of regulatory processes and interactions is obvious. Because of the incomplete resolution of the problems in this area, particularly regarding the dedifferentiation of muscle, it will be necessary for the most part to treat dedifferentiation as an overall reaction of the distal tissues of the limb stump rather than as a specific cellular reaction.

A superficial perusal of the literature would indicate a number of apparent inconsistencies in descriptions of factors leading to and controlling the process of dedifferentiation. Upon closer examination, however, it becomes readily apparent that the factors regulating this phenomenon are different and quite consistent when the studies on amphibians are divided into two groups—those conducted upon larval forms (mainly *Ambystoma*) and those conducted upon adults. Axolotls tend to fall into the latter group. As a general rule, it is quite easy to induce dedifferentiation in larval urodeles. Not only amputation, but a number of other experimental manipulations can foster profound dedifferentiative changes which may lead to regression of the entire limb. In adult urodeles, on the other hand, there is a greater tendency for the internal tissues of the limb to remain stable in the fully differentiated state. This tendency is even more pronounced in adult anura. There is little question that damage of some sort is normally required to elicit the phase of dedifferentiation. The intensity of the dedifferentiative response depends not only upon the type and amount of damage, but it can also be modified to a great degree by supplementary experimental manipulations performed upon the limb.

In larval *Ambystoma,* early experiments by BUTLER [1933] showed that localized X-irradiation of limbs caused a profound prolongation of the dedifferentiative phase after amputation. This often led to regression up to the shoulder. BUTLER hypothesized that the primary effect of irradiation was to block differentiation, and that the lack of differentiative activity in an X-rayed amputated limb permitted the dedifferentiative process to continue unabated. Even in the absence of amputation, regression of X-rayed limbs is the rule – particularly if minor trauma is inflicted upon the distal limb segment. Similarly, ultraviolet irradiation [BUTLER and PUCKETT, 1940; BUTLER and BLUM, 1955] was found to cause an exaggerated dedifferentiative phase in amputated larval limbs as well as pronounced regression in other apparently undamaged limbs. It is of considerable interest that in the latter case of regression after irradiation of the elbow region, a substantial percentage of these limbs went on to form accessory structures at the site of the radiation-induced dedifferentiation. THORNTON

[1943] found that after limb amputation, extensive dedifferentiation occurred in *Ambystoma* larvae immersed in a 1:1500 solution of colchicine. Experiments by SCHOTTÉ and BUTLER [1941] and BUTLER and SCHOTTÉ [1941] on both *Ambystoma* and *Triturus* larvae showed that denervated limbs not only underwent dedifferentiation at the appropriate interval after amputation, but that the dedifferentiative phase was exaggerated to the point of nearly complete resorption of the experimental limbs. In contrast, no gross regression or histological dedifferentiation was noted in denervated, but otherwise uninjured limbs. SCHOTTÉ and HARLAND [1943a] observed a similar phenomenon of regression in amputated denervated limbs of anuran tadpoles. In a further investigation of this phenomenon THORNTON and KRAEMER [1951] and THORNTON [1953] found that after denervation, extreme regression could be stimulated by relatively minor pinching or piercing injuries. A participating or stimulatory role by the larval wound epidermis was ruled out in two ways. First, after crushing, dedifferentiation began despite the lack of a superficial wound and a lack of epidermal-mesodermal contact. Second, dedifferentiation did not occur after the creation of a skin wound in the absence of trauma to the underlying tissues. THORNTON [1953] considered mesodermal trauma to be the main factor in stimulating dedifferentiation in larval limbs and observed that cartilage was affected first. As the dedifferentiation of cartilage progressed and muscle attachments were loosened, dedifferentiative changes appeared in muscle. GOSS [1968, pp. 157–8] also viewed the disintegration of cartilage as a key factor in larval dedifferentiation and stressed that in several systems the absence of a nerve supply is correlated with considerable instability of cartilage. The adult limb, on the other hand, contains bone and is not so unstable in the denervated state.

In general, it appears difficult to inhibit regeneration in limbs of larval urodeles without producing the condition of regression or excessive dedifferentiation. One exception to this is the inhibition of regeneration in *Ambystoma* larvae by beryllium [THORNTON, 1949, 1950, 1951]. Although dedifferentiation began normally in beryllium-treated limbs, it generally halted without leading to extreme regression. Another example of an inhibition of regeneration without extensive regression also came from the laboratory of THORNTON [1958], who found that after ultraviolet irradiation of the wound epidermis dedifferentiation did occur, but a blastema never formed. Although a slight degree of regression was noted, the limbs stabilized and new cartilage was laid down upon former areas of dedifferentiation. Dedifferentiation was found to occur in larvae after daily removal of the apical epidermal cap [THORNTON, 1957], but its extent did not substantially exceed that of normally regenerating limbs. Its limitation was morphologically associated with the presence of an apically situated connective tissue scar.

In adult urodeles (as well as other adult species capable of regenerating limbs) it is a general rule that the dedifferentiative phase of regeneration is much easier to stop than to maintain. Many of the experimental manipulations which lead to regressive activity in larval forms simply stop the dedifferentiative process in adults. Thus, local X-irradiation inhibits regeneration in adults, but instead of leading to excessive dedifferentiation as in larval limbs, the overall phase of dedifferentiation is blocked [TUCHKOVA, 1964; ROSE and ROSE, 1965; POLEZHAEV, 1966a, b; CARLSON, 1970a]. It is interesting to note that after the inclusion into irradiated limbs of various normal tissues, such as skin [UMANSKY, 1937; TRAMPUSCH, 1951], cartilage [EGGERT, 1966; DESSELLE, 1968], muscle [THORNTON, 1942; TRAMPUSCH, 1951; POLEZHAEV and TUCHKOVA, 1968] and intact nerve [TRAMPUSCH, 1964], dedifferentiation begins and regeneration takes place. However, the origin of the dedifferentiated cells is still uncertain. It is likely that many of the dedifferentiated cells arose from the normal implants, but neither the recuperation of dedifferentiative capacity of irradiated tissues nor the migration of normal cells into the irradiated limb has been ruled out. Until this impasse is resolved, it is unwise to do more than mention these experiments in the context of this chapter.

In denervated limbs of adult newts, ROSE [1948a] stressed the continued lack of dedifferentiative activity following amputation. He noted that during several months of observation, regression was not found.

HALL and SCHOTTÉ [1951] found that prior hypophysectomy almost totally inhibits dedifferentiation in the amputated limb of the adult newt. Lack of dedifferentiation was closely correlated with the premature invasion of differentiated connective tissue elements beneath a greatly thickened wound epidermis. In later work, these authors [SCHOTTÉ and HALL, 1952] found that by delaying hypophysectomy with respect to the time of amputation, various degrees of regeneration, from abortive to complete, resulted. They considered the first 3-5 days after amputation to be the critical period. According to these authors' interpretation, the effects of hypophysectomy act directly upon the cells at the end of the limb stump and cause them to be incapable of dedifferentiation. Because of the inability of these cells to undergo dedifferentiation, a secondary process—formation of a pad of connective tissue—occurs instead. Hypophysectomy does not alter normal regeneration in larvae, but SCHOTTÉ [1961] reported that if an adult pituitary is transplanted into a larva and it is removed at a later date, regeneration of the larval limb does not occur. Contrary to the usual reaction in larvae, these inhibited limbs do not regress, but rather react like limbs of adults. Dedifferentiation is minimal, and the end of the limb stump is capped by a pad of connective tissue. Recent experiments of TASSAVA *et al.*, [1968] have not supported SCHOTTÉ'S earlier findings. These authors found no inhibition of limb regeneration after removal of adult pituitary grafts from *Ambystoma* larvae. Although it is presently somewhat difficult to reconcile the differences in the results of these two experiments, the histological results of SCHOTTE's inhibited limbs are of interest with respect to the inhibition of dedifferentiation.

Systemically applied actinomycin D also inhibits dedifferentiation in adult newts although the earlier processes of wound healing and demolition are indistinguishable from normal [CARLSON, 1966, 1967a]. The difficulties of interpreting the effects of systemically applied metabolic inhibitors at the molecular level do not allow one to conclude that the inhibition of RNA formation alone is the sole reason for the inhibition of dedifferentiation, but that remains one of the alternatives. In a continuation of this work, CARLSON [1969] reported an inhibition of limb regeneration in the axolotl after treatment of the skin (epidermis + dermis) with actinomycin D. Despite the presence of an epidermal-mesodermal interface, dedifferentiation was substantially reduced. This result is in contrast to that obtained by THORNTON [1958] in larvae on ultraviolet irradiation of the wound epidermis. As previously noted, he found that dedifferentiation occurred although this case was somewhat exceptional for larvae in that extensive regression did not occur.

In adult amphibians, at least, the apposition of a normal wound epidermis to the damaged mesodermal tissues seems to be a prerequisite for dedifferentiation. It has been shown by GODLEWSKI [1928] and others that dedifferentiation is blocked when the wound surface in the axolotl is sealed off by full thickness skin. Unpublished experiments in my laboratory have confirmed this observation in the adult newt as well as in the axolotl. POLEŽAJEW and FAWORINA [1935] considered the wound epidermis in the axolotl to be a positive factor in the stimulation of dedifferentiation of the underlying tissues. TABAN [1955] reported that the wound epithelium of *Triton cristatus* could liquify a blood clot. This represents a confirmation of the earlier results of ADOVA and FELDT [1939], who found proteolytic activity in the wound epithelium. Nevertheless, it must be conceded that despite demonstrations of several ways by which the dedifferentiative phase can be blocked, our knowledge of positive events leading to dedifferentiation is slight, at best. Most of the positive information which has been obtained has come from studies on the stimulation of supernumerary limb formation and the stimulation of regeneration in normally non-regenerating limbs.

Concerning events leading to the cessation of the dedifferentiative phase, most of the experimental work has been done on larval urodeles. The first major hypothesis in this area was proposed by BUTLER and PUCKETT [1940], who noted the close temporal relationship between the establishment of a blastema and the end of dedifferentiation. In both irradiation and in denervation experiments BUTLER [1933, 1935], BUTLER and SCHOTTÉ [1941] and SCHOTTÉ and BUTLER [1941] observed that in the absence of a blastema, regression of the amputated limb was the rule. However, if a limb already possessing a blastema was denervated [SCHOTTÉ and BUTLER, [1944 or X-rayed [LITSCHKO, 1937, regression did not occur SCHOTTÉ *et al.* [1941] tested this hypothesis directly by denervating limbs of *Ambystoma* larvae and transplanting regeneration blastemas to the cut ends of the limb stumps. They found that while older blastemas showing signs of differentiation were swept up into a massive dedifferentiative reaction, younger undifferentiated blastemas effectively limited the extent of dedifferentiation, and resorption did not occur. Later work by THORNTON and KRAEMER [1951] on the regression of denervated larval *Ambystoma* limbs following relatively minor internal injury showed that under certain circumstances the dedifferentiative process could be stopped in the absence of a regeneration blastema. This was particularly true when certain digits in the limbs did not completely resorb. These authors found a consistent correlation between the reinnervation of a limb and the cessation of dedifferentiation. Although nerve stains were not employed in SCHOTTÉ *et al.*'s [1941] study, one can assume that these limbs were truly denervated at the time of inhibition by the transplanted regeneration blastema. One possible explanation which would embrace both experimental results given above is based upon SINGER's [1965] theory of the production of the 'trophic substance' by other tissues than nerve. It is quite likely that the regeneration blastema represents a fairly concentrated source of this as of yet hypothetical substance. That such might be the case is indicated in the experiments of WOOLFITT [1968], who implanted pieces of dorsal iris into the regeneration blastema and obtained lens regeneration from them in the absence of neural retina. It is not unlikely that the neural retina also represents a rich source of such a trophic factor and that the regeneration blastema was able to act as a surrogate retina in this experiment. If this reasoning be referred back to the experiments of SCHOTTE *et al.* [1941], it is possible that the young blastemas which they transplanted served as an alternate source of trophic substance and that this compensated for the lack of nerve endings. The recent work of DECK and FUTCH [1969] tends to support this viewpoint. They infused blastemal extracts into denervated limbs of newts and observed the subsequent formation of early blastemas. These, however, exhibited little tendency to differentiate. An interesting test for this concept of alternate sources for the trophic substance would be to implant pieces of neural retina into denervated limbs of *Ambystoma* larvae and see whether post-amputational regression could be arrested. There seems to be little doubt that dedifferentiation is normally regulated by some nervous influence and that in the absence of a threshold nerve supply the process continues unchecked in larvae, but comes to a halt in adults.

In normal regeneration the dedifferentiated cells migrate toward the distal end of the limb stump and accumulate under the wound epidermis to form the regeneration blastema. This marks the end of what is often called the overall regressive phase of the regenerative process, and with the establishment of the blastema the progressive phase begins. Factors concerned with the establishment of the regeneration blastema will be taken up in the next section.

C. Blastema Formation

Dedifferentiation as a process does not simply cease to exist, but rather it becomes, in a sense, swept into and almost overwhelmed by the next major phase in the chain of regenerative events. This phase is the establishment of the blastema. The step from the peak of dedifferentiative activity to the fully established apical blastema actually represents a culmination of a number of interacting lesser processes, the full extent of which is surely not known at this time. As presently recognized, two major processes are primarily concerned with the building up of the blastema. One is an apical migration of the mesenchymatous appearing cells resulting from the dedifferentiative phase. The other is the proliferation of these same cells. Almost completely unknown is whether any segregative processes occur during the establishment of the blastema, and if so, whether they occur during the earliest accumulative phase or during some later stage at which time the cells would have acquired and/or expressed a greater amount of information relative to their ultimate fate within the limb regenerate. The recent work of STEEN [1970] certainly lends support to the likelihood of segregative reactions during the period of the early blastema.

Another facet which has attracted considerably greater attention, but which is still quite incompletely resolved, concerns the possible transfer of morphogenetic information from the limb stump to the blastema. When does this occur and how might it, if at all, affect the early establishment of the blastema? A closely related and interlocking problem is whether the regeneration blastema establishes an apical proliferation center as FABER [1959, 1965] suggests. If so, when is this center set up and what causes it to come into being? Elaboration on many of these latter questions would more logically be placed in a discussion of the realization of proliferative, differentiative and morphogenetic potentials of the regeneration blastema and will not be treated in any depth in this review. It must be kept in mind that it is not at all known how early many of these properties of the blastema are acquired or established and that some of them may play a significant role in organizing the early blastema. This section of the review will concern itself primarily with the problem of what factors in the regenerating limb bring about the transition from the dedifferentiative phase to the accumulative phase (the establishment of a blastema).

Although many descriptive aspects of dedifferentiation still remain in doubt, the bulk of recent evidence [summarized by STEEN, 1970] suggests that dedifferentiation involves essentially all the internal tissues of the limb except for endothelial cells of the vasculature, neurons and probably pigment cells. The mitotic counts of CHALKLEY [1954, 1959] and the autoradiographic evidence of HAY and FISCHMAN [1961], O'STEEN and WALKER [1961] and CARLSON [unpublished] have revealed intense and early proliferative activity of mononuclear cells in the areas of tissue dedifferentiation. Although the need for an increase in the number of potential blastemal cells is an obvious one, there is good reason for suspecting that early mitotic activity allows the dedifferentiating cells the opportunity for a 'scrambling of the genes', which may be necessary for reprogramming the synthetically active sites on the informational macromolecules. Functionally, this may be similar to the observations of bursts of mitotic activity in germinal centers of antigenically stimulated lymph nodes. Whatever the significance of the mitotic activity, the fact remains that there is established a substantial population of cells which can potentially be used in building up the blastema.

Thanks to the outstanding experiments of THORNTON, it can be considered an established fact that the wound epidermis plays a major role in causing the dedifferentiated cells to aggregate into a terminal blastema. Pronounced thickenings of the apical part of the wound epidermis had been commonly noted in histological sections of regenerating limbs in the earlier literature [e.g., METTETAL, 1939, figs. 10 and 17], and it was recognized by SAUNDERS [1948] that a thickened ridge of apical ectodermal cells has a positive influence on the outgrowth of limb buds in the chicken embryo. The importance of a thickened wound epidermis in the regenerating amphibian limb was first recognized by THORNTON [1954], who noted that regressing denervated limb stumps of *Ambystoma* larvae regenerated only if, upon reinnervation, the nerve fibers entered an epidermal area not underlain by dermis. When epidermal invasion by nerve fibers occurred, the epidermis thickened into a cap-like structure, beneath which the mesenchymatous cells liberated by the regressing limb soon accumulated. In the absence of epidermal innervation, an epidermal cap did not form, and in the absence of an apical epidermal cap a blastema did not appear. THORNTON postulated that the thickened apical epidermis somehow attracted the blastemal cells to aggregate beneath it. This hypothesis was verified in a number of ways. A blastema is not established following daily removal of the apical cap [THORNTON, 1957]. Treating the skin with actinomycin D [CARLSON, 1969] also prevents blastema formation. Additional support for this concept was obtained when THORNTON [1960a] experimentally produced asymmetrical epidermal caps and found that the subsequent blastemas were correspondingly asymmetrical. Aside from the necessity of the apical epidermal cap for blastemal cell aggregation, very little is known about this structure. The stimulus for its formation, how it produces its effect and what causes it to recede are almost completely unknown.

It has long been known that the wound epithelium is formed by a migration of epidermal cells arising from the region just proximal to the amputation surface. As is the rule for most migrating sheets of cells, the migratory cells themselves do not divide. Rather, the source of new cellular material resides in a very active zone of cell proliferation in the epithelium bordering the wound. The autoradiographic studies of HAY and FISCHMAN [1961] and HAY [1965] further revealed that significant DNA synthesis is not begun until well after the apical epidermal cap is established and a blastema has already been formed. ROSE [1948b] noted from 0.0–1.7% mitoses in the wound epithelium of newts in the epidermal mound stage (16–18 days), but no earlier counts were made. In contrast to the relative lack of mitotic activity, autoradiographic studies have shown that uptake of H^3-uridine in the apical cap is moderately increased over normal [BODEMER, 1962] whereas protein synthesis, as indicated by incorporation of thio-amino acids and leucine [BODEMER and EVERETT, 1959; ANTON, 1965], is intense, particularly in the basal layers of the apical cap [ANTON, 1968].

Most of the information concerning the nature of the stimulus which leads to the establishment of the apical epidermal cap is indirect. After the demonstrations by SINGER [1949] and TABAN [1949] and others of the penetration of the wound epithelium by nerve fibers, it was generally conceded that some influence from the nerve directly stimulates the epidermis to thicken (and presumably, to express special characteristics). Subsequent work, however, indicated that epidermal innervation was not the *sine qua non* which was once believed. YNTEMA [1959a, b] demonstrated that larval limbs, aneurogenic from birth, maintain the capacity to regenerate. SIDMAN and SINGER [1968] found that partially denervated limbs, with only the motor components remaining, did regenerate. Both SINGER [1959] and THORNTON [1960b] have reported that motor fibers do not penetrate the wound epidermis. Working with partially denervated limbs, SINGER and INOUE [1964] observed the appearance of an apical cap not followed by blastema formation. Thus, we know that conditions compatible with the appearance of an apical cap in a distal wound epithelium are the presence of an adequate number of nerve fibers either within the epidermis or in the mesodermal tissues beneath, or the total absence of nerves throughout the history of the limb.

In contrast, an apical cap does not form in denervated limbs [THORNTON, 1954] or in aneurogenic limbs to which previously innervated skin is transplanted [STEEN and THORNTON, 1963]. ROSE and ROSE [1967] have reported that in X-irradiated limbs of newts, epidermal innervation is not present, but that after covering the x-rayed mesodermal portion of the limb with normal epidermis, invasion of the epidermis by nerves occurs and regeneration takes place. Recently, THORNTON and THORNTON [1970] orthotopically transplanted aneurogenic forelimbs to normal larvae. If the previously aneurogenic limbs were allowed to become innervated by the host for at least 13 days and were then denervated and amputated, apical cap formation and subsequent regeneration failed. However, if such previously aneurogenic limbs were denervated and maintained in the denervated state for at least 30 days before amputation, they were once again able to form apical caps and to regenerate despite the lack of innervation.

It must also be kept in mind that apical cap formation fails in the absence of an intimate epidermal-mesodermal interface. Whether this is due to the absence of epidermal-mesodermal contact or to the accompanying lack of innervation or to both has not been directly tested.

To summarize conditions leading to apical cap formation, experimentation to date has shown that in an amputated limb both mesodermal contact and innervation are normally required. It is widely believed that the nerve supplies a trophic substance which acts (directly or indirectly?) upon the epidermis, thus imparting to the epidermis the competence to direct the next phase of the regenerative process [see ROSE, 1962, 1964; SINGER, 1965; THORNTON, 1968]. The ability of an aneurogenic epidermis to circumvent the need for activation by nerves has been attributed to the production of the 'trophic substance' by non-neural tissues in increased quantities when the dominating influence of the nerve is absent [SINGER, 1965; THORNTON and THORNTON, 1970]. In this latter case the tissue source of trophic substance has not been determined. If the epidermis be postulated, one would be faced with the rather unusual situation of a tissue stimulating itself to undergo an apparently qualitatively different developmental process. Another possibility within the limits of present experimental data would be that the mesoderm supplies the actual stimulus for apical cap formation and that the neural-epidermal relationship may permit the wound epidermis to become receptive to a stimulus emanating from the underlying mesoderm. An early experiment of POLEZHAEV [1933b], however, indicates that the nerve may act directly upon epidermis to produce thickening (and possibly the apical cap). In this experiment POLEZHAEV removed all of the structures from a hind limb of an axolotl except for the nerve and a flap of skin. The nerve was tied to a suture and was deviated under the skin at the base of the tail. A hole was made through the skin with a needle, and the end of the nerve was pulled through the hole. The wound created by removal of the limb was covered with the skin flap and the portion of the nerve which had been pulled through the hole in the skin at the base of the tail was cut off even with the skin surface. Epithelialization occurred over the cut end of the nerve. Since there was essentially no damage to the underlying mesodermal tissues during the deviation procedure and since the end of the nerve was effectively separated from underlying areas by a tight cuff of dermis, the net result was an almost pure nerve-epidermal interface. Histological preparations 3 months after the initial operation revealed that the epidermis overlying the end of the nerve was several times thicker than normal. It was not stated whether the thickening was due to mitotic activity. It is of interest to compare POLEZHAEV's results with those of OVERTON [1950], who observed the stimulation of mitoses in the epidermis of the dorsal fin of *Ambystoma* after implanting grafts of spinal cord beneath. A difference between these results and the autoradiographic demonstration by HAY and FISCHMAN [1961] that apical cap formation in the newt is not preceded by significant mitotic activity will have to be resolved before a satisfactory resolution of the mechanism of nerve stimulated apical cap formation is possible.

If the exact set of reactions which cause the apical epidermal cap to appear is still quite obscure, the means by which the apical cap exerts its effects are even more so. The early knowledge of epidermal innervation led to the tentative hypothesis that the nerve fibers extending toward the apical cap might act as guide wires for directing blastemal cells toward their destination. YNTEMA's [1959a, b] studies of regeneration in aneurogenic limbs effectively eliminated that as a viable alternative.

It is commonly hypothesized that the apical cap exerts its influence upon blastemal cell aggregation by chemical means. Direct evidence for this hypothesis is totally lacking, but the heightened RNA [BODEMER, 1962] and protein [BODEMER and EVERETT, 1959; ANTON, 1965, 1968] synthetic activity of the apical epidermis provides reasonable grounds for such an assumption. Since any chemical activity would have to be directed toward the underlying tissues and would likely involve the exportation of the product, it would be of interest to determine whether the Golgi apparatus of the cells in the epidermal cap is located basal to the nuclei. Such is the case in the kidney where the macula densa abuts upon the afferent arteriole in the juxtaglomerular apparatus, and in a recent report TRELSTAD [1970] has shown a similar situation in the corneal epithelium of the chick. The likelihood of the secretion of basement lamellar material by the epidermis [HAY and REVEL, 1963; HAY, 1964] makes it difficult to interpret the real significance of any cytological or cytochemical findings related to the wound epidermis. Attempts in my laboratory to produce a blastema in a limb with a Millipore filter inserted between the wound epidermis and the underlying tissues have thus far been unsuccessful. However, cellulose acetate is a poor substrate for epithelial migration, and in most cases obvious disturbances in the wound epidermis have been present.

The basal cells of the apical epidermal cap are in intimate physical contact with the underlying mesoderm [SCHMIDT, 1968, p. 76], but little is known about the anatomical nature or functional significance of this arrangement. The recent studies of intercellular communication as a morphogenetic mechanism [LOWENSTEIN and PENN, 1967] have not been extended to the regenerating limb.

Another factor which may relate to the epidermally directed distal migration of blastemal cells is the presence of differences of electrical potential (30 mv) between the midline (+) and the limb tip (−) of normal animals [BECKER, 1961]. Following amputation, the potentials change in a characteristic pattern, but it has not yet been possible to demonstrate a direct causal relationship between this phenomenon and any of the processes leading to blastema formation. ROSE [1964] has discussed some possible ways in which BECKER's findings might be related to tissue interactions in the early regenerating limb.

As elongation of the blastema begins and the digital primordia appear, the apical epidermal cap gradually loses prominence until the terminal epidermis may become somewhat thinner than that along the lateral surfaces of the regenerate. Virtually nothing is known about factors causing the apical cap to recede. Not only has essentially no experimental work been conducted on this phase of the life history of the apical cap, but there are no extensive published descriptions of the morphology of the receding apical cap or of its correlation with the exact developmental phases occurring in the mesodermal tissues beneath.

In normal limbs, the presence of nerves is essential for blastema formation. The role of nerves has been documented in many reviews [ROSE, 1948a; SINGER, 1952, 1959; THORNTON, 1968] and material covering the importance of nerves will not be repeated here. However, recent work is beginning to accumulate data on the mode of action of nerves. Aside from the possible direct or indirect interaction with the wound epidermis, nerves seem to have a direct action upon the newly forming blastema. It has been generally recognized that nerves exert a stimulatory effect upon cell division [SINGER and CRAVEN, 1948]. OVERTON [1950] showed that implants of spinal cord stimulated cell division in the overlying epidermis in *Ambystoma* larvae, and SINGER *et al.* [1964] have presented evidence compatible with localized areas of mitotic activity around the ends of nerves in early blastemas. In recent work, DRESDEN [1969] has demonstrated a general decrease in macromolecular synthesis (DNA, RNA, protein) in denervated regenerates in the early palette stage. LEBOWITZ and SINGER [1970] found a decrease in protein synthesis in the denervated early bud blastema and were also able to restore partially the level of protein synthesis by infusing nerve homogenates into the denervated limbs. The nerve is generally assumed to exert its effect by liberating a trophic substance, but despite intensive effort, its isolation has not yet been reported. Nevertheless, experimental evidence favors this interpretation of nerve action upon the early blastema [THORNTON, 1970; THORNTON and THORNTON, 1970]. VAN ARSDALL and LENTZ [1968] have described membrane-bound granules in the nerve fibers of newt limbs during the early post-amputational period. They have interpreted their findings as being compatible with a secretory function of the nerves. It is still too early to determine which of the supposedly neurotrophic effects on the regenerating limb are mediated directly by nerves and which are possibly secondary effects, but the application of biochemical methods to the problem may hasten its solution.

IV. Supernumerary Limb Formation

The study of supernumerary limb formation allows one to view the early regenerative process from a slightly different vantage point. Certain characteristics of early post-amputational regeneration (such as gross loss of structure) assume a lesser degree of importance whereas the need for other factors (such as nerves) is reinforced upon the observer.

A fundamental question relating to supernumerary limb formation concerns the presence or absence of a morphogenetic negative feedback system in the intact limb. Does a given experimental manipulation actually trick a limb into replacing a part which it 'thinks' is missing because of a break in communication or does the manipulation cause a set of positive morphological reactions after which the limb cannot help but undergo a regenerative response? The first instance assumes that a condition of morphological homeostasis (i.e., the intact limb) exists and is communicated to the tissues of the limb as a negative message. When something (e.g., amputation or any of the procedures listed in table I) disrupts the homeostatic condition by either removing a part of the structure or by disrupting the lines of communication, the remaining tissues of the limb are permitted to mobilize themselves to restore homeostasis by either forming a new structure or by re-establishing a satisfactory form of communication. The second instance assumes that in a fully developed limb a continuous negative feedback is no longer operative or necessary to maintain normal structure, but that when certain conditions (e.g., the combination of mesodermal damage, a wound epidermis and adequate innervation) are present, normally quiescent cells are in a sense drawn into a regenerative reaction.

Table 1. Asummary of major methods used to stimulate supernumerary extremities in post-embryonic amphibians

Method	Selected references
A. Severe trauma	
1. Ligatures around limbs	DELLA VALLE, 1913; NASSONOV, 1930
2. Chewing injuries	PRZIBRAM, 1921; BRUNST, 1961
3. Skeletal injury	TORNIER, 1897, 1898; NASSONOV, 1936a
B. Nerve deviation	LOCATELLI, 1929; GUYENOT *et al.,* 1948; KIORTSIS, 1953; BODEMER, 1958
C. Tissue implants	BREEDIS, 1952; RUBEN, 1955
1. Neoplastic	NASSONOV, 1934a, b, 1938a, b, c, 1941; RUBEN and FROTHINGHAM, 1958; RUBEN and STEVENS, 1963; BALLS and RUBEN, 1964; STEVENS *et al.,* 1965; CARLSON, 1967b, 1968, 1971
2. Normal	
D. Celloidin implant	RUBEN, 1957
E. Nerve deviation + tissue implant	BODEMER, 1959, 1960
F. Implants of carcinogens	BREEDIS, 1952; RUBEN and BALLS, 1964
G. Irradiations	
1. Ultraviolet	BUTLER and BLUM, 1955, 1963
2. X-ray	BRUNST, 1950a, b; BRUNST and FIGGE, 1951
H. Apical epidermal cap transplantation	THORNTON and THORNTON, 1965
I. Limb transplantation	CARPENTER, 1932; SWETT, 1932; LECAMP, 1935; YNTEMA, 1962; THORNTON and TASSAVA, 1969
J. Skin transplantation	GLADE, 1957; DROIN, 1959
K. Alterations to limb skeleton	
1. Removal of proximal portion of humerus	STUDITSKY, 1948
2. Separation of distal epiphysis from shaft in humerus or femur	PURDY, 1967

The extent of the reaction would depend upon the number of cells directly involved and the amount of support (morphogenetic, trophic, etc.) supplied by the limb. As will be seen in this section, the evidence seems to favor the latter viewpoint.

As a prelude to this discussion, histological studies have established rather conclusively that the basic morphological features of supernumerary limb development are those which occur in the regenerating limb [KASANŻEFF, 1930; BODEMER, 1958; BUTLER and BLUM, 1963; CARLSON, 1967b]. In addition, PEREDELSKY [1940a] found that the ability of anuran tadpoles to produce implant-induced accessory structures was lost at the same stage as the ability to regenerate amputated limbs.

From the histological descriptions present in the literature, a number of host reactions seem to be common to most systems of experimentally stimulated supernumerary limb formation. Most consistently reported is a traumatic reaction of the underlying mesodermal tissues leading to dedifferentiative changes. In some cases direct mechanical trauma appears to be the primary stimulus leading to dedifferentiation. This is particularly evident in the cases of supernumerary limb formation following the application of a ligature to a limb [DELLA VALLE, 1913; NASSONOV, 1930; KASANZEFF, 1930]. BODEMER [1958, 1960] found it necessary to traumatize severely the muscle underlying the ends of deviated nerves, or supernumerary limbs would not form in the adult newt. This trauma led to massive dedifferentiation of the damaged muscles. In a recent report, THORNTON and TASSAVA [1969] described in several species of *Ambystoma* the formation of supernumerary limbs from both aneurogenic and normal limbs transplanted as homografts or heterografts to normal hosts. In their interpretation of the mechanism, they considered the aggregation of cells derived from traumatized mesodermal tissues to be a primary step in the process.

After the injection of carcinogens into the limbs of adult newts, BREEDIS [1952] observed non-specific inflammation of long duration associated with the foreign material. In some cases degenerative changes resembling dedifferentiation occurred in muscle and bone. These changes were followed by the appearance of a regeneration blastema. Such regenerative changes were produced much more frequently in association with certain carcinogens (e.g., coal tar fractions) than with others (acetylaminofluorene or scarlet red in olive oil) and were not correlated with the amount of tissue destruction.

Following ultraviolet irradiation of larval *Ambystoma* limbs, BUTLER and BLUM [1955, 1963] noted extensive cytolysis and dedifferentiation in the irradiated areas. Since epidermal changes occurred first, it is not possible to state with certainty that the mesodermal cytolysis and dedifferentiation were direct sequelae of the ultraviolet irradiation. After irradiating larval axolotl tails with low doses of x-radiation, BRUNST [BRUNST, 1950a, b 1952; BRUNST and FIGGE, 1951] obtained a substantial incidence of supernumerary limbs and tails, which arose at the border between the irradiated and unirradiated tissues. This border region, called by BRUNST the 'zone of stimulation', is characterized by high mitotic activity of not only mesodermal tissues, but also of the neural tube. BRUNST [1952] postulated that necrohormones or injury products from irradiated cells act upon the healthy tissues in the zone of stimulation to initiate the regenerative changes which bring about the formation of accessory structures.

By far the greatest amount of work has been done in the area of implant-induced supernumerary limb formation. Although a number of general interpretations of the mechanisms operative in this process have been proposed, there is general agreement among investigators that tissue destruction is necessary for the initiation of the regenerative process. In contrast to the situations described above, the important factor in implant experiments seems to be destruction of the implant rather than a destruction of the mesodermal tissues of the limb because there is no positive correlation between the amount of tissue destruction in the host and the incidence of supernumerary limb formation [BREEDIS, 1952; RUBEN and STEVENS, 1963]. It has also been noted that with the exception of bone and cartilage implants, an implant which 'takes' is not likely to stimulate accessory growth [NASSONOV, 1938c; RUBEN, 1960]. Exactly what the implant does has been a matter of conjecture for years. NASSONOV [1938c] propounded a very elaborate theory in which specific organizers (in the embryonic sense) were released from inactivators during degeneration of the implant. The organizers were said to act upon the mesodermal tissues of the host whereas a 'coriocide', also released from the degenerating implant, causes a decomposition of the dermis. NASSONOV also postulated that epidermal invagination and evagination factors are present in the implanted tissues.

In 1960 RUBEN proposed an immunobiological model of implant-induced supernumerary limb formation according to which a tissue implant elicits from the host a foreign body response which proceeds to destroy the implant. Then cytolytic products from the degenerating implant act directly upon the peripheral nerves of the host. As did NASSONOV [1938c], RUBEN stressed the importance of cytolytic products, working either alone or through peripheral nerve effects, in bringing about a dissociation of the dermis as well as other tissues of the host. BODEMER [1960] stressed the relationship between a cytolyzing implant and the appearance of mesodermal dedifferentiation and also indicated that the effect might be either direct or mediated through the peripheral nerves.

Despite varying opinions concerning the mechanism of implant-induced supernumerary limb formation, preliminary work by several investigators on a number of implant systems has suggested that some type of chemical factors(s) is involved. Inductive activity persists after freezing [STEVENS *et al.*, 1965] and lyophilization [CARLSON and MORGAN, 1967] as well as in homogenates of newt muscle [BODEMER, 1959] and frog kidney [CARLSON and MORGAN, 1967]. In every case tested the stimulatory ability of tissues has been abolished by boiling (axolotl cartilage [NASSONOV, 1936b]; newt bone and cartilage [BODEMER, 1959]; frog kidney [STEVENS *et al.*, 1965; CARLSON and MORGAN, 1967]). In further work on axolotl cartilage, KUZMINA [1940] found that inducing activity is lost after exposure to temperatures of 50 °C and higher whereas it is present after heating in temperatures of 45 ° or less. Workers in NASSONOV'S laboratory [reviewed by FEDOTOV, 1946; ZELINSKY, 1946] found that a number of acid hydrolysates of cartilage were quite active, and polypeptides derived from implant proteins were suspected to be the inductive agents.

Recently, CARLSON [1971] has determined the anatomical distribution and relative potency of inductive capacity in 20 tissues and organs of *Rana pipiens*. It ranged from 100% in lung and over 90% in the urinary system to 0% in brain, nerve, skeletal muscle and skin. Moderate inductive powers were found in the digestive and female reproductive tracts, in spleen and in the fat body. It is also known that inductive ability in frog kidney increases from a relatively low level in the tadpole to adult levels early in metamorphosis [CARLSON, 1968]. These findings, coupled with the poor stimulatory effects of implanted tissues from many other animals including homografted tissues, strongly indicate that there are chemical entities released from certain normal or degenerating cells which are specifically capable of stimulating dissociative or dedifferentiative reactions in mesodermal tissues. It is noteworthy that ATTARDI *et al.* [1965] have recently described a mesodermal dedifferentiating factor which was isolated from mouse submaxillary glands. If this concept is valid and can be related to implant-induced supernumerary limbs, it would be most interesting to find out whether dedifferentiation could be stimulated in the absence of gross mechanical trauma even though such a mechanism may not be at all related to events in the amputated extremity.

Although there is little doubt that damage to mesodermal tissues is a common feature of almost all supernumerary limb-forming systems studied to date, there is also considerable evidence that damage alone does not constitute an adequate stimulus. Thus, BREEDIS [1952] was unable to stimulate accessory growth in adult newts by fracturing the humerus or by deep burns. CARLSON [unpublished] injected trypsin or pronase into forelimbs of newts and failed to elicit a regenerative response even though a violent inflammatory reaction followed injection of the enzymes. Neither BREEDIS [1952] nor RUBEN and STEVENS [1963] found a positive correlation between the amount of mesodermal destruction and supernumerary limb production in their experiments. Finally, CARLSON [1970b] found that mincing the entire pubo-ischio-tibialis muscle in the axolotl hind limb was not sufficient to elicit an epimorphic regenerative response. Thus, even though most methods used to produce supernumerary limbs exert a rather profound effect upon mesodermal tissues, there must be a more widespread response of other limb components if a full supernumerary regenerative response is to occur.

Almost all investigations of supernumerary limb formation have stressed the importance of alterations in the normal skin morphology in order for regenerative changes to occur. In most cases, however, it is difficult to say whether the initial stimulus acts upon the skin directly or whether this reaction is secondary to others which have occurred in the underlying tissues. In the case of supernumerary limb formation after ultraviolet irradiation of *Ambystoma* larvae, there can be no doubt that the skin was the primary target of the stimulus. Within 24 hours after irradiation, the epidermis in the irradiated area had blistered and sloughed off [BUTLER and BLUM, 1955, 1963]. Within a few days the epidermal defect was filled in with a wound epidermis presumably derived from protected skin. The wound epidermis, not underlain by a basement membrane, was thicker than normal and eventually became innervated. One week after irradiation, regression of underlying tissues began, and in the third week a blastema had become established. From the information given in the reports, it was not possible to estimate accurately the depth of penetration by the ultraviolet rays. Thus it is not known whether the mesodermal reaction was a direct result of the irradiation or if it was a secondary effect mediated through the damaged epithelium. In the work of THORNTON [1958] it was determined that ultraviolet rays did not penetrate past the epidermis, but the reaction of his animals does not seem to have been so severe as that elicited in the experiments of BUTLER and BLUM.

BALINSKY [1956] stressed the importance of direct epidermal-mesodermal contact in embryonic limb development and related this concept to supernumerary limb formation. In axolotls, both PURDY [1967] and CARLSON [unpublished] have found that the epidermal-mesodermal contact following skin removal alone did not prove to be a sufficient stimulus for accessory limb formation. Nevertheless, direct epidermal-mesodermal contact has been stressed by many as an indispensable condition for supernumerary limb formation. In some experimental models a wound epidermis has been a concomitant of the stimulating procedure (e.g., in implantation experiments). In the absence of a skin wound close to a tissue implant, dissolution of the dermis is a prerequisite for a regenerative response [PEREDELSKY, 1940b; RUBEN and FROTHINGHAM, 1958]. This reaction is probably not the primary stimulus for regeneration although NASSONOV [1938c] placed great emphasis upon it. In the case of supernumerary limbs forming at the site of a tight ligature placed around a limb, dissolution of the dermis is one of the primary morphological events [KASANZEFF, 1930].

Another major method of stimulating supernumerary limb formation is the deviation of peripheral nerves to some site other than their normal course [LOCATELLI, 1929; GUYÉNOT *et al.*, 1948; KIORTSIS, 1953; BODEMER, 1958]. Unfortunately, this method does not provide as straightforward a means of analyzing the problem of the stimulus for supernumerary limb formation as it may appear because deviation of a nerve alone does not constitute an adequate regenerative stimulus [BODEMER, 1959]. Two other conditions are necessary for consistent formation of supernumerary limbs. First, a wound epidermis must be present over the end of the denervated nerve [LOCATELLI, 1929; POLEZHAEV, 1933b; BODEMER, 1958, 1959, 1960]. If a layer of full thickness skin is left or secondarily heals over the nerve, regeneration does not occur [GUYÉNOT *et al.*, 1948]. Along with an overlying wound epidermis, damage to the underlying mesodermal tissues is normally needed. In adult newts, BODEMER [1958, 1959] found that extensive mechanical trauma to the musculature around the end of the deviated nerve was essential. In the absence of trauma the wound epidermis thickened and became invaded by nerve fibers, but only limited dedifferentiation of the soft tissues occurred. He also found [BODEMER, 1959] that implantation of certain types of homografted tissues at the tip of the nerve could substitute for the mechanical traumatization of muscle.

According to most reports, the skeletal tissues play a minor role in the initiation of a supernumerary regenerative response [BODEMER, 1958,

CARLSON, 1967b]. There are, however, reports of accessory structures arising after alterations confined primarily to skeletal tissues. In the axolotl, STUDITSKY [1948] removed the proximal part of the humerus and described the structure of an accessory limb containing both bone and muscle which arose in the region of the removed bone. It was connected to the cartilaginous callus which had formed around the cut end of the bone. Because portions of certain muscles in the area were cut off in the initial operation and because developmental stages were not described, it is difficult to determine the role of the removed skeletal part as a stimulatory factor. Recently, PURDY [1967] produced a low percentage of accessory structures in young axolotls by separating the distal articulating heads from the shafts in either the humerus or the femur, but this procedure was ineffective if the skin overlying the end of the bone was unbroken. When dorsal skin wounds were created over the distal end of the bones, supernumerary regenerative responses occasionally occurred.

Analysis of the various methods used to stimulate supernumerary limb formation reveals no obvious common features in the stimuli themselves which might explain their effectiveness. With respect to their influence upon the responding system, a feature common to all which were extensively studied was an effect on the soft mesodermal tissues, leading to dedifferentiation. In some cases this was preceded by mechanical trauma; in other cases by an apparently non-specific local cytolysis; and in still others the introduction of a fairly specific dedifferentiation-promoting stimulus remains a possibility. Damage to mesodermal tissues, however, is not enough to elicit a regenerative response. The presence of a wound epidermis and an adequate nerve supply are indispensable 'co-factors' which must accompany the mesodermal damage. Thus, RUBEN [1959] found that in a limb bearing a frog kidney implant and possessing a wound epidermis, withdrawal of innervation before a critical time inhibited blastema formation even though dissociation of soft mesodermal tissues had occurred. Conversely, both NASSONOV [1938c] and RUBEN and FROTHINGHAM [1958] stressed the need for a wound epidermis to be present in limbs bearing tissue implants and possessing normal innervation. With the possible exception of late embryonic stages [BALINSKY, 1956], a wound epidermis alone is not an adequate stimulus. In an experiment conducted for a different purpose, ROSE and ROSE [1965] allowed limb stumps of adult newts to be covered entirely with a wound epidermis derived from the shoulder region. They did not note the formation of accessory limbs under these conditions. A wound epidermis plus a nerve in the absence of mesodermal trauma is likewise ineffective [BODEMER, 1959] as is the combination of a wound epidermis plus severe muscle trauma without the presence of a deviated nerve [CARLSON, unpublished observations on mature axolotls —200 mm]. The experiments involving the deviation of a nerve as a primary stimulus also reveal a dependence upon other factors, for a deviated nerve in the absence of a wound epidermis does not lead to regeneration [GUYÉNOT *et al.*, 1948] nor does a deviated nerve plus a wound epidermis in the absence of trauma [BODEMER, 1959].

From the work done in the field of supernumerary limb formation so far, it is apparent that no single stimulus which involves only one component of a limb is adequate to bring about a regenerative response. At least three components of the limb (the nerves, the epidermis and the soft mesodermal tissues) must react in concert with one another. Of these components a relative excess of a stimulus to soft tissues (e.g., mechanical damage or tissue implants) can to a certain extent compensate for a quantitatively insufficient nerve supply [BODEMER, 1960], whereas under some circumstances (particularly in the axolotl) an excess nerve supply appears to be able to overcome a low level mesodermal reaction. The work of BODEMER [1960] shows that certain types of tissue implants can effectively replace the need for direct mesodermal trauma, and that of RUBEN and STEVENS [1963] clearly shows that the amount of trauma is not correlated with the success of supernumerary limb induction. These authors [RUBEN and STEVENS, 1963] also demonstrated that an implanted tissue does not bring about an increase in the number of local nerves. These findings, along with the recent results of CARLSON [1968, 1971], strongly indicate that in certain types of tissues there are chemical factors emanating from normal or degenerating cells which can specifically supplant direct mechanical trauma in stimulating a mesodermal dedifferentiative response. No report giving quantitative reactions between the amount of wound epidermis and other components of the system exists in the field of supernumerary limb formation.

It is difficult to compare the effectiveness of many of the stimulatory methods which have been reported because of inherent differences in experimental animals. In larval *Ambystoma* dissociative and degenerative changes in the mesodermal tissues occur much more readily and in response to apparently more minor stimulatory influences than is the case in adult newts. In axolotls the nerve is apparently able to provide a relatively greater cellular contribution to supernumerary growth, reducing the quantitative need for mesodermal dissociation. In adult newts the differentiated state of the mesodermal tissues is relatively stable and requires a greater stimulus as well as the cooperation of other tissues to bring about a dedifferentiative reaction.

Despite the difficulties in comparing some of the experimental models described above, several generalizations can be made. In no case, except for some of the relatively pure carcinogens, is the exact stimulatory agent known. In most cases the target tissue seems to be the soft mesodermal tissues. In several cases the epidermis seems to be another primary or at least a secondary target. Although the role of the peripheral nerve as a target organ has been more difficult to demonstrate, this possibility is not unlikely. It is quite clear that no supernumerary regenerative response is possible (except for aneurogenic limbs) without the interplay of a wound epidermis, the peripheral nerves and a dissociating population of mesodermal cells.

V. Stimulation of Regeneration in Higher Vertebrates

A. The Histology of Non-Regenerating Limbs

The major histological features of non-regenerating limbs are remarkably similar whether the limb be one which does not normally regenerate (e.g., post-metamorphic frogs, mammals) or one in which the natural regenerative capacity has been inhibited (e.g., limbs in newts after x-irradiation, denervation or hypophysectomy).

The early phases of epithelial healing and removal of cellular debris at the amputation surface are normally quite similar to those stages in a regenerating limb. From this point on, the non-regenerating limb is characterized by a relatively static morphological picture in which the damaged tissues are rather quickly repaired by tissue regenerative processes after a minimal amount of dedifferentiative activity. The epidermis is not invaded by appreciable numbers of nerve fibers, nor does it thicken to form an apical cap [THORNTON, 1956]. Epithelial tongues may be directed toward damaged tissues lying beneath the wound epidermis. In comparison with regenerating limbs, the wound epidermis becomes quickly underlain by a newly differentiated dermis which effectively seals off an epidermal-mesodermal contact. During the time when the mesodermal tissues in a regenerating limb are undergoing an extensive loss of structure or dedifferentiation, these same tissues in a non-regenerating limb are undergoing individual healing or tissue regenerative responses. Within a week or two the ends of the bones in the limb stump of a frog or a mammal are surrounded by a rapidly differentiating cap of cartilage, which is formed in the same manner as a callus during fracture healing [ROSE, 1944; POLEZHAYEV, 1946d; SCHOTTE and SMITH, 1959]. The injured muscle fibers undergo a local regenerative response which is well underway during the period when massive regressive changes are occurring in the muscles of regenerating limbs. Even the connective tissue in a non-regenerating limb produces a characteristic dense cap-like scar between the end of the bone and the wound epithelium. Several authors [SCHOTTÉ and HARLAND, 1943b; ROSE, 1944; THORNTON, 1956; SCHOTTÉ and SMITH, 1961] have noted the presence of fluid-filled cavities between the wound epidermis and the underlying tissues of the limb stump. They have speculated that these cavities may interfere with normal epidermal-mesodermal interactions.

The general description given above applies to limbs with almost no epimorphic regenerative activity. Some potentially instructive examples of incomplete regeneration of metamorphosing tadpole limbs have been described by ROSE [1948a] and THORNTON [1956]. These authors have described cases in which regeneration has proceeded fairly normally up to a certain point and then progressed no further. THORNTON [1956] has even described a limb in which an elongating blastema had formed, but which had failed to differentiate.

B. Methods of Stimulating Limb Regeneration

A summary of most of the techniques which have been used in attempts to stimulate limb regeneration in higher vertebrates is given in table II. This field has also been reviewed by POLEZHAEV [1946a, 1968], SINGER [1954] and ROSE [1964].

In terms of understanding the stimulatory events in limb regeneration, many of these experiments are quite instructive, but in others no satisfactory explanation of the mechanisms involved can yet be offered. As a prelude to this discussion, one must be very careful in comparing the results of different investigations. It is apparent that some methods have succeeded in eliciting true epimorphic regenerative responses in which the cells of the regenerate arise from a blastema, whereas in other experiments the regeneration seems to have been of the tissue-specific variety, which may not be subject to the same controls and releases as the former type.

Despite the apparent diversity of the stimulatory methods listed in

Table II. Summary of major methods used to stimulate regeneration in limbs normally not capable of regeneration

Method	Selected references
I. Anuran amphibians	
1. Prolongation of regenerative capacity	
A. Repeated amputation	MARCUCCI, 1915; POLEZHAEV, 1936b; POLEZHAEV and MOROSOV, 1941
B. Transplantation	LIOSNER, 1931; YAKOVLEVA, 1938; POLEZAJEW and GINZBURG, 1939
C. Re-regeneration	POLEZHAEV, 1933a
2. Stimulating regeneration	
A. Replacing skin with tadpole skin	GIDGE and ROSE, 1944
B. Repeated skin removal	ROSE, 1944
C. Mechanical trauma	POLEZHAEV, 1939, 1946b
D. Chemical irritation	
a) NaCl	ROSE, 1942, 1944, 1945; POLEZHAYEV, 1946c; SINGER *et al.*, 1957
b) Na_2CO_3 and lactose	POLEZHAEV, 1946b
c) Glucose, iodine, nitric acid and colchicine	POLEZHAEV, 1945b, 1946c
E. Augmentation of nerve supply	SINGER, 1954; KONIECZNA-MARCZYNSKA and SKOWRON-CENDRZAK, 1958; KUDOKOTSEV, 1965
F. Adrenal transplants	SCHOTTÉ and WILBER, 1958; MIZELL, 1963
G. Ultraviolet irradiation	ROGAL, 1951b
H. Tissue implants or tissue extracts	POLEZHAEV and RAMYENSKAYA, 1950; MALININ and DECK, 1958; MALININ, 1960; KUDOKOTSEV, 1960; CARLSON, unpuplished
I. Vitamin D deficiency	ROSE, 1944
J. Electrical stimulation	BODEMER, 1964; SMITH, 1967
K. Electrical stimulation and NaCl	BODEMER, 1964
L. Heat shock	ROSE and ROSE, 1947
M. Removal of a bone	GOSS, 1953
N. RNP injection	SMITH and CRAWFORD, 1969
II. Lizards and mammals	
A. Repeated skin removal	UMANSKY and KUDOKOTSEV, 1948
B. Nerve deviation	SIMPSON, 1961; SINGER, 1961; KUDOKOTSEV, 1962
C. Parathyroid hormone	UMANSKY and KUDOKOTSEV, 1951
D. ACTH and cortisone	SCHOTTÉ and SMITH, 1961
E. Trypsin and $CaCl_2$ applied to wound surface	SCHARF, 1961, 1963; KUDOKOTSEV and KUNTSEVICH, 1965; KUDOKOTSEV, 1966
F. Implants of brain and spinal ganglia	MIZELL, 1968; MIZELL and ISAACS, 1970
G. Avitaminosis A and D	ROGAL, 1951a
H. Injections of vitreous body	POLEZHAEV, 1968

table II, a closer examination of the experiments indicates that most of them can be assigned to one of four categories with respect to their effects upon the responding system. These categories are: 1) manipulations affecting epidermal-mesodermal relationships; 2) attempts to increase the nerve supply; 3) traumatic stimuli to mesodermal tissues, and 4) attempts to reduce the formation of scar tissue.

The first of these general methods is to establish a sufficiently great area of epidermal-mesodermal contact to support a regenerative response. Long ago it was recognized by MORGAN [1908] and later by SCHOTTÉ and HARLAND [1943b] and ROSE [1944] that in postmetamorphic frogs the relatively loose skin of the limb tends to pinch off the amputation surface at the expense of the wound epithelium. ROSE [1944] and UMANSKY and KUDOKOTSEV [1948] attacked this problem directly by repeatedly removing the old skin which encroached upon the wound surface. This procedure led to regeneration of limbs with defective distal parts in frogs [ROSE, 1944], whereas UMANSKY and KUDOKOTSEV [1948] described the stimulation of tail-like regenerates containing cartilage, muscle and nerves in the limb of the lizard, *Lacerta.* Also working on *Lacerta,* POLEZHAYEV [1946b] was unable to induce regeneration by cutting off the edges of the old skin or by transplanting skin of the tail (which does regenerate) to the limb. Immersion of limb stumps in saturated solutions of NaCl appears to stimulate regeneration to a large extent by means of its effect upon the wound epidermis [ROSE, 1942, 1944, 1945; POLEZHAYEV, 1946c; SINGER *et al.,* 1957], but a substantial mesodermal inflammatory response also occurs. SINGER *et al.* [1957] placed greater emphasis upon the mesodermal effects and related this to the lowering of the threshold of nerve fibers required for limb regeneration. Even the somewhat unlikely condition of vitamin D deficiency in frogs appears to allow regeneration by preventing the encroachment of full thickness skin over the amputation surface [ROSE, 1944].

Striking improvements in the regenerative response of anuran limbs have been effected by augmentation of the nerve supply to the limb stump (SINGER [1954] in *Rana;* KONIECZNA-MARCZYNSKA and SKOWRON-CENDRZAK [1958] in *Xenopus;* and KUDOKOTSEV [1965] in *Bombina*). This procedure has been extended to lizards with some success [SIMPSON, 1961; SINGER, 1961; KUDOKOTSEV, 1962], but its application to mammalian extremities has produced somewhat disappointing results [BAR-MAOR and GITLIN, 1961; KUDOKOTSEV and DANCHENKO, 1966]. SIMPSON [1970] has obtained better results in the limb of *Lygosoma* by implanting ependymal tissue into the limb. Although histological data are scanty in these experiments, SINGER [1954] stated that the growth of nerve-induced regenerates in the frog appeared to occur largely by cellular proliferation of a blastema, which seemed to be better established than in salt-induced regenerates.

BODEMER [1964] augmented a component of the normal function of nerves in the anuran limb stump by electrically stimulating the nerves at periodic intervals following amputation. This procedure produced the dedifferentiation of mesodermal tissues and the formation of a blastema beneath a wound epidermis, but the regenerates were quite small. In a somewhat similar experiment in frogs, SMITH [1967] attempted to simulate the local bioelectric fields present in urodeles by the implantation of bimetallic rods into the amputated fore-limbs. Instead of the typical healing response of a non-regenerating limb, he observed histological changes (thickened wound epidermis, mesodermal destruction and accumulation of cells) more typical of regenerating limbs. Grossly, the outgrowths tended to be conical in shape.

The recent work of MIZELL [MIZELL, 1968; MIZELL and ISAACS, 1970] shows considerable promise. Working with newborn opossums, he found that implantation of a core of brain tissue or spinal ganglia into the amputated hind limb elicited a considerable regenerative response leading to the restoration of an imperfectly formed, but recognizable, hind foot. In contrast, implants of kidney, liver or adrenal were ineffective. Histological studies indicated that in regenerating limbs considerable dedifferentiative and osteoclastic activity occurred. At this point, the mechanism of the stimulation is still somewhat obscure, but it is assumed to be related to the augmentation of the normal nerve supply of the limb.

Many of the means used to stimulate limb regeneration produce an increase in the destruction of mesodermal tissues which, in turn, leads to the establishment of a population of dedifferentiated cells. POLEZHAYEV [1968], in particular, stresses the great importance of mesodermal trauma as a stimulatory event in epimorphic regeneration, and many of his experimental manipulations have been performed with the aim of increasing the amount of destruction in non-regenerating limbs. Thus, he [POLEZHAYEV, 1936b, 1939] repeatedly traumatized the limb stumps of metamorphosing tadpoles in order to bring about dedifferentiation. In *Bombina* he [POLEZJAYEV, 1946b] increased destructive processes, by applying ligatures to limbs two weeks before amputation. Various forms of chemical irritation, particularly 5% glucose, but to a lesser extent 10% iodine, 5% lactose, $NaHCO_3$ at pH 8.0 and 10% nitric acid also stimulated regeneration by increasing mesodermal trauma [POLEZHAYEV, 1946b, c], whereas

NaCl exerted a relatively greater effect upon the epidermis. In both lizards and in mammals, UMANSKY and KUDOKOTSEV [1951, 1952] ascribed the inability to regenerate to the lack of harmony between rates of destruction and regeneration of bone and muscle. Attempting to bring these processes in bone and in muscle more in phase, they injected parathyroid hormone into lizards and young rats in order to increase the amount and rate of skeletal destruction. They succeeded in stimulating regeneration to some extent. Others have directly manipulated the skeleton of non-regenerating limbs and have reported contradictory results. SAMAROVA [1950] found that removal of the skeleton in anuran amphibians resulted in the further inhibition of regeneration, whereas GOSS [1953] reported a stimulation. Although few details on either of these experiments are available, it may be that the inhibitory effects were due to the collapse of old skin over a limb stump lacking skeletal support. This has been frequently noted in similar experiments in my laboratory.

Taking their cue from NASSONOV's work on the induction of supernumerary limbs by hydrolysates of cartilage, both POLEZHAYEV and RAMYENSKAYA [1950] and KUDOKOTSEV [1960] have attempted to stimulate limb repair with injections of hydrolysates and extracts of cartilaginous and skeletal elements of the axolotl. KUDOKOTSEV was trying to supply something not provided by the relatively intact skeleton of the limb stump. In both cases, the extracts appeared to act primarily upon the mesodermal tissues, and a moderate stimulation of regeneration was noted. MALININ and DECK [1958] and then MALININ [1960] implanted early embryos or limb tissues of the tadpole into frogs' limbs and obtained imperfect regeneration. The tissues primarily affected were mesodermal, and the authors concluded that a chemical released from the implant was the stimulatory agent. One interesting finding was that the stimulatory influence was not inactivated by boiling the implants. This is in marked contrast to the effects of boiling tissue implants in supernumerary limb-forming systems [BODEMER, 1959; STEVENS *et al.*, 1965; CARLSON and MORGAN, 1967].

The other general technique used in attempts to stimulate regeneration is to depress the formation of the connective tissue scar which interposes itself between the wound epidermis and the underlying tissues of the limb stump. SCHOTTÉ and WILBER [1958] implanted additional adrenal tissue into frogs and obtained a considerable enhancement of regenerative ability. It was assumed that an increased release of adrenal glucocorticoids played a role in the inhibition of scar tissue formation. In an attempt to induce regeneration of mouse digits, SCHOTTÉ and SMITH [1961] injected ACTH or cortisone into the animals. Although epidermal healing was delayed and dermal healing was sometimes completely inhibited, neither dedifferentiation nor the formation of a blastema occurred. The experiments of SCHARF [1961, 1963] and KUDOKOTSEV and KUNTSEVICH [1965] fall into this category. It appears that treatment of the amputation surface with trypsin and calcium chloride does to some extent retard certain of the healing and scarring processes, but this treatment does not seem to have stimulated a convincing degree of epimorphic regeneration. Much more work on the role of scarring in non-regenerating limbs is needed for although the scar obviously represents a formidable barrier to limb regeneration, it may be looked upon as the end product of an overall healing response not compatible with limb regeneration rather than as a primary inhibitory factor.

An examination of many of the experiments designed to stimulate limb regeneration reveals that most of them affect more than one component of the limb or that they are only effective in the presence of other components or processes which are normally required in regenerating systems.

POLEZHAEV [1968] stresses the importance of destructive processes and dedifferentiation, and he ascribes the loss of regenerative capacity in higher vertebrates to a lessened capacity for destruction and dedifferentiation of the tissues in the limb. In many of his experiments, the application of destructive influences to mesodermal tissues led to dedifferentiation, and, subsequently, regeneration continued after a population of dedifferentiated cells had been established. SINGER *et al.* [1957] found that trauma alone was not sufficient to stimulate mesodermal dedifferentiation for in the absence of innervation, treatment with hypertonic NaCl resulted in smooth healing of amputated limbs in *Rana.* These authors stated, however, that tissue regeneration of limb components was not inhibited by denervation. They concluded that injury lowers the threshold of responsiveness by the mesodermal tissues to the nerve. In his nerve deviation experiments, SINGER [1954] established that following a given amount of trauma (amputation), the amount of regeneration was roughly correlated with the amount of innervation. Yet, BODEMER [1964] found in his experiments on the electrical stimulation of frog limbs that the positive effects were considerably enhanced by supplementary mesodermal trauma.

From the work cited above, it appears that in the stimulation of limb regeneration a relative increase in either the amount of mesodermal trauma or in the nervous influence suffices to compensate for a relative deficiency of the other. Aside from the fact that a wound epidermis must be present in order for limb regeneration to occur in higher vertebrates, it is not known whether an absolute epidermal threshold exists beyond which an increase in the epidermal influence is ineffectual. It is also not known whether augmentation of some other component of the total regenerating system could effectively replace a quantitative epidermal deficiency. Thus, as in normal limb regeneration and in supernumerary limb formation, interactions between a thickened wound epidermis not underlain by dermis, damaged mesodermal tissues and an adequate nerve supply are needed for regeneration, and the fact that stimulatory methods have been successfully directed at each of these components suggests that no single factor is responsible for the loss of regenerative capacity in higher vertebrates. Superimposed upon this is the modulating influence of the endocrine system. Experiments to date strongly indicate that although endocrines may not be primary effectors in the basic regenerative process, the absence of a proper hormonal environment leads to physiological reactions to amputation which are clearly inhibitory to regeneration.

Mention must also be made of the concept that regeneration of limbs in higher vertebrates is prevented by the rapid healing responses of individual tissues within the limb stump. There is no question that non-regenerating limbs of frogs or mammals are characterized by a rapid (*sic* 'premature') differentiation of skeletal elements (in the form of a cartilaginous callus), of dermis, of connective tissue (as the distal fibrocellular scar) and of muscle. Although it has often been stated or implied, it has not been proven that strong tissue regenerative responses are inhibitory to an overall epimorphic response. Conversely, it may even be possible that epimorphic regeneration has an inhibitory effect upon tissue regeneration. It is common knowledge that a variety of inhibitory influences (such as denervation, hypophysectomy, x-irradiation and skin flaps), when applied to the amputated limb of a newt, results in a healed limb stump in which prominent tissue regenerative responses predominate. Unless these inhibitory procedures are all considered to act by directly stimulating tissue regenerative processes, an alternative explanation lies in the removal of an inhibition superimposed by the epimorphic response. Much more work on interactions between different types of restorative processes will be needed before this question can be satisfactorily resolved [CARLSON, 1970c].

VI. Conclusions

When normal limb regeneration, supernumerary limb formation and experimentally stimulated limb regeneration are viewed together, a number of troubling aspects of the early regenerative process become more clearly outlined.

To date there is essentially no evidence for a discrete all-or-none stimulus for regeneration. Instead, the evidence favors a cascading series of events and interactions which, ultimately, builds up to a full-blown epimorphic regenerative process. Exactly what the 'cascade' consists of is by no means fully characterized, but a rough outline of some fairly well established steps can be constructed.

In a temporal sense, the two primary components of a regenerating system appear to be damaged mesodermal tissues in close association with a wound epidermis. It is not known whether anything leading directly to a regenerative response is accomplished by this combination alone during the initial period of their association. A nervous influence during this early period is unlikely because of the short degenerative phase of nerves following amputation and because of insufficient time for sprouting in the case of supernumerary limb formation. This, however, has not been experimentally tested. Within a few days after the initial stimulatory event, numerous regenerating nerve fibres become associated with the wound epithelium. Experimental evidence points to a positive interaction between nerve or nerve product and epidermis, but its exact nature remains unknown. Shortly after this association of nerve and epidermis, an exaggerated wave of histolysis sweeps through the mesodermal tissues of the limb stump. In adults, this phase seems to be dependent upon the presence of an activated (by nerves and/or mesodermal trauma) wound epidermis while in larvae, the wound epidermis does not seem to play a strongly positive role. As histolysis continues, dedifferentiated cells appear in the areas of destruction. It is extremely important to determine, first of all, exactly what constitutes dedifferentiation at the cellular level. Until then, it is very difficult to do more than generalize about causative factors relating to this phenomenon. It has been experimentally determined that both nerve and epidermis can profoundly affect the dedifferentiative phase, but it is still not known whether the effects are primary or secondary. At this stage more than any other, there seems to be a great difference between adult and larval tissues. As dedifferentiation is occurring, the wound epithelium (again under the apparent influence of nerves) thickens into the characteristic apical epidermal cap. The dedifferentiated cells are attracted to the apical cap and accumulate beneath it to form the regeneration blastema. At least partially under the influence of the nerve (direct?) the blastemal cells are stimulated to divide, thus increasing the mass of the blastema.

When all of the experimental evidence is taken into account, each of the major tissue components of the regenerating limb seems to play a number of discrete roles during different phases of the early regenerative process. For instance, the following properties have been attributed to the wound epidermis: phagocytosis, promotion of histolysis, promotion of dedifferentiation, attraction of blastemal cells as well as other less well established functions. Since the roles proposed for various tissues have been established upon different experimental systems, it is important to determine if they are really distinct or whether secondary effects have been mistaken for primary actions.

Although every step in the early regenerative process is necessary, special importance must be attached to the dedifferentiative reaction, for in adults, at least, the presence of dedifferentiated cells is almost an expression of the intent to regenerate. Yet, less is known about dedifferentiation than almost any other phase of the process. Surely, some changes preparatory to dedifferentiation must be occurring during the period just preceding the overt morphological manifestation of the process, but except for inferences made from studies with metabolic inhibitors or autoradiographic methods, there is very little concrete information with which to work. The removal of established cellular and intercellular structure by histolysis seems to be important in allowing dedifferentiation to occur, but it is quite possible that this is a permissive event and that a positive stimulus is also required. Further studies of experimental means of producing dedifferentiation may be quite helpful in attacking this problem. The possibility of natural chemical factors which may promote dedifferentiation is likely in view of the results of investigations of implant-induced supernumerary limb formation and the reported isolation of a mesodermal dedifferentiating factor from the submaxillary gland [ATTARDI *et al.*, 1965]. It is quite likely that if a natural dedifferentiation stimulating factor is found, it would exert a maximum effect only in conjunction with other normal concomitant factors, such as nerve action. Another potentially important approach might be to extend to amphibians the work of HESS and ROSNER [1970], who reported an increase in satellite cell numbers in denervated mammalian skeletal muscle.

It is apparent from this review that early limb regeneration is a very complex process involving many interacting systems. Interruption of a single critical step can bring the entire process to a halt, but logic does not allow one to conclude that the interrupted step is the all important one. Some appreciation for the probable complexity of limb regeneration might be gained by considering the phenomenon of blood clotting. The absence or malfunction of any of over a dozen factors results in defective clotting, yet each of these factors represents only a small part of the total picture. When one considers the extent of the morphological alterations in a regenerating limb and compares that with the morphology of clotting blood, it staggers the imagination to ponder the complexity of intricacy of interplay between cells, tissues and their environment which must be required to cause a complex structure like a limb to undergo the necessary preparations for reproducing itself.

References

ADOVA, A. N. and FELDT, A. M.: C. R. Acad. Sci. URSS *25:* 43 (1939).
ANTON, H.J.: In V. KIORTSIS and H.A.L. TRAMPUSCH Regeneration in animals and related problems, pp. 377–395 (North-Holland Publ. Co., Amsterdam 1965).
ANTON, H. J.: Roux' Arch. Entw. Mech. *161:* 49 (1968).
ATTARDI, D. G.; LEVI-MONTALCINI, R.; WENGER, B. S. and ANGELETTI, P. U.: Science *150:* 1307 (1965).
BALINSKY, B. I.: Proc. nat. Acad. Sci., U.S. *42:* 781 (1956).
BALLS, M. and RUBEN, L. N.: Develop. Biol. *10:* 92 (1964).
BAR-MAOR, J. A. and GITLIN, G.: Transplant. Bull. *27:* 460 (1961).
BECKER, R. O.: J. Bone J. Surg. *43A:* 643 (1961).
BODEMER, C. W.: J. Morph. *102:* 555 (1958).
BODEMER, C. W.: J. exp. Zool. *140:* 79 (1959).
BODEMER, C. W.: J. Morph. *107:* 47 (1960).
BODEMER, C. W.: Anat. Rec. *142:* 457 (1962).
BODEMER, C. W.: Anat. Rec. *148:* 441 (1964).
BODEMER, C. W. and EVERETT, N. B.: Develop. Biol. *1:* 327 (1959).
BREEDIS, C.: Cancer Res. *12:* 861 (1952).
BRUNST, V. V.: J. exp. Zool. *114:* 1 (1950a).
BRUNST, V. V.: J. Morph. *86:* 115 (1950b).
BRUNST, V. V.: Amer. J. Roentgenol. *68:* 281 (1952).
BRUNST, V. V.: Quart. Rev. Biol. *36:* 178 (1961).
BRUNST, V. V. and FIGGE, F. H. J.: J. Morph. *89:* 111 (1951).
BUTLER, E. G.: J. exp. Zool. *65:* 271 (1933).
BUTLER, E. G.: Anat. Rec. *62:* 295 (1935).
BUTLER, E. G. and BLUM, H. F.: J. nat. Cancer Inst. *15:* 877 (1955).
BUTLER, E. G. and BLUM, H. F.: Develop. Biol. *7:* 218 (1963).
BUTLER, E. G. and PUCKETT, W. O.: J. exp. Zool. *84:* 223 (1940).
BUTLER, E. G. and SCHOTTE, O. E.: J. exp. Zool. *88:* 307 (1941).
CARLSON, B. M.: Doklady Akad. Nauk SSSR *171:* 229 (1966).
CARLSON, B. M.: J. Morph. *122:* 249 (1967a).
CARLSON, B. M.: J. exp. Zool. *164:* 227 (1967b).

CARLSON, B. M.: Experientia *24:* 1064 (1968).
CARLSON, B. M.: Anat. Rec. *163:* 389 (1969).
CARLSON, B. M.: Anat. Rec. *166:* 423 (1970b).
CARLSON, B. M.: Amer. Zool. *10:* 175 (1970c).
CARLSON, B. M.: Oncology *25:* 365 (1971).
CARLSON, B. M. and C. F. MORGAN: J. exp. Zool. *164:* 243 (1967).
CARPENTER, R. L.: J. exp. Zool. *61:* 149 (1932).
CHALKLEY, D. T.: J. Morph. *94:* 21 (1954).
CHALKLEY, D. T.: In C.S. THORNTON Regeneration in vertebrates, pp. 34–58 (Univ. of Chicago Press, Chicago 1959).
DECK, J.D. and FUTCH, C.B.: Develop. Biol. *20:* 332 (1969).
DELLA VALLE, P.: Bull. Soc. Nat. Napoli *25:* 95 (1913).
DESSELLE, J-C.: C. R. Acad. Sci., Paris *267:* 1642 (1968).
DRESDEN, M. H.: Develop. Biol. *19:* 311 (1969).
DRESDEN, M. H. and GROSS, J.: Develop. Biol. *22:* 129 (1970).
DROIN, A.: Rev. suisse Zool. *66:* 641 (1959).
EFIMOV, M. I.: Zh. exp. Biol. *7:* 352 (1931).
EGGERT, R. C.: J. exp. Zool. *161:* 369 (1966).
FABER, J.: Arch. Biol. *71:* 1 (1959).
FABER, J.: In V. KIORTSIS and H.A.L. TRAMPUSCH Regeneration in animals and related problems, pp. 404–419 (North Holland Publ. Co., Amsterdam 1965).
FEDOTOV, D. M.: Nature *158:* 367 ((946).
GIDGE, N. M. and ROSE, S. M.: J. exp. Zool. *97:* 71 (1944).
GLADE, R.: J. Morph. *101:* 477 (1957).
GODLEWSKI, E.: Roux' Arch. Entw. Mech. *114:* 108 (1928).
GOSS, R. J.: Anat. Rec. *115:* 311 (1953).
GOSS, R. J.: Principles of regeneration (Academic Press, New York 1968).
GRILLO, H. C.; LAPIERE, C. M.; DRESDEN, M. H. and GROSS, J.: Develop. Biol. *17:* 571 (1968).
GUYENOT, E.; DINICHERT-FAVARGER, J. and GALLAND, M.: Rev. suisse Zool. *55* Suppl. *2:* 1 (1948).
HALL, A. B. and SCHOTTE, O. E.: J. exp. Zool. *118:* 366 (1951).
HAY, E. D.: Develop. Biol. *1:* 555 (1959).
HAY, E. D.: In D. RUDNICK Regeneration, pp. 177–210, (Ronald Press, New York 1962).
HAY, E. D.: In W. MONTAGNA and W. C. LOBITZ The epidermis, pp. 97–116 (Academic Press, New York 1964).
HAY, E. D.: In R.L. DE HAAN and H. URSPRUNG Organogenesis, pp. 315–336 (Holt, Rinehart & Winston, New York 1965).
HAY, E. D.: Regeneration (Holt, Rinehart & Winston, New York 1966).
HAY, E. D. and FISCHMAN, D. A.: Develop. Biol. *3:* 26 (1961).
HAY, E. D. and REVEL, J. P.: Develop. Biol. *7:* 152 (1963).
HESS, A. and ROSNER, S.: Amer. J. Anat. *129:* 21 (1970).
KARCZMAR, A. G.: J. exp. Zool. *103:* 401 (1946).
KASANZEFF, W.: Roux' Arch. Entw. Mech. *121:* 658 (1930).
KIORTSIS, V.: Rev. suisse Zool. *60:* 301 (1953).
KONIECZNA-MARCZYNSKA, B. and SKOWRON-CENDRZAK, A.: Folia biol., Warsaw *6:* 37 (1958).
KUDOKOTSEV, V. P.: Doklady Akad. Nauk SSSR *132:* 715 (1960).
KUDOKOTSEV, V. P.: Doklady Akad. Nauk SSSR *142:* 233 (1962).

KUDOKOTSEV, V. P.: Biol. Nauki No. *3:* 40 (1964).
KUDOKOTSEV, V. P.: Vestnik Kharkov Univ. Ser. Biol. No. *1:* 119 (1965).
KUDOKOTSEV, V. P.: In (Committee, ed.) Conditions of regeneration of organs and tissues in animals (in Russian), pp. 137–140 (Moscow 1966).
KUDOKOTSEV, V. P. and DANCHENKO, L. K.: Biol. Nauki No. *2:* 37 (1966).
KUDOKOTSEV, V. P. and KUNTSEVICH, V. A.: Byull. exp. Biol. Med. No. *9:* 106 (1965).
KUZMINA, N. A.: C. R. (Doklady) Acad. Sci. URSS *26:* 504 (1940).
LASH, J. W.: J. exp. Zool. *128:* 13 (1955).
LASH, J. W.: J. exp. Zool. *131:* 239 (1956).
LEBOWITZ, P. and SINGER, M.: Nature *225:* 824 (1970).
LE CAMP, M.: Bull. Biol. France Belg., Suppl. *19:* 1 (1935).
LENTZ, T. L.: Am. J. Anat. *124:* 447 (1969).
LIOSNER, L. D.: Roux' Arch. Entw. Mech. *124:* 571 (1931).
LITSCHKO, E. Ya.: In To Academician N.V. Nassonov (in Russian), pp. 365–388 (Izdatel. Akad. Nauk SSSR, Moscow 1937).
LOCATELLI, P.: Roux' Arch. Entw. Mech. *114:* 686 (1929).
LOEWENSTEIN, W. R. and PENN, R. D.: J. Cell Biol. *33:* 235 (1967).
MALININ, T. I.: J. exp. Zool. *143:* 1 (1960).
MALININ, T. and DECK, J. D.: J. exp. Zool. *139:* 307 (1958).
MANNER, H. W.: J. exp. Zool. *122:* 222 (1953).
MARCUCCI, E.: Arch. Zool. *8:* 89 (1915).
MAURO, A.: J. biophys. biochem. Cytol. *9:* 493 (1961).
METTETAL, C.: Arch. Anat. Histol. Embryol. *28:* 1 (1939).
MIZELL, M.: Amer. Zool. *3:* 510 (1963).
MIZELL, M.: Science *161:* 283 (1968).
MIZELL, M. and ISAACS, J. J.: Amer. Zool. *10:* 141 (1970).
MORGAN, T. H.: Amer. Naturalist *42:* 1 (1908).
NASSONOV, N. V.: Roux' Arch. Entw. Mech. *121:* 639 (1930).
NASSONOV, N. V.: C. R. Akad. Nauk URSS, N.S. *3:* 261 (1934a).
NASSONOV, N. V.: C. R. Akad. Nauk URSS, N.S. *3:* 328 (1934b).
NASSONOV, N. V.: C. R. (Doklady) Acad. Sci. URSS *2(XI):* 207 (1936a).
NASSONOV, N. V.: C. R. Akad. Nauk URSS, N.S. *3:* 261 (1934a).
NASSONOV, N. V.: C. R. (Doklady) Acad. Sci. URSS, *19:* 127 (1938a).
NASSONOV, N. V.: C. R. (Doklady) Acad. Sci. URSS *19:* 133 (1938b).
NASSONOV, N. V.: C. R. (Doklady) Acad. Sci. URSS *19:* 137 (1938c).
NASSONOV, N.V.: Accessory formations, developing after implantation of cartilage under the skin of tailed amphibians (in Russian) (Izdatel. Akad. Nauk. SSSR, Leningrad 1941).
NEEDHAM, A. E.: Proc. zool. Soc. Lond., Ser. A *111:* 59 (1941).
NICHOLAS, J. S.: In B.H. WILLIER, P.A. WEISS and V. HAMBURGER Analysis of development, pp. 674–698 (Saunders, Philadelphia 1955).
NORMAN, W. P. and SCHMIDT, A. J.: J. Morph. *123:* 271 (1967).
ORECHOWITSCH, W. N. and BROMLEY, N. W.: Biol. Zbl. *54:* 524 (1934).
O'STEEN, W. K. and WALKER, B. E.: Anat. Rec. *139:* 547 (1961).
OVERTON, J.: J. exp. Zool. *115:* 521 (1950).
PEREDELSKY, A. A.: C.R. (Doklady) Acad. Sci. URSS *26:* 499 (1940a).
PEREDELSKY, A. A.: C.R. (Doklady) Acad. Sci. URSS *26:* 495 (1940b).

POLEZHAEV, L. V.[2]: Biol. Zh. *2:* 357 (1933a).
POLEZHAEV, L. V.: Biol. Zh. *2:* 368 (1933b).
POLEZHAEV, L. V.: Zool. Zh. *15:* 277 (1936a).
POLEJAIEV, L. W.: Arch. Anat. micr. *32:* 437 (1936b).
POLEZHAEV, L. V.: C.R. (Doklady) Acad. Sci. URSS *22:* 648 (1939).
POLEZHAEV, L. V.: Doklady Akad. Nauk SSSR, *48:* 232 (1945).
POLEZHAEV, L. V.: Biol. Rev. *21:* 141 (1946a).
POLEZHAYEV, L. W.: C. R. (Doklady) Acad. Sci. URSS *54:* 461 (1946b).
POLEZHAYEV, L. W.: C. R. (Doklady) Acad. Sci. URSS *54:* 281 (1946c).
POLEZHAYEV, L. W.: C. R. (Doklady) Acad. Sci. URSS *54:* 653 (1946d).
POLEZHAEV, L. V.: Izvestia Akad. Nauk SSSR, Ser. Biol. *(1):* 37 (1966a).
POLEZHAEV, L. V.: Izvestia Akad. Nauk SSSR, Ser. Biol. *(2):* 254 (1966b).
POLEZHAEV, L. V.: The loss and restoration of regenerative capacity in the tissues and organs of animals (in Russian) (Izdatel. Nauka, Moscow 1968).
POLEŽAJEW, L. W. and FAWORINA, W. N.: Roux' Arch. Entw. Mech. *133:* 701 (1935).
POLEŽAJEW, L.W. and GINZBURG, G. I.: C. R. (Doklady) Acad. Sci. URSS *23:* 733 (1939).
POLEZHAEV, L. V. and MOROSOV, I. I.: C. R. (Doklady) Acad. Sci. URSS *30:* 675 (1941).
POLEZHAEV, L. V. and RAMYENSKAYA, G. P.: Doklady Akad. Nauk SSSR *70:* 141 (1950).
POLEZHAEV, L. V. and TUCHOVA, S. Ya.: Doklady Akad. Nauk SSSR *180:* 754 (1968).
PRZIBRAM, H.: Roux' Arch. Entw. Mech. *48:* 205 (1921).
PURDY, S. G.: M. S. thesis, Michigan State University, East Lansing, Michigan (1967).
ROGAL, I. G.: Doklady Akad. Nauk SSSR *78:* 161 (1951a).
ROGAL, I. G.: Doklady Akad. Nauk SSSR *81:* 953 (1951b).
ROSE, F. C. and ROSE, S. M.: Anat. Rec. *99:* 653 (1947).
ROSE, F. C. and ROSE, S. M.: Growth *29:* 653 (1947).
ROSE, F. C. and ROSE, S. M.: Growth *29:* 361 (1965).
ROSE, F. C. and ROSE, S. M.: Growth *31:* 375 (1967).
ROSE, S. M.: Proc. Soc. exp. Biol. Med. *49:* 408 (1942).
ROSE, S. M.: J. exp. Zool. *95:* 149 (1944).
ROSE, S. M.: J. Morph. *77:* 119 (1945).
ROSE, S. M.: Ann. N.Y. Acad. Sci. *49:* 818 (1948a).
ROSE, S. M.: J. exp. Zool. *108:* 337 (1948b).
ROSE, S. M.: In D. RUDNICK Regeneration, pp. 153–176 (Ronald Press, New York 1962).
ROSE, S. M.: In J. A. MOORE Physiology of the amphibia, pp. 545–622 (Academic Press, New York 1964).
RUBEN, L. N.: J. exp. Zool. *128:* 29 (1955).
RUBEN, L. N.: Anat. Rec. *128:* 613 (1957).
RUBEN, L. N.: Nature *183:* 765 (1959).
RUBEN, L. N.: Amer. Naturalist *94:* 427 (1960).
RUBEN, L. N. and BALLS, M.: J. Morph. *115:* 239 (1964).
RUBEN, L. N. and FROTHINGHAM, M. L.: J. Morph. *102:* 91 (1958).
RUBEN, L. N. and STEVENS, J. M.: J. Morph. *112:* 279 (1963).
SALPETER, M. M. and SINGER, M.: Develop. Biol. *2:* 516 (1960).

2 The names POLEZHAEV, POLEJAIEV, POLEZHAYEV and POLEZAJEW are all transliterations of the same Russian surname. Because of this, the papers by this author are listed chronologically rather than strictly alphabetically.

SAMAROVA, V. A.: Trudy Inst. Biol. Kharkov Univ., p. 14 (1950); quot. UMANSKY and KUDOKOTSEV [1952].
SAUNDERS, J. W.: J. exp. Zool. *108:* 363 (1948).
SCHARF, A.: Growth *25:* 7 (1961).
SCHARF, A.: Growth *27:* 255 (1963).
SCHMIDT, A. J.: Cellular biology of vertebrate regeneration and repair (Univ. Chicago Press, Chicago 1968).
SCHOTTÉ, O. E.: In D. RUDNICK Synthesis of molecular and cellular structure, pp. 161–192 (Ronald Press, New York 1961).
SCHOTTÉ, O. E. and BUTLER, E. G.: J. exp. Zool. *87:* 279 (1941).
SCHOTTÉ, O. E. and BUTLER, E. G.: J. exp. Zool. *97:* 95 (1944).
SCHOTTÉ, O. E.; BUTLER, E. G. and HOOD, R. T.: Proc. Soc. exp. Biol. Med. *48:* 500 (1941).
SCHOTTÉ, O. E. and HALL, A. B.: J. exp. Zool. *121:* 521 (1952).
SCHOTTÉ, O. E. and HARLAND, M.: J. exp. Zool. *93:* 453 (1943a).
SCHOTTÉ, O. E. and HARLAND, M.: J. Morph. *73:* 329 (1943b).
SCHOTTÉ, O. E. and SMITH, C. B.: Biol. Bull. *117:* 546 (1959).
SCHOTTÉ, O. E. and SMITH, C. B.: J. exp. Zool. *146:* 209 (1961).
SCHOTTÉ, O. E. and WILBER, J. F.: J. Embryol. exp. Morph. *6:* 247 (1958).
SIDMAN, R. L. and SINGER, M.: J. exp. Zool. *144:* 105 (1960).
SIMPSON, S. B.: Proc. Soc. exp. Biol. Med. *107:* 108 (1961).
SIMPSON, S. B.: Amer. Zool. *10:* 157 (1970).
SINGER, M.: J. exp. Zool. *111:* 189 (1949).
SINGER, M.: Quart. Rev. Biol. *27:* 169 (1952).
SINGER, M.: J. exp. Zool. *126:* 419 (1954).
SINGER, M.: In C. S. THORNTON Regeneration in vertebrates, pp. 59–80 (Univ. Chicago Press, Chicago 1959).
SINGER, M.: Proc. Soc. exp. Biol. Med. *107:* 106 (1961).
SINGER, M.: In V. KIORTSIS and H. A. L. TRAMPUSCH Regeneration in animals and related problems, pp. 20–32 (North-Holland Publ. Co., Amsterdam 1965).
SINGER, M. and CRAVEN, L.: J. exp. Zool. *108:* 279 (1948).
SINGER, M. and INOUE, S.: J. exp. Zool. *155:* 105 (1964).
SINGER, M.; KAMRIN, R.P. and ASHBAUGH, A.: J. exp. Zool. *136:* 35 (1957).
SINGER, M.; RAY, E. and PEADON, A. M.: Folia biol., Krakow *12:* 347 (1964).
SINGER, M. and SALPETER, M. M.: In M. X. ZARROW Growth in living systems, pp. 277–311 (Basic Books, New York 1961).
SMITH, S. D.: Anat. Rec. *158:* 89 (1967).
SMITH, S. D. and CRAWFORD, G. L.: Oncology *23:* 299 (1969).
STEEN, T. P.: Amer. Zool. *10:* 119 (1970).
STEEN, T. P. and THORNTON, C. S.: J. exp. Zool. *154:* 207 (1963).
STEVENS, J.; RUBEN, L. N.; LOCKWOOD, D. and ROSE, H.: J. Morph. *117:* 213 (1965).
STUDITSKY, A. N.: Doklady Akad. Nauk SSSR *60:* 309 (1948).
SWETT, F. H.: J. exp. Zool. *61:* 129 (1932).
TABAN, C.: Arch. Sci. *2:* 553 (1949).
TABAN, C.: Rev. suisse Zool. *62:* 387 (1955).
TASSAVA, R. A.; CHLAPOWSKI, F. J. and THORNTON, C. S.: J. exp. Zool. *167:* 157 (1968).
TAUBE, E.: Arch. Entw. Mech. *49:* 269 (1921).
THORNTON, C. S.: J. Morph. *62:* 17 (1938).

THORNTON, C. S.: J. exp. Zool. *89:* 375 (1942).
THORNTON, C. S.: J. exp. Zool. *92:* 281 (1943).
THORNTON, C. S.: J. Morph. *84:* 459 (1949).
THORNTON, C. S.: J. exp. Zool. *114:* 305 (1950).
THORNTON, C. S.: J. exp. Zool. *118:* 467 (1951).
THORNTON, C. S.: J. exp. Zool. *122:* 119 (1953).
THORNTON, C. S.: J. exp. Zool. *127:* 577 (1954).
THORNTON, C. S.: J. exp. Zool. *131:* 373 (1956).
THORNTON, C. S.: J. exp. Zool. *134:* 357 (1957).
THORNTON, C. S.: J. exp. Zool. *137:* 153 (1958).
THORNTON, C. S.: Develop. Biol. *2:* 551 (1960a).
THORNTON, C. S.: Copeia No. *4:* 371 (1960b).
THORNTON, C. S.: Adv. Morphogenes. *7:* 205 (1968).
THORNTON, C. S.: Amer. Zool. *10:* 113 (1970).
THORNTON, C. S. and KRAEMER, D. W.: J. exp. Zool. *117:* 415 (1951).
THORNTON, C. S. and TASSAVA, R. A.: J. Morph. *127:* 225 (1969).
THORNTON, C. S. and THORNTON, M. T.: Experientia *21:* 146 (1965).
THORNTON, C. S. and THORNTON, M. C.: J. exp. Zool. *173:* 293 (1970).
TORNIER, G.: Zool. Anz. *20:* 356 (1897).
TORNIER, G.: Zool. Anz. *21:* 372 (1898).
TORNIER, G.: Roux' Arch. Entw. Mech. *22:* 348 (1906).
TOTO, P. D. and ANNONI, J. D.: J. dent. Res. *44:* 71 (1965).
TOWLE, E. W.: Biol. Bull. *2:* 289 (1901).
TRAMPUSCH, H. A. L.: Proc. Kon. Ned. Akad. Wetensch. *54:* 373 (1951).
TRAMPUSCH, H. A. L.: In M. SINGER and J. P. SCHADE Mechanisms of neural regeneration, pp. 214–227 (Elsevier, Amsterdam 1964).
TRAMPUSCH, H. A. L. and HARREBOMEE, A. E.: In V. KIORTSIS and H. A. L. TRAMPUSCH Regeneration in animals and related problems, pp. 341–376 (North Holland Publ. Co., Amsterdam 1965).
TRELSTAD, R. L.: J. Cell Biol. *45:* 34 (1970).
TUCHKOVA, S. Ya.: Doklady Akad. Nauk SSSR *158:* 1420 (1964).
UMANSKY, E.: Biol. Zh. *6:* 739 (1937).
UMANSKY, E. E. and KUDOKOTSEV, V. P.: Doklady Akad. Nauk SSSR *61:* 757 (1948).
UMANSKY, E. E. and KUDOKOTSEV, V. P.: Doklady Akad. Nauk SSSR *77:* 533 (1951).
UMANSKY, E. E. and KUDOKOTSEV, V. P.: Doklady Akad. Nauk SSSR *86:* 437 (1952).
VAN ARSDALL, C. B. and LENTZ, T. L.: Science *162:* 1296 (1968).
WEISS, P.: Principles of development (Holt, New York 1939).
WEISS, P.: In The Harvey lecture series, vol. 55, pp. 13–42 (Academic Press, New York 1961).
WOOLFIT, R.: W. Virginia med. J. *64:* 387 (1968).
YAKOVLEVA, T. M.: Biol. Zh. *7:* 489 (1938).
YNTEMA, C. L.: J. exp. Zool. *140:* 101 (1959a).
YNTEMA, C. L.: J. exp. Zool. *142:* 423 (1959b).
YNTEMA, C. L.: J. exp. Zool. *149:* 127 (1962).
ZELINSKY, N. D.: Zh. Obsch. Biol. *7:* 161 (1946).

Author's address: Dr. BRUCE M. CARLSON, Department of Anatomy, University of Michigan, *Ann Arbor, Mich.* (USA)

Neoplasia and Cell Differentiation, pp. 106–152
(Karger, Basel 1974)

Genome Control and the Genetic Potentialities of the Nuclei of Dedifferentiated Regeneration Blastema Cells

ANN M. C. BURGESS
The London Hospital Medical College, London

Contents

I. The Activation and Repression of the Bacterial Genome

One of the most tantalizing problems in embryogenesis is the means by which the genes are controlled and the way in which the differential pattern of gene activity emerges in the developing organism. In a metazoan, not all the cells are identical, yet they all originated from a common gene complex which, during the course of development, is given a wide variety of expression. This type of system presumably incorporates a very large number of specific controls and a very elaborate and highly integrated mechanism for their action. Even in more simple systems like bacteria, new enzymes may be required if changes take place in the substrate, while in the viruses there is a temporal sequence of events which would seem to indicate the timed switching on and off of the genes. While not many people would disagree with the concept of genetic regulation, there has been and indeed still is, disagreement over the nature of the mechanism particularly with regard to the Metazoa.

A. The Lac System of *Escherichia coli*

The concept of a group of genes being controlled by a regulator gene had long received considerable support but was first put forward in 1961 by the French workers JACOB and MONOD as an integral part of the operon theory. They investigated the lac system of *Escherichia coli* and by using the experimental methods of genetics, showed that the control of β-galactosidase is negative. In effect this means that the product of the regulator gene is a repressor, the lac repressor, which binds to a site on the lac gene complex, or lac operon and prevents the expression of the structural genes of the operon. They postulated that as long as the regulator gene remained in the active state, the structural genes would remain in the inactive or repressed state. The method of activation of the operon is by the inactivation of the repressor which is then thought to lose its capacity for binding to the operator and so release the structural genes for synthesis of RNA by RNA polymerase which starts at the 'promoter'. The mechanism for the inactivation of the repressor is not known but may be concerned with slight alterations in shape brought about by the inducer molecule. As a consequence of the changes in shape, the repressor molecule is thought to lose its ability to bind to the operator. In not all systems has this account been found to be accurate, in some cases the repressor is only able to bind to the operator in the presence of a smaller molecule which is known as the corepressor.

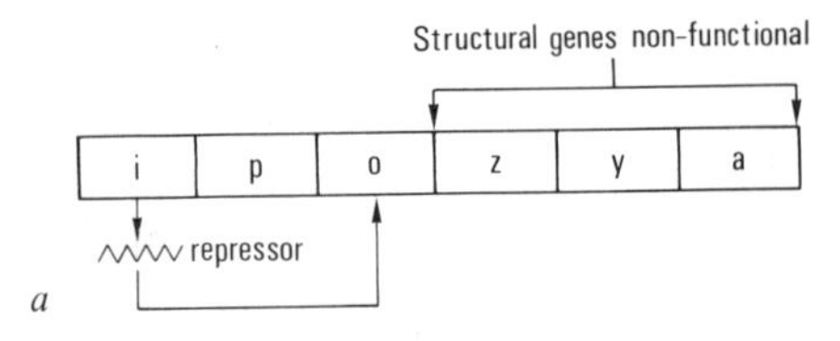

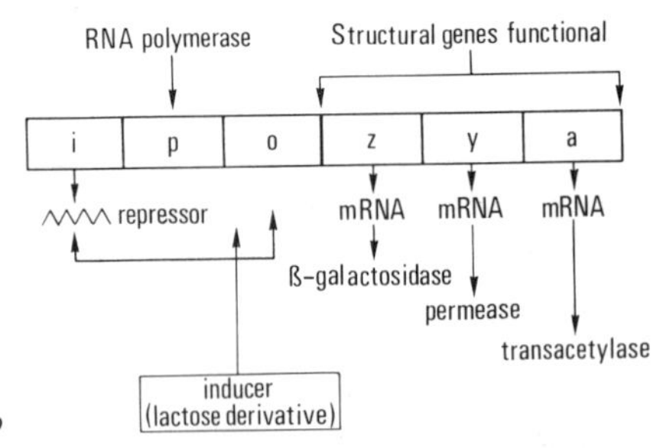

Fig. 1a, b. The lac repressor system of *Escherichia coli. a* The *i* gene produces repressor which binds to the operator (o) and prevents the transcription of the structural genes (z, y and a). *b* The inducer (lactose derivative) interacts with the repressor molecule which then becomes unable to bind to the operator. RNA polymerase then proceeds to transcribe the structural genes. p = Promoter.

In the case of the lac system it is postulated that the lac repressor prevents the expression of the three structural genes under the control of the lac operator (fig. 1a). These three genes have been found to be concerned only with the structure of the proteins which they produce and have no effect on the regulation of their synthesis. This is most clearly illustrated when mutations occur in any of the three genes: the *z* gene that codes for β-galactosidase, the *y* gene that is responsible for concentrating lactose in the cell and the *a* gene that produces a transacetylase enzyme, the function of which is as yet unknown. Mutations occurring in any of these genes will produce changes in the corresponding enzymes but will not affect their regulation. The only gene that appears to affect regulation is the *i* gene. JACOB and MONOD proposed that it is the product of the *i* gene that is responsible for the production of the repressor. It has been shown recently that it is almost certainly not lactose itself which functions as the inducer of lactose metabolism but that a derivative of lactose is probably responsible [BURSTEIN *et al.*, 1965]. By inactivating the repressor it allows the transcription of the genes of the lac operon to proceed (fig. 1 b).

B. The λ-Phage System

Another interesting example of genetic repression is shown by phage λ which infects *E. coli*. This is particularly interesting because it shows how a single gene, the C_1 gene, can turn off the entire phage genome. KAISER and JACOB [1957] had already shown that the C_1 gene is responsible for a single repressor, the phage repressor. It is now known that the phage repressor operates not at one, but at two different sites on the phage genome. Both sites are adjacent to the C_1 gene and the two different operons are transcribed in opposite directions. An interesting fact has also been observed about these two genes which could throw some light on the situation as it exists in higher animals. It has been noted that some of the products of the genes which are turned on earlier, are responsible for the turning on of later genes. Furthermore, it was demonstrated by EBISUZAKI in 1963, that if defects are produced in the earlier formed proteins of phage T4 by the action of 7-azatryptophan, 5-methyl tryptophan or fluorophenylalanine, the synthesis of later structural proteins is inhibited. The extrapolation of the results of experiments of this kind to the developing embryo, especially in situations where induction is known to occur, is very tempting.

Similar work on the l-arabinose operon of *E. coli* which was done by ENGLESBERG *et al.* [1969] has shown that the product of the regulator gene *(ara c)* can exist in two forms: the repressor and the activator. In the absence of the inducer and in the presence of an operator site the *c* gene product takes on a configuration which results in it functioning as a repressor. Small quantities of the activator state are detectable and the two states are presumably in equilibrium. However, when the inducer is present the repressor configuration is no longer detectable and the operon becomes active.

C. The Isolation of the Lac Repressor

There are many other instances of a group of genes being controlled by a regulator gene but until 1966, when GILBERT and MÜLLER-HILL succeeded in isolating the first repressor, all the evidence had come from genetic experiments, which although very elegant and in some cases brilliant, gave no indication of the molecular mechanism underlying the system.

It had been postulated that the genetic repressors were the histone proteins that are found in association with the chromosomes of the higher animals or that they contained RNA and, in fact, evidence has been presented in support of both these hypotheses but the few repressors that have been isolated have been shown to be large proteins. This does not mean that it has been demonstrated that the nuclear histones play no part in genetic regulation, especially in the higher animals where the system of genetic regulation has not been described in detail, but that they play no part in genetic regulation in terms of the repressor genes which have so far been analyzed in bacterial and phage systems. The idea that repressors are composed, at least in part, of RNA was founded on experimental data which has subsequently proved to be incorrect.

Even with the isolation of the lac repressor by GILBERT and MÜLLER-HILL and the isolation of the λ phage repressor by PTASHNE [1967a], the level at which the repressor operated was still open to speculation. It was postulated either that it acted at the level of RNA and prevented the transcription of the genetic message from RNA to protein, or that the repressor prevented the transcription of regions of the DNA by binding directly to it. The latter hypothesis was most commonly believed to be true and this belief was vindicated by the later experiments of GILBERT and MÜLLER-HILL and of PTASHNE.

In approaching the problem of the isolation of the lac repressor, GILBERT and MÜLLER-HILL reasoned that *E. coli* that had not been previously induced with lactose ought to contain some molecules of the lac repressor but they suspected that it would only be present in very small amounts. The problem of isolating a specific substance in minute quantities was overcome by a method which must be admired for its direct approach and simplicity of concept. It was based on the assumption that the repressor and the inducer would in some way interact and that if the inducer was labelled with a radioactive tracer, it should be possible to isolate the repressor by collecting the radioactive fraction. Basically, they dialysed a highly concentrated preparation of *E. coli* against iso-propyl-thio-galactoside which is an analogue of lactose and the most vigorous known inducer of the lac genes. After a great deal of experiment they eventually found an extract of *E. coli* that would bind small quantities of IPIG and which gave a slightly higher count than the background of the dialysis medium. It was then just a matter of painstakingly fractionating the total protein and finding which fraction was bound to the labelled IPIG for the lac repressor to be identified. Further purification was achieved with ammonium sulphate precipitation and DEAE chromatography until a final fraction was obtained that contained around 1% repressor. In such an apparently simple and yet brilliant way was the first genetic repressor isolated. The isolation of the repressor was all the more remarkable because it is only present in extremely small amounts. It is estimated that only 0.002% of the total protein is repressor, in molecular terms this means that there are only 10–20 molecules per cell [Gilbert and Müller-Hill, 1967]. Further pruification was achieved by Riggs and Bourgeois [1968] using gel filtration with G200 Sephadex and phosphocellulose chromatography.

As further proof that the substance they had isolated was really the lac repressor, two further experiments were undertaken. In the first, a strain of *E. coli*, which had been shown by genetic experiment not to manufacture lac repressor, was dialysed against labelled IPIG but the number of counts in the dialysis bag never rose above the background level indicating that no IPIG had been bound, and in the second experiment, which is the reverse of the first, a strain was used containing a mutant *i* gene. The results of the second experiment showed that the IPIG had been more tightly bound. Thus, the evidence that for the first time a genetic repressor had been isolated was complete.

D. Characteristics of Repressors

A short while later, Ptashne [1967a] succeeded in isolating the λ phage repressor. Although it is too early to make many generalizations about repressors, both the lac and λ phage repressors turn out to have certain features in common. First they are both quite large oligomeric proteins, each composed of 4 sub-units. The lac repressor has a molecular weight of about 150,000 [Riggs and Bourgeois, 1968] and the λ phage repressor 112,000. At neutral pH they both have an overall negative charge which makes them slightly acidic, this contrasts with the histone proteins which are basic and carry an overall high positive charge at neutral pH; the histones also have low molecular weights. However, not all genetic repressors are acidic proteins as may be seen by the work of PIROTTA and PTASHNE [1969] who have isolated another repressor, the 434 phage repressor which has been found to be slightly basic.

E. The Binding of Repressors to DNA

Although repressors had been isolated, the problem of the level at which they acted was still unsolved. In 1967(b), PTASHNE published a paper in which he showed that the λ phage repressor was specifically bound to λ DNA. His method consisted of the co-centrifugation of DNA and labelled repressor in a glycerol gradient. DNA is much heavier than the repressor and unless they were in some way bound together, would sediment much faster under centrifugation. He added radioactively labelled λ phage repressor to λ phage DNA and found that much of the repressor sedimented with the DNA, indicating that they were complexed. In a further experiment, the DNA from a different phage was used but in this case the repressor did not sediment with the λ phage DNA and so was assumed not to be complexed with it. Similar results were obtained by GILBERT and MÜLLER-HILL [1967] with the lac repressor.

Confirmation was obtained by RIGGS *et al.* [1968] using a completely different method. RIGGS and BOURGEOIS [1968] has already observed that the lac repressor would bind very strongly to Millipore filters and furthermore, that IPIG would bind with it. RIGGS *et al.* also showed that the DNA-repressor complex would also become attached to a cellulose nitrate filter, again illustrating that the repressor binds directly to the DNA.

F. The Complexing of Repressors with Specific Regions of the DNA

Two facts were now reasonably certain: the inducer binds to the repressor and the repressor binds not to RNA but to the DNA itself. This last fact established what had hitherto been strongly suspected, that the operator is actually a part of the DNA molecule. It now remained to be determined whether or not the binding of the repressor was to a specific site on the DNA molecule or whether it was random and unselective. The solution to this problem was immensely important as on it ultimately rested the whole structure of the specific repressor theory put forward by JACOB and MONOD. If it was found that the repressor complexed with the DNA unspecifically at almost any point along the molecule, then the specific repressors predicted by the theory had in fact not been separated and the method of GILBERT and MÜLLER-HILL had been founded on the fallaceous assumption that the interaction between the repressor and inducer was a specific one or the whole concept of the specific repressor theory was incorrect despite all the evidence that had accumulated which indicated that the theory was not only brilliant but also sound.

The problem was approached by the use of phages and bacteria con-training mutant DNA in the region of the operator. If *E. coli* carries a mutation in the lac operator, it can be shown by the co-centrifugation technique that the lac repressor will not complex with the DNA. However, even more impressive is the evidence from λ phage which shows that both the operators have to carry mutations to prevent the binding of the repres-sor. If it is borne in mind that the repressor operates at two different sites on the genome, then the evidence for the specific binding of the repressor by DNA becomes almost irrefutable. Confirmation was obtained by RIGGS *et al.* [1968] using the cellulose nitrate filter technique on the lac system. They also concluded that not only does the repressor complex with the DNA but that the reaction is quite specific. Their evidence included the fact that IPIG prevents the DNA binding of the repressor whereas non-inducing galactosides do not. Furthermore, mutant DNA without a lac region will bind in only insignificant amounts to the repressor trapped on the filter. It was now known that DNA and repressor complexes are specific but the question still remained of whether this would prevent transcription.

G. The Mechanism of Repression in Microorganisms

It has recently been shown that when λ phage repressor is bound to the two operators, RNA polymerase is not able to initiate transcription of the two operons. It has not yet been shown that the lac repressor specifically prevents the transcription of the lac operator. The mechanism which allows the repressor to recognize the particular sequence of base pairs known as the operator is unknown. One of the more obvious theories, which postulates that repressors are composed in part of RNA which enables them to recognize specific regions of the DNA molecule, is attractive but seems to be quite unfounded. Neither the lac nor the λ phage repressors seem to be associated in any way with RNA. Hardly anything is known regarding the nature of the mechanism of the repressor-operator complex. It is known, however, that the repressor will only bind to native DNA; it will not bind to denatured, single-stranded DNA [GILBERT and MÜLLER HILL, 1967]. It is sometimes thought that it is the negatively charged phosphates in the DNA chain which to some extent attract the repressor–this may be so but it does not explain the specificity of the repressor. It is interesting to note though that the repressors that have been isolated, so far, are large protein molecules, the tertiary structure of which may play a part in the recognition of specific operators.

II. The Differential Function of Genes

The mechanism of genetic control is as yet unknown in the Metazoa although the large volume of evidence which has accumulated round the subject does permit speculative guesses to be made as to the type of system which is likely to be involved. Although fairly recently the situation seems to have been largely clarified in the phage and bacteria, there are many fundamental differences between these organisms and the Metazoa. The systems of gene control which have been found to operate in microorganisms do not necessarily explain the situation existing in higher animals where the whole system is undoubtedly far more complex. This is not meant to imply that the mechanism of gene control as elucidated in the phage and bacteria is irrelevant to the Metazoa but rather that it is almost certainly not the whole story.

A. Metazoan and Bacterial Systems Contrasted

In the first place, phage and bacteria do not undergo a period of development leading to the formation of a large number of cell types. The entire genome or at least the major part of the entire genome, is expressed during one life cycle of a microorganism although there may be temporal differences between the expression of various regions with the result that the total genetic complement does not function simultaneously. This situation is to be contrasted with that found in the Metazoa where the entire genome is almost certainly never expressed by any one particular set of chromosomes but instead various cells become specialized to perform certain specific functions. Any part of the genome that is not concerned with the specialized function or with producing proteins which are basic to the operation of any cell, the 'irrelevant genes', are presumably repressed or, as it may be preferable to say, 'non-functional' since we do not know them to be repressed in the bacterial sense. Furthermore, with a few exceptions, all the somatic cells contain the diploid number of chromosomes and, therefore, contain twice the haploid quantity of DNA that is present in the gametes [BOIVIN *et al.*, 1948; MIRSKY and RIS, 1949].

The regulatory system in bacteria is essentially one that is adapted to respond rapidly to changes in the substrate and there is a continuous and rapid triggering of the regulator genes which are producing the messenger RNA for the repressor proteins and a rapid production of the short-lived messenger RNA of the structural genes. For instance, it has been calculated by KAEMPFER and MAGASANIK [1967] that in *E. coli,* messenger RNA for β-galactosidase is being synthesized within 2.5 min of induction.

Many instances may be cited in the Metazoa where such a system would be of limited use. The most obvious example is found in embryonic induction where long-term and far reaching effects may result from a brief exposure of particular cells to an inductive stimulus. The idea of a succession of operon systems, each being triggered off by the previous one, is feasible, though somewhat unlikely because of the enormous number of repressor regulator genes which would need to be functioning continuously. Add to this the fact that it has been estimated by various workers that about 80–95% of the genome is repressed in a differentiated vertebrate cell nucleus, then, even allowing for a high degree of redundancy, the number of repressor genes would be prodigious. In the bacterial systems so far analyzed, the number of regulator genes for a system tends to be similar to the number of structural genes and the situation is almost certain to be considerably more complex in the Metazoa.

This is not intended to imply that more complex organisms are incapable of a rapid response at the chromosome level and there are examples showing the speed with which such a response can be effected, but most of the work concerns the effect of hormones on the mammalian genome, cell-free systems [e.g. KIDSON and KIRBY, 1964; KIDSON, 1965; BARKER and WARREN, 1966; BREUER and FLORINI, 1966; KIM and COHEN, 1966; MEANS and HAMILTON, 1966] and investigations carried out on plants [ROYCHOUDHURY *et al.,* 1965; JOHRI and VARNER, 1968]. All these examples are concerned with activation and it is well known that the activation of genes in the Metazoa is far more rapid than gene repression because of the comparatively long life of the mRNAs.

B. Activation or Repression

If the basic system of differential gene function in multicellular organisms is one of an activation of particular sites on the genome rather than a system of the selective repression of all but relevant genes, then this difficulty becomes less formidable and genomic repression does not have to be specific and may just be an overall effect; the onus then falls on the factors which activate specific regions of the genome.

C. The Metazoan Nucleus

Metazoan cells have a definite cell nucleus which is surrounded by a nuclear membrane which effectively separates the chromosomes from the rest of the cell. It is possible that one of its functions is to act as a barrier to keep 'gene regulatory substances' in a high concentration around the chromosomes. The only time the nuclear membrane breaks down in the life history of a normal cell is at mitosis, at which time the chromosomes may be assumed to be in more intimate contact with the general cell cytoplasm than at any other time.

Various nuclear proteins have been identified in the nuclei of vertebrate cells. Among them are the DNA and RNA polymerases, various enzymes of metabolism, the acidic nuclear proteins and the basic nuclear proteins. The last group may be subdivided into the protamines and the histones. The latter have had a great deal of attention focused on them because of their suggested role as agents in the differential function of the genome.

D. The Role of Nucleoproteins in Gene Regulation

The involvement of proteins in gene regulation seems to be a fact and many investigators have shown that if the proteins are removed from chromatin, the result is the derepression of the genome [GEORGIEV *et al.*, 1966; MARUSHIGE and BONNER, 1966; PAUL and GILMOUR, 1966a, b; BONNER *et al.*, 1968; PAUL and GILMOUR, 1968]. That this effect is due to a change in the types of genes which are expressed to the extent of a factor of about twenty which makes it unlikely that the phenomenon is due to an increase in the activity of RNA polymerase resulting in a quantitative increase in the products of the same genes. The question of which of the chromatin proteins are responsible for gene repression is less easy to answer and the mechanism by which they operate is, at the moment, completely unknown although there are some hypothetical ideas which have been put forward and which, in the course of time, may prove to be correct.

E. Non-Histone Nucleoproteins

Besides the role that histone proteins are thought to play, some observations have been made on the non-histone nuclear proteins. The experiments of GEORGIEV *et al.* [1966] show that the removal of some non-histone proteins will increase template activity and PAUL and GILMOUR [1968] have also produced evidence to show that the template range increases with the removal of non-histone protein from chromatin. It has been reported by HNILICA and KAPPLER [1965] that in liver cells, the rate at which the non-histone proteins are synthesized is related to the degree of informational RNA synthesis that the cell is engaged in. Non-histone proteins, in any case, show a high level of metabolic activity as a number of studies have shown [HNILICA and KAPPLER, 1965; WANG, 1965, 1967]. Some investigators have shown that non-histone proteins are found in close association with chromatin DNA and when reacted *in vitro,* will complex with DNA [DOUNCE and HILGARTNER, 1964; LEVESON and PEACOCKE, 1966, 1967; WANG, 1967]. WANG has also presented data showing that nuclear histones will complex with the acid nuclear proteins and it has been suggested [PATEL and WANG, 1965; WANG, 1965, 1967] that the synthesis of the non-histone nuclear proteins is DNA dependent. The acidic nuclear proteins are also present in large amounts but have been poorly studied, largely because their extreme insolubility makes their characterization very difficult.

F. Occurrence of Nucleohistones

Despite the widespread occurrence of histones in plant as well as in animal cells [KOSSEL, 1921; SETTERFIELD *et al.,* 1960; PHILLIPS, 1961; IWAI, 1964], their presence in bacteria is still disputed. CRUFT and LEAVER [1961] for example, have published experimental evidence to show that they have isolated histones from *Staphylococcus aureus,* whereas ZUBAY and WATSON [1958] have been unable to identify histones in *Escherichia coli.* Whatever the general situation may be, it seems fairly certain either that histones are absent from bacteria or that they are present in extremely small and probably insignificant amounts. In this respect they differ a great deal from most metazoan cells where histones are present in large quantities. In fact, on a dry weight basis, DNA and the nuclear histones are present in approximately equal amounts.

G. Nucleohistones and Gene Regulation

It was first suggested by STEDMAN and STEDMAN [1943] that the nuclear histones had a function in gene modification. At the time it was not appreciated that nucleic acids were the major component of chromosomes and the STEDMANS considered that they were concerned with the formation of the spindle at mitosis. Their theory of the histones as modifiers of the genome was based on evidence that the quantity of nuclear histones varied from one cell type to another. Later evidence, including some of STEDMANS, indicates that there is a lack of variation in nuclear histones from different cell types and also in various tissues of the same species [CRUFT *et al.,* 1957 and 1958; BUSCH and DAVIS, 1958; HNILICA *et al.,* 1962; BUSCH *et al.,* 1963; HNILICA and BUSCH, 1963]. On the other hand, it has been reported by BLOCH [1962a, b] that in some lower species, the histones vary according to the stage of development and the degree of specialization.

It has now been found by various workers [HUANG and BONNER, 1962; ALLFREY *et al.,* 1963; BONNER and HUANG, 1963] that in a cell free system, the addition of histones will inhibit DNA-primed RNA synthesis. The experiments of HUANG and BONNER on the *in vitro* synthesis of RNA from the DNA of pea seedling nuclei showed that a stoichiometric relationship exists between the blocking of RNA synthesis and the prevailing histone concentration. It was also found possible to reform the nucleohistones by the addition of histones to the isolated DNA. A year later HUANG *et al.* [1964] found that whole nucleohistone from thymus was either unable to prime RNA synthesis or could do so only at an extremely low level. The very lysine-rich histones appeared to completely inhibit synthesis but the slightly lysine-rich and arginine-rich histones were less efficient in the role. However, ALLFREY *et al.* [1963] had already published the results of work done on isolated calf thymus nuclei which showed a completely opposite effect. As ALLFREY *et al.* themselves pointed out, histones are also able to prevent ATP synthesis and it might be this effect rather than a true blocking of the genes which gave this result with cell nuclei in contrast to the cell free system used by HUANG *et al.* Recent experiments by CLARK and FELSENFELD [1971] indicating that only about half the DNA of calf thymus is protected by its associated proteins might be held to favour this view although as the authors admit, results obtained by the action of staphylococcal nuclease on mammalian chromatin must be of dubious biological validity.

It was also noticed by ALLFREY *et al.* that isolated nuclei treated with trypsin were able to synthesize RNA at three to four times the rate of those that were untreated. Trypsin hydrolyses about 80% of the nucleohistones but leaves 95% of the non-histone proteins almost unharmed; this means that the nucleohistones can be preferentially removed [BONNER and HUANG, 1966]. As RNA polymerase appears to be unaffected by this method, the explanation of the effect cannot be that trypsin increases the activity of the polymerase and so quantitatively rather than qualitatively increases RNA synthesis. The readdition of histones to the system completely reverses the effect.

In a series of experiments HINDLEY [1964] has shown that the gradual extraction of histones produces an effect similar to that observed by ALLFREY *et al.* As the very lysine-rich fraction was removed, no significant effect was noted in the rate of RNA synthesis but with the withdrawal of the slightly lysine-rich and the arginine-rich fractions, the rate of synthesis increased appreciably. The results of HUANG *et al.* are similar to those of BARR and BUTLER [1963] who found that the very lysine-rich histones produced the greatest inhibition of RNA synthesis.

In view of these conflicting results it should be appreciated that the quantities of histones used in this type of experiment vary greatly between different investigators and also that the activity of DNA polymerase itself can be influenced by the concentration of histones [BAZILL and PHILPOT, 1963].

The evidence does make it at least possible that the histones are concerned in differential gene function. The fact that the bacterial repressors so far isolated have not turned out to be histones does not eliminate them as potential candidates for vertebrate gene regulation because, as has already been noted, there is a lack or paucity of histones in microorganisms. In the salivary glands of *Drosophila,* HORN and WARD [1957] and SWIFT [1963, 1964] have shown that there is a close spatial relationship between nucleohistones and DNA in the chromatin and according to SWIFT, in the interchromomeric regions, both DNA and the nucleohistones are only present in very small amounts. More recently BOUBLIK *et al.* [1971] have even produced some evidence indicating how certain individual histones interact with DNA. MARKERT and URSPRUNG [MARKERT and URSPRUNG, 1963; URSPRUNG and MARKERT, 1963] tried injecting histones and albumens into the nuclei of frogs eggs and reported that this treatment resulted in all development being arrested after late blastula stages. Nuclei from these arrested blastulae were transplanted into enucleate frogs eggs but were never able to support development beyond late blastula stages. This result may have been brought about by the complete and irreversible blocking of the genome by the injected histones and albumens or it could possibly have resulted from chromosome damage at the time of injection.

H. The Mechanism of Repression in the Metazoa

Although it is well known that histones are able to bind to DNA, there is so far no evidence that they are able to bind to specific sites in the way that bacterial and phage repressors have been found to operate and as was originally predicted by JACOB and MONOD [1961]. In fact, the binding of nucleohistones to DNA is thought to be mediated through the phosphate groups of the DNA which would tally with the non-specific relationship. Histones have been shown, nonetheless, to have an effect on the structure of the chromosome and it has been known for some time that they are able to affect the loops on the lampbrush chromosomes of Amphibia [IZAWA *et al.*, 1963]. LITTAU *et al.* [1965] have noted that removal of the lysine-rich histones results in the chromosomes taking up the reticulate form, while readdition of those histones will reverse the process and the chromosomes will adopt the clumped or condensed form. It is obvious that the critical experiment showing the role of the nucleohistones in gene regulation has not yet been performed.

III. The Regeneration Blastema as a Dedifferentiating System

There are two well-known environments in which fully differentiated cells may be induced to undergo the process generally known as dedifferentiation. First, the conditions of some tissue culture media will stimulate cells grown *in vitro* to dedifferentiate and it has often been observed that this property is augmented by frequent sub-culturing. The second milieu favouring cellular dedifferentiation is that which is produced during epimorphic regeneration. The regeneration of naturally or experimentally extirpated parts has been studied by many workers using a wide variety of animal species and the Amphibia have proved to be the vertebrate group most favourable for this type of experiment. The faculty for regeneration is far more highly developed in the urodeles than in the anurans where the ability is generally either lost or greatly diminished at metamorphosis.

The regeneration blastema of late larval amphibians constitutes a good experimental system for the investigation of the phenomenon of cellular dedifferentiation and for attempting to determine the genetic potentialities of dedifferentiated cells. Late larvae still retain their regenerative ability, albeit reduced, and at the same time cellular differentiation is complete and only a relatively small proportion of fibroblasts is present.

A. Regeneration in Amphibia

There is close agreement among biologists concerning the gross pattern of regeneration in the class Amphibia and while it is not intended that a detailed description of amphibian regeneration be given in this chapter, it is felt that a résumé of the main features of amphibian and especially early anuran regeneration is, at this point, appropriate and may be summarized as follows. Within a few hours of amputation, the cut surface is quickly covered by a layer of epithelium which migrates distally from the epithelium of the stump. This is followed by a period of phagocytosis which removes the cellular debris which has accumulated as a consequence of the wounding. Two to three days following amputation, dedifferentiated cells start to accumulate at the transected surface. These cells constitute the regeneration blastema which grows by proliferation and eventually gives rise to the regenerated part.

B. The Local Origin of Blastema Cells

Concerning the accumulation of blastema cells at the site of transection, there would now seem to be very little doubt that they arise from some of the fully differentiated tissues of the stump and not from a reserve of undifferentiated cells as is observed in some invertebrate types. BUTLER [1933] experimented with *Ambystoma* and found that if only the limb was irradiated with x-rays prior to amputation, then the ability of that limb to regenerate was lost. If, however, the limb was shielded and a large part of the body was irradiated, then the limb was able to regenerate. He also found that an unirradiated limb grafted into a fully irradiated host would regenerate normally. More recently, work by CARLSON [1967] has shown that if actinomycin D is given to the adult newt one day before amputation, dedifferentiation of the stump tissues is inhibited and a blastema fails to form. These two types of evidence would seem to show, beyond all reasonable doubt, that the blastema cells arise at the site of amputation and are not transported there from other sites in the body as had been previously postulated. There has been in the past a great deal of argument concerning the origin of the blastema cells and a certain amount of confusion seems to have arisen from the fact that most of the invertebrates that have been investigated seem to possess a reserve of undifferentiated cells which form the regeneration blastema, whereas in the vertebrates most investigators hold the view that the blastema cells arise by the process of dedifferentiation of the stump tissues. Hence, all attempts to formulate a general theory for the origin of blastema cells that would embrace the whole animal kingdom, have only clouded the issue and made particular instances more difficult to accommodate within the general theoretical structure.

C. The Role of the Epidermis

Not all the stump tissues form blastema cells. The idea that blastema cells were largely of epidermal origin was first put forward by GODLEWSKI [1928] and was later championed by ROSE [1948] using *Triturus viridescens.* This theory has largely fallen into disrepute following the work of KARCZMAR and BERG [1951], CHALKLEY [1954] HAY and FISCHMAN [1961]. KARCZMAR and BERG followed the localization of alkaline phophatase during regeneration in *Ambystoma opacum* and *A. punctatum* and observed that although blastema cells gave positive results when treated with phosphatase stains, the epidermis would only give a positive reaction after the early stages of regeneration had been completed. CHALKLEY showed that there was no need to postulate that any role was played by the epidermis in the formation of blastema cells as the rate and time of cell division in the inner stump tissues was sufficient to account for blastema formation. However, the most convincing evidence has been presented by HAY and FISCHMAN who used tritiated thymidine autoradiography on the regenerating forelimb of *Triturus viridescens.* They found that the stump tissues in the process of dedifferentiating incorporated large quantities of tritiated thymidine but that the apical epithelium did not, in spite of the fact that epidermis from other sites became labelled. As almost all the blastema cells were later found to be labelled, the evidence against the epidermis contributing to the blastema, at least to any significant extent, appears to be very strong.

D. The Dedifferentiation of the Stump Tissues

Most investigators hold the view that the blastema cells arise by the process of dedifferentiation of the muscle, cartilage, endomysium, epimysium and Schwann cells of the stump tissues and that as well as these, contributions are made by the fibroblasts (fig. 2a and b). THORNTON [1938a, b] carried out a most extensive and lucid study of the formation of blastema cells from the stump tissues in the forelimb of *Ambystoma punctatum* and concluded that the principal source of undifferentiated cells was the dedifferentiation of skeletal muscle. MANNER [1953] investigated the regenerating forelimbs of adult *Triturus viridescens* and on the basis of observations with the light microscope, concluded that the process was actually one of the disintegration of the mesodermal tissues and that this phenomenon had wrongly been described as dedifferentiation. He described large numbers of pycnotic nuclei that were present during the accumulation of the blastema. He concluded that the blastema was formed essentially of fibroblasts and that it contained the disintegrating products of cartilage and muscle. However, BODEMER and EVERETT [1959] have studied protein synthesis in regenerating newt limbs and using methionine S^{35} as a label, have produced autoradiographs which show that protein is being synthesized by the blastema cells. This evidence, combined with CHALKLEY's studies in 1954, which showed that active mitosis occurred in blastemal cells, and HAY and FISCHMAN's excellent work [1960] which produced evidence of considerable synthesis of DNA in the distal inner stump tissues of *Triturus viridescens* limbs at 5 to 10 days post-amputation, all argue strongly against cellular degeneration. HAY and FISCHMAN also reported that after the first few post-operative days, there were very few pycnotic nuclei to be seen. It would seem likely that those that were present initially were probably due to the operative trauma and direct physical cellular damage. It seems, therefore, as if HAY and FISCHMAN were correct in suggesting that MANNER had confused the process of muscle dedifferentiation and degeneration.

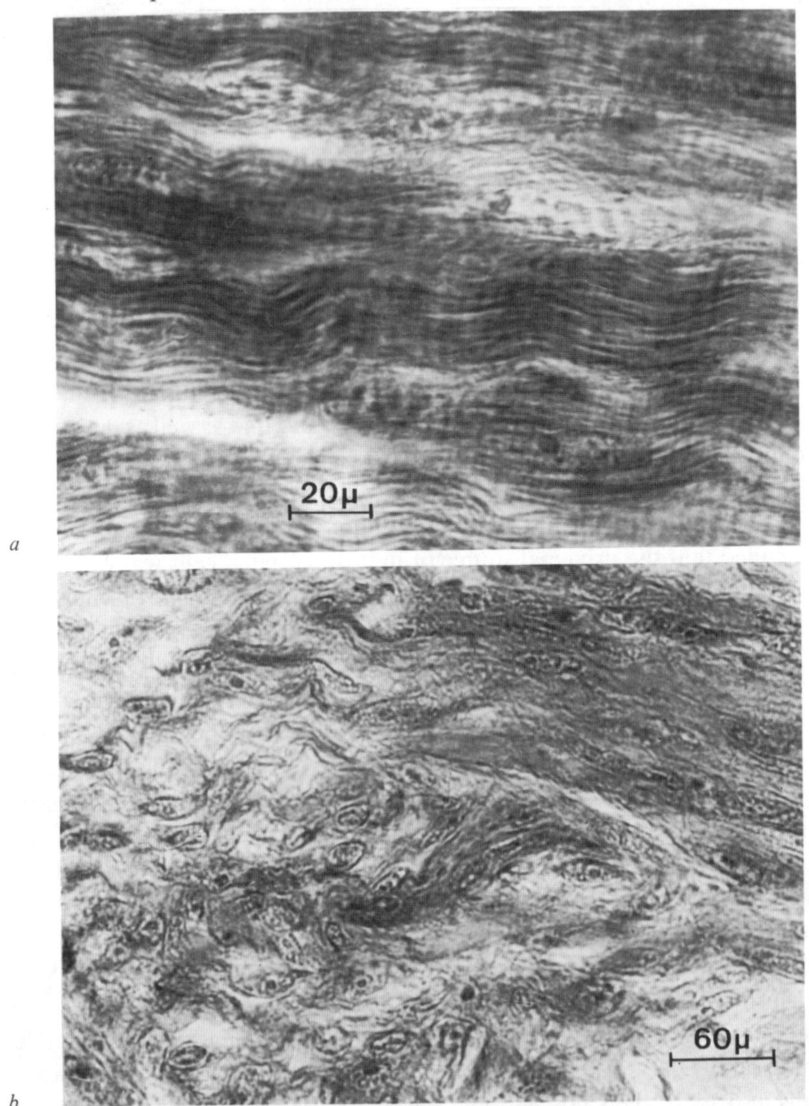

Fig. 2a, b. a Fully differentiated skeletal muscle from a *Xenopus* larva at stage 59. *b* Muscle dedifferentiation following amputation of the hind limb; same stage as *a.* Both photographed from sections stained by a modified method of Masson's triple stain [BARKER, personal communication].

IV. The Significance of the Dedifferentiated State

THORNTON's conclusions [1938a, b] on the formation of the blastema are confirmed and extended in a series of observations made by HAY [1962] on *Ambystoma* using both the light and electron microscopes. HAY describes the blastema cells as 'resembling mesenchyme cells' and in another paper, HAY and FISCHMAN [1961] refer to 'undifferentiated appearing blastema cells'. HAY [1959] observed that following the breaking up of the myofibrils, the myofilaments and Z-bands gradually disappear from the sarcoplasm and the sarcoplasmic reticulum breaks up into small vesicles and finally disappears. This follows HAY's observations [1958] that the endoplasmic reticulum, which in the differentiated chondrocyte forms a continuous network of membrane bound cavities, becomes discontinuous in the dedifferentiated cell and the flattened vesicles of the Golgi apparatus become predominantly vesicular in the dedifferentiated state.

In the 1959 paper, HAY states that the blastema cells derived from muscle cannot be distinguished from those derived from cartilage or other connective tissue. She claims that not only do the dedifferentiated tissues of the stump lose the cytological specializations which were characteristic of their once fully differentiated state, but they also gain some embryonic qualities and she quotes as examples the rounded nuclei with prominent nucleoli, an increased nucleocytoplasmic ratio and an increased number of free ribosomes. She considers that the complete reversal of muscle and cartilage to a common basophilic cell type with all the ultrastructural characters of embryonic cells must imply that the process of dedifferentiation involves more than just the loss of specialized characteristics and that there must be some underlying return to a genomically more versatile state such as is found in undifferentiated embryonic cells. This raises an extremely controversial point: does the process of dedifferentiation entail the regaining of the pleuripotency associated with embryonic cells or is it simply a loss of the cytological specializations which once defined it as a particular type of differentiated cell (fig. 3).

There is a certain amount of disagreement over the use of the appropriate terms in this field and some definition is necessary so the reader will know in exactly what sense the following terms are used in this chapter. A differentiated cell is defined as one that exhibits specific morphological and biochemical characteristics which distinguish it not only from undifferentiated 'embryological' cells but also from other differentiated cell types exhibiting a different mode of specialization. The converse of the differentiated state is that in which the cell is described as 'undifferentiated'. An undifferentiated cell may be defined as one which does not and *never has* exhibited any signs of morphological or biochemical specialization

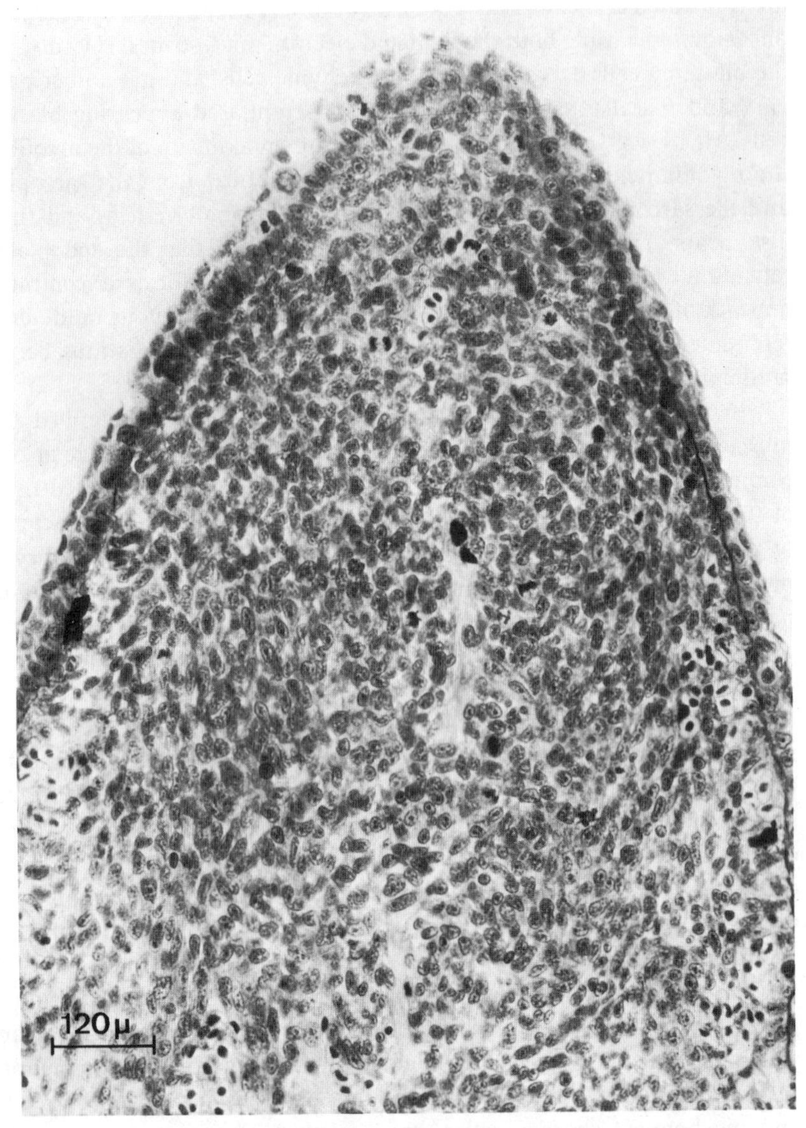

Fig. 3. The regeneration blastema at 7 days post-amputation. By this dedifferentiation is complete and the blastema is composed of a mass of cells all more or less similar in appearance. (Stain as in fig. 2.)

which would lead to its identification as a particular type of specialized cell. This state is characteristic of early embryonic cells which are assumed to be totipotent. It will, of course, be appreciated that such terms as 'differentiated' and 'undifferentiated' are relative and represent points on a continuous spectrum. The failure to acknowledge this fact has often led to confusion and misunderstanding. A dedifferentiated cell is one that at one time exhibited morphological and biochemical specializations that enabled it to be defined as a particular type of cell (i.e., 'differentiated' in the most extreme sense) but which has since lost its specialized cytoplasmic structure so that it has come to *resemble* an undifferentiated cell.

A. The Difficulty of Cell Identification

The problem of the dedifferentiation and redifferentiation of cells is not only an extremely vexed question, it is also one that is very difficult to prove or disprove experimentally. The principal difficulty lies in finding an utterly reliable method of labelling cells which will allow them to be identified right through the processes of dedifferentiation and redifferentiation. As the cells lose their cytological specializations during dedifferentiation it is only of limited use to be able to identify them by the specialized proteins which they characteristically produced in the differentiated state. HOLTZER [1961] tried to trace dedifferentiating and redifferentiating muscle by using fluorescein labelled anti-myosin and although he showed that muscle would bind the anti-myosin up until the 7th to 8th day of regeneration and again on the 12th to 13th day, there was a period of 5 to 7 days during which it was not bound and it is therefore not possible to be certain whether they were in fact the same cells that had originally constituted the fully differentiated muscle or whether a certain amount of metaplasia had taken place.

Two other types of label have been widely used in attempts to solve this problem but both of them have inherent difficulties. There is the method of identifying the cells by some characteristic of their morphology, such as ploidy or the number of nucleoli, which is not lost during dedifferentiation. Alternatively there is the method of radioactively labelling the DNA of the nucleus. The disparity in size between haploid and diploid nuclei was used by HERTWIG [1927] in an attempt to show that blastema cells were derived from stump tissues. Unfortunately, the criterion of nuclear size is not always sufficiently accurate to provide a definite answer. HAY [1952] used polyploid grafts to trace the formation of the blastema and relied on the nucleolar number as a marker. Nucleolar number is not always constant, however, due either to spontaneous ploidy or to fusion of the nucleoli [FRANKHAUSER and HUMPHREY, 1943; WALLACE, 1963] and in any case, at some stages of differentiation they can be extremely difficult to find. Against these disadvantages must be set the fact that the label does not become diluted and neither can it be transferred to neighbouring cells.

The other method which has been commonly employed is that of labelling with radioisotopes which generally involves autoradiography as a means of identifying the site of the label. RIDDIFORD [1960], HAY and FISCHMAN [1961], O'STEEN and WALKER [1961] and STEEN and THORNTON [1963] have used thymidine-^{3}H but have been unable to reach any unequivocal conclusions about the stability of differentiation because the method has several drawbacks which make the positive identification of cells throughout the entire process impossible. First, the tritium is to some extent labile so it is possible for it to be transferred to surrounding cells. The use of a ^{14}C label would be more stable and less inclined to spread from one cell to another but, unfortunately, the rather higher β emission of ^{14}C makes the resolution of the autoradiographs less accurate than the low energy β emission of ^{3}H. The second difficulty with a radioactive label is that it becomes progressively more diluted as the cells undergo mitosis until in a rapidly proliferating tissue such as the blastema, it can become very difficult to identify the label with any certainty. The final and ubiquitous difficulty in using a radioactive label in any biological system is the possibility of radiation damage, the mild effects of which may be very subtle and difficult to recognize and may therefore pass unnoticed unless the experiment is very carefully controlled. This may result in the wrong interpretation being put on the observed facts. There is a certain amount of evidence that thymidine-^{3}H is toxic to developing chick embryos [SAUER and WALKER, 1961] and to mammalian cells in tissue culture [DREW and PAINTER, 1959]. There is also a report by XEROS [1962] that cold thymidine is able to inhibit cell division and STEEN [1967] has found that it will inhibit the rate of regeneration very slightly. He also observes that thymidine-^{3}H given in 15 C/gm doses (6.7 c/mM specific activity) causes a quite definite slowing down of regeneration. Although some gross abnormalities were observed in these experiments, STEEN notes that the tissues appeared to be histologically normal. The most successful of the labelling methods is to use a double label of both ploidy and radioisotopes as STEEN did, although even this technique is unable to give an unequivocal answer to the problem.

B. The Dedifferentiated State

Two things can be said with certainty about dedifferentiating cells. First that there is a loss of the specialized morphological structures which once defined the cell as a particular type. This has been reported many times both in histological studies [e.g., BUTLER, 1933; THORNTON, 1938a, b; TRAMPUSCH and HARREBOMEE, 1965] and also in cytological studies [e.g., HAY, 1958, 1959, 1962; SALPETER and SINGER, 1960]. Immunochemical studies of muscle by DE HAAN [1956] and LAUFER [1959] have confirmed, at a more sensitive level, the earlier observations of BUTLER [1933]. There seems to be no doubt that dedifferentiation involves the suspension of the synthesis of proteins highly characteristic of a particular differentiated cell type, while equally, there is good evidence that there is a certain amount of protein synthesis [BODEMER and EVERETT, 1959] presumably at a much more basic level. The second well established fact about dedifferentiation is that the nucleo-cytoplasmic ratio increases. This is probably due to two causes: the loss of a certain amount of cytoplasm and the increase in size of the nucleus and nucleoli. This latter phenomenon is generally regarded as being associated with the start of the active synthesis of nucleic acids and so is not necessarily an 'embryonic' characteristic although, of course, many embryonic cells are actively dividing and synthesizing DNA. HAY's report that there is an increase in the number of ribosomes in the dedifferentiated cell may also partly be the result of the swelling of the nucleus; possibly, the same number of ribosomes occupy a smaller space and thus appear more numerous.

SALPETER and SINGER [1960] made a particularly interesting observation on dedifferentiated cells when they examined the regenerating forelimb of the adult newt, *Triturus.* They showed that the blastema cells of the adult have a rich granular endoplasmic reticulum. This differs from the report of HAY in 1958 which states that in differentiating chondrocytes the endoplasmic reticulum becomes discontinuous, and in 1959, that the sarcoplasmic reticulum of muscle cells breaks up into small vesicles and gradually disappears. This divergence in findings between adult and larva is extremely interesting. As SALPETER and SINGER have pointed out, if rounded nuclei with prominent nucleoli, increased nucleo-cytoplasmic ratio, increased number of free cytoplasmic ribonucleoprotein granules and a decreased amount of endoplasmic reticulum are used as criteria for indicating the degree of dedifferentiation a cell has undergone, then the cells of the adult blastema are less dedifferentiated than those of the larva and according to the criteria used by HAY, are not developmentally equivalent. However, it has often been observed that cells undergoing intense mitotic activity tend to have less endoplasmic reticulum than those that are not [PORTER, 1954; PALADE and PORTER, 1954; PALADE, 1955]. SALPETER and SINGER put forward the wholly plausible explanation that as the larval blastema is larger than the adult blastema, a higher mitotic rate is to be anticipated in the larva and, consequently, this might be expected to lead to a paucity of endoplasmic reticulum. PORTER and MACHADO [1960] have expressed the view that the endoplasmic reticulum is not an end product of cellular differentiation but, rather, that it is a means of achieving such differentiation and that it is the cells secreting large quantities of protein that have a complex endoplasmic reticulum [e.g., PALADE, 1955; EKHOLM and SJÖSTRAND, 1957; NILSSON, 1958]. Thus, it appears that the criteria used by HAY to support the idea that all dedifferentiating cells in the blastema become identical, except in the strictly morphological sense, have been an unfortunate choice. To extrapolate from this situation and draw conclusions concerning the genetic versatility of such cells is not justified.

HAY's work on the morphological similarity of the blastema cells has been substantiated at the histochemical level by several workers, amongst them DE HAAN [1956], LAUFER [1959] and HOLTZER *et al.* [1960]. The latter studied the dedifferentiation of cartilage from 10-day chick vertebrae. After releasing the chondrocytes by digesting the matrix with trypsin, they cultured them under conditions promoting rapid cell division and after a varying number of mitoses (1–60 days), the cells were centrifuged into pellets. Thionin failed to give the positive reaction for chondroitin sulphate in all pellets that had undergone mitosis for more than 9 days. HOLTZER [1961] also studied the dedifferentiation of muscle using fluorescein labelled antibodies. In mature muscle the fluorescent antibody reacts with antigens confined to the A-bands of cross-striated muscle. Using regenerating salamander tail and limbs, HOLTZER found that from 40–80% of the blastema cells at 3–4 days post-amputation bind labelled anti-myosin but that by 7 days the fluorescent antibody is no longer bound. This work is extensive and has been extremely carefully carried out and it cannot have levelled at it many of the criticisms which are often applied to immunological work. Thus, it is fairly certain that by the 7th day of regeneration, if there is any myosin present in the blastema cells, it is present in such minute quantities that it cannot be detected morphologically by the electron microscope nor by the very sensitive technique of immunological reactions. While it seems almost irrefutable that it is the dedifferentiation of stump tissues which gives rise to the cells of the regeneration blastema and that during the process the cells lose all the cytological specializations which characterized them in the differentiated state, it is far more difficult to determine the genetic potentialities of these simple-looking and apparently identical dedifferentiated cells. To deduce that dedifferentiated cells actually return to an embryonic state and regain the wide spectrum of genetic potentialities associated with the undifferentiated state is not justifiable on the evidence which has, so far, been put forward.

There is no doubt that 'embryonic' and blastemal cells resemble each other extremely closely as has been noted and remarked on by numerous investigators but when the following facts are considered, this resemblance does not seem so remarkable or so significant. Blastema cells in the process of dedifferentiation, lose all the signs of morphological specialization which once defined them as a particular type of differentiated cell. Embryonic cells are undifferentiated and have not yet developed any specialized cytoplasmic apparatus which will characterize them in their differentiated state. In the absence of any marked specialized morphological differences which would enable them to be distinguished, it is hardly surprising that blastema and embryonic cells should show a high degree of resemblance.

V. How Stable Is Differentiation?

The stability of differentiation is a problem which has fascinated biologists for many years and will probably prove to be a subject for discussion and experiment for some time to come. The basic interest lies in an understanding of genome control and the phenomena of metaplasia and neoplasia which are obviously closely related to the stability of differentiation. To expect that there would be one simple, clear-cut and comprehensive theory that would account for the phenomenon of differentiation throughout the whole animal kingdom, would be to ignore the complexity and variety of living organisms and to deny to them that very adaptability that has been so characteristic of their evolution.

The problem of whether metaplasia normally occurs in the regenerating system is as yet unsolved. It has not yet proved possible to design a wholly convincing experiment to give a definite answer to this problem although ingenious experiments have been devised that have yielded data both for and against metaplasia. Perhaps this is only to be expected in a system as complex as that of regeneration and it would not seem unreasonable to suppose that different tissues vary in their ability to undergo metaplasia in the same way that tissues vary in their ability to dedifferentiate and take part in blastema formation.

A. Metaplasia in Regenerating Muscle

HOLTZER [1961] nearly solved the problem for dedifferentiating muscle and STEEN's investigations [1967] which were carefully carried out, yielded illuminating results relevant to this topic. STEEN used a double label of nucleolar number (heat induced triploidy) and thymidine-^{3}H in his study of limb regeneration in the axolotl *(Siredon mexicanum).* He was able to trace the transplanted cells of labelled muscle in the exarticulated stump into the newly regenerated cartilage. Furthermore, he showed that even when cartilage was present in the stump, the transplanted, labelled muscle still contributed to the cartilage of the regenerate. This work confirms the opinion of a large number of earlier investigators who showed by various methods, that the cartilage of the regenerate does not necessarily originate from the cartilage of the stump [MORRILL, 1918; WEISS, 1925; BISCHLER, 1926; BISCHLER and GUYENOT, 1926; THORNTON, 1938b]. More recent evidence along these lines has been published by GOSS [1958]. However, by contrast KONIGSBERG [1963] found that clonal cultures of chick skeletal muscle were stable but, as STEEN points out, muscle is a mixture of cell types having associated with it, among other things, connective tissue so that the possibility of fibroblasts contributing to the regenerating cartilage cannot be excluded. This may explain why KONIGSBERG found that his clonal cultures were stable; he was using a pure cell line instead of a mixture of cell types.

B. Metaplasia in Regenerating Cartilage

Cartilage appears to be far more stable in its differentiation than muscle, though again, the lack of both fibroblasts and any indication of metaplasia in cartilage may be significant. PATRICK and BRIGGS [1964] experimented with regenerating axolotl limbs and grafts of triploid cartilage and found that triploid cells were only localized in the cartilage of the regenèrate. They concluded that chondrocytes are stable and that they do not dedifferentiate into a pleuripotent cell type. This last statement is barely justified by their results although they do produce good evidence for the stability of chondrocytes. EGGERT [1966] used a rather different approach and grafted cartilage into x-irradiated limb stumps. He also came to the conclusion that chondrocytes are stable in their differentiation. STEEN's experiments quoted above also throw light on this problem. Cartilage was removed from diploid host limbs and replaced by diploid donor cartilage labelled with thymidine-^{3}H. The regenerates that formed subsequent to amputation only contained labelled cells in the regenerated cartilage. STEEN took a great deal of care to ensure that only cartilage was transplanted. All attached muscle, connective tissue and perichondrium were carefully excised and if there had been any periosteal ossification, this too was removed. It was observed that during the course of dedifferentiation, the donor chondrocytes became mixed with host cells, yet showed no tendency to metaplasia. Thus in the axolotl limb, chondrocytes appear to be stable in their differentiation. Not only this but the stability appears to be inherited, as STEEN claims that dilution of the tritium label indicated that the cells had undergone at least 5 mitoses. COON [1966] has studied clonal cultures of chick cartilage cells and finds that even after 35 generations involving 4 successive clonal passages, the phenotype is stable and the dedifferentiated chondrocytes will still redifferentiate according to their original phenotype. It has been observed by CAHN and CAHN [1966] that retinal pigment cells grown in mass monolayer cultures or in culture media containing a high molecular weight fraction from embryo extract, all lose their pigmentation by 6–8 days but that by 10–14 days they begin to repigment. These results may indicate that differentiation in retinal pigment cells is stable and heritable through several generations in an unpigmented state because of some stable repression of the genome, but they may equally well be interpreted as the effect of the medium influencing what regions of the genome shall be repressed and what transcribed.

VI. The Potentialities of Blastema Cells

All the methods which have been mentioned so far, have been used in an attempt to analyze the stability of differentiation in a regenerating system, and have relied on various morphological or biochemical characteristics to demonstrate the functional state of the cell. The results obtained by these methods have been used to draw conclusions concerning the underlying state of the genome. It has frequently been assumed that dedifferentiated blastema cells, besides becoming simple and 'embryonic' in appearance, also regain all the genetic potentialities which early embryonic cells are assumed to possess. As it has already been pointed out, this assumption is not justified because, although these methods can show, at least to some extent, what a cell is or is not producing, they do not indicate what the cell is capable of producing. This is the basic problem in trying to discover the genetic potentialities of blastema cells and in seeking to establish whether, under the conditions of epimorphic regeneration, differentiation is reversible.

A. The Reversibility of Gene Function

There is no doubt that in bacteria, the repression and activation of the genes is completely reversible as, for example, ITO [1968] has shown for the tryptophan operon in *E. coli,* or that the cessation of transcription is extremely rapid after repression of the operon [IMAMOTO, 1968, 1970], but as it has already been pointed out, the system of gene control in bacteria is almost certainly different from that found in the Metazoa. It is axiomatic to the theory of epigenesis that there must exist a mechanism to control cellular differentiation. According to the variable gene activity theory, which was first considered by such workers as MORGAN [1934] but was not really given formal expression until 1950 by STEDMAN and STEDMAN and in 1951 by MIRSKY, cellular differentiation is based on the principal of the functioning of different parts of an identical genome in the various types of specialized cell. The mechanism of differential gene control is unknown although various theories have been put forward, each with its own degree of credibility.

There are many instances in which cells appear to be induced to change from one differentiated state to another as has recently been demonstrated by LOPASHOV and SOLOGUB [1972]. STONE [1950] showed that the retina pigment cells of the adult salamander eye were capable of redifferentiating into neural retinal cells and in 1955 he showed that both iris and lens could be regenerated from retina pigment cells. He has also noted [1952, 1953] that lens regeneration from the dorsal iris of the newt eye is inhibited if a living fully differentiated lens, even of another species, is present, so perhaps the regeneration of lens from retinal pigment cells is not too surprising. There is also a report by PRADA [1946] who has evidence that the lens normally produces an inhibitory lens forming factor. Furthermore, the neural retina lies over the retinal pigment layer and as in all experiments of this nature, it is extremely difficult to be certain that the layers have been completely separated. Moreover, all these cells are of the neuro-epithelial type. The experiments of WEISS [1925] are interesting and showed that bone was able to regenerate in *Triton* limb regenerates after the bone had been excised from the limb previous to amputation. This is some of the best and also earliest evidence of metaplasia during regeneration in vertebrates. However, it has been shown by LEVANDER [1945] that bone or muscle can be formed in connective tissue if it is treated with extracts of the respective tissue. Ossification and chondrification in any case, occur quite frequently in connective tissue, as has been recorded many times in pathology. WEISS [1939] investigated the changes from columnar to stratified epithelium which poor vitamin A conditions produce and adopted the term 'modulation' to describe situations in which a cell alters its morphological appearance in response to altered circumstances. He implied that there is also a loss of some functions without necessarily a change in genetic potentialities. Since the term 'dedifferentiation' has become so emotive, in that to some people it implies a widening of the genetic spectrum while to others simply a loss of specialized morphological characteristics, the term 'modulation' may perhaps best be used to describe these examples.

The approach to the problem of the dedifferentiated cells of the regeneration blastema had been almost exclusively morphological and biochemical until fairly recently, when an attempt was made to determine the genetic potentialities of the blastema cells by means of the technique of nuclear transplantation into enucleated eggs [BURGESS, 1967]. The amputated hind limbs of *Xenopus laevis* larvae at stage 59–60 [NIEUWKOOP and FABER, 1956] were used as a source of blastema cells.

B. *Xenopus* Larvae as a Source of Blastema Cells for Nuclear Transplantation

By stage 59–60 (fig. 4), the musculature of the hind limb is fully differentiated; the tibio-fibula is fully cartilaginous [NIEUWKOOP and FABER, 1956] and fibroblasts are only sparsely distributed [DENT, 1962]. At this age, the larvae will develop a satisfactory regeneration blastema whose general pattern of formation by dedifferentiation is essentially the same as that found by other workers in different amphibian species [e.g., THORNTON, 1938a, b; SCHOTTE and HARLAND, 1943; FORSYTH, 1946; HAY,

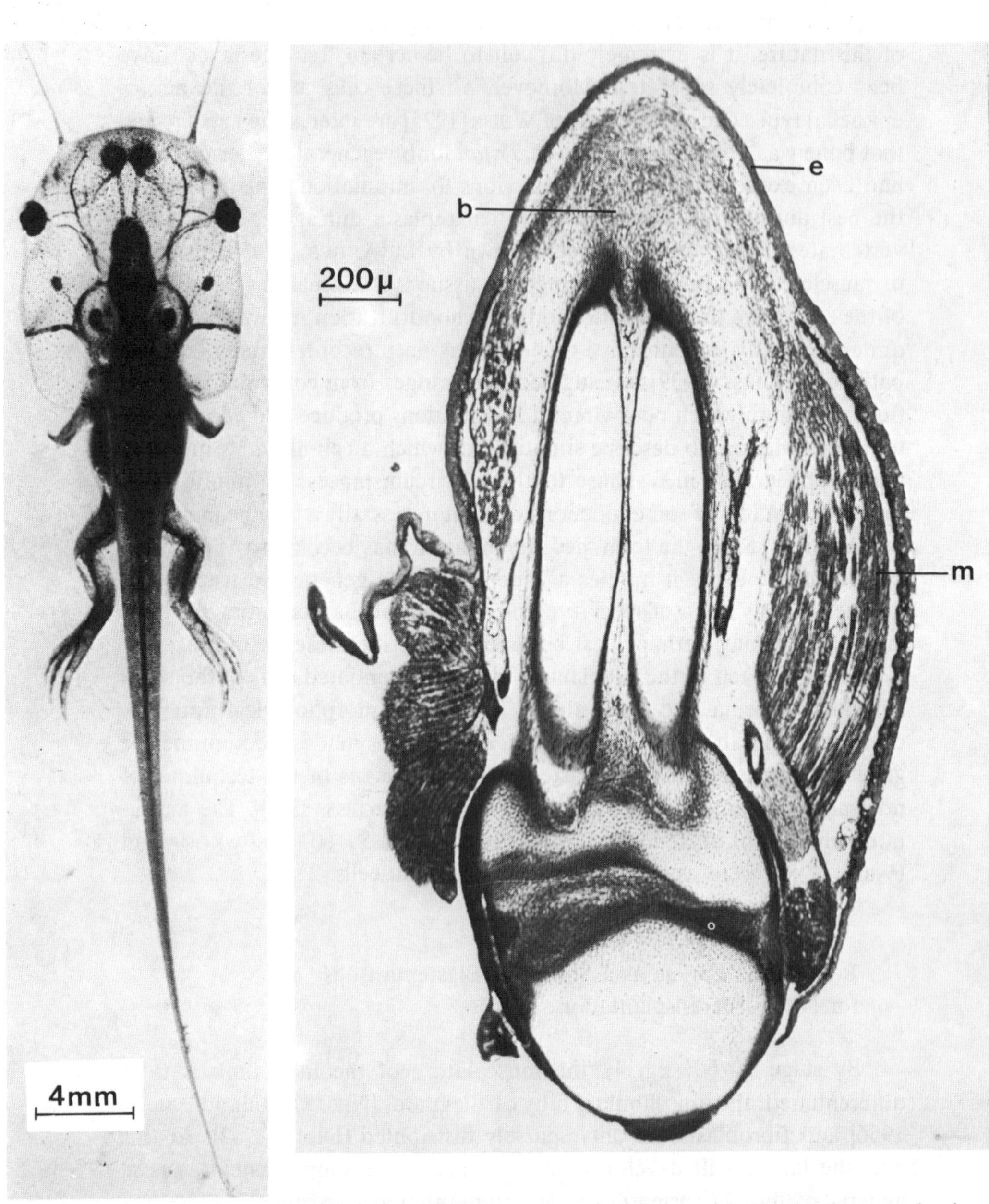

Fig. 4. Dorsal view of a normal *Xenopus* larva at stage 59.

Fig. 5. Longitudinal section through the regenerating hind limb stump showing the regeneration blastema 5 days post-amputation. b = Developing blastema; m = muscle; e = epidermis.

1959], although they are not so prolific as blastemata developed after amputation at earlier stages. However, in the limbs of younger larvae, the tissues are not fully differentiated and there is a large number of fibroblasts present which makes them unsuitable for use in an experiment designed to test the genomic potentialities of dedifferentiated cells.

In the hind limbs of *Xenopus* larvae at stage 59–60, all the various cell types to be found in the definitive adult limb may be identified (fig. 5). The tibio-fibula, while remaining mainly cartilaginous, shows signs of ossification and all the muscles to be found in the adult limb are present and fully differentiated. The perimysia at this stage are considerably thinner than in the adult [NIEUWKOOP and FABER, 1956]. Other potential sources of blastema cells are the intermuscular sheaths, the perineural sheaths (but not the associated nerves), the perichondrium and, possibly, the subcutaneous connective tissue. A large volume of evidence shows that it is most unlikely that either the blood vessels or epidermis dedifferentiate.

Three types of control were used in these experiments [BURGESS, 1967]; dead blastema cell nuclei, gastrula endoderm nuclei and epidermal cell nuclei. Due to the paucity of cytoplasm in the epidermal cells and in the newly formed blastema cells (fig. 6), it is not practical to transplant the nuclei singly, as was done in the original method devised by BRIGGS and KING [1952] and later adapted for use with *Xenopus* by ELSDALE, GURDON and FISCHBERG [1960]. Instead, clumps of nuclei were transplanted according to the method of KING and MCKINNEL [1960] in their experiments on the developmental potentialities of the renal adenocarcinoma cells of *Rana*.

The basic comparison was between the developmental potentialities of the epidermal cell nuclei, which do not undergo dedifferentiation, and gastrula endoderm cell nuclei which are relatively undifferentiated. It is not practical to use either differentiated muscle, from which most of the dedifferentiated cells of the blastema appear to originate, or differentiated cartilage, which is another major source of these cells and which would, therefore, form the obvious control group. However, the syncytial nature of muscle renders it very difficult to handle with the micropipette used in nuclear transplantation and in the case of chondrocytes, difficulty is experienced in releasing them from the matrix in which they are embedded. Epidermal cells form a suitable differentiated control cell type and, what is more, the work of HAY and FISCHMAN [1961] among others, has clearly shown that the epidermis does not normally contribute to the formation of the blastema and does not dedifferentiate. The use of gastrula endoderm cells as the undifferentiated cell type is usual in this type of experiment [e.g., BRIGGS and KING, 1953, 1957] and permits comparisons to be made with the results of other investigators.

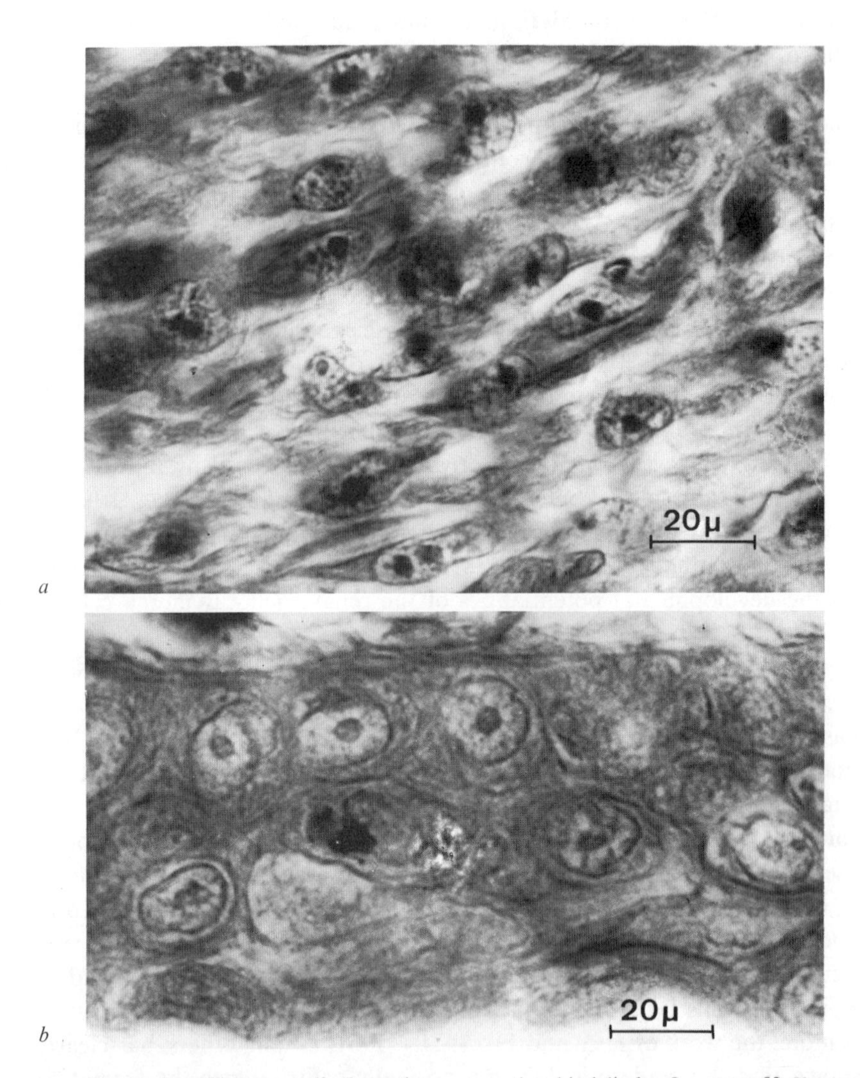

Fig. 6a, b. a Blastema cells from the regenerating hind limb of a stage 59 *Xenopus* larva. *b* Epidermis from the hind limb of a larva at the same stage as *a.*

C. The Developmental Potentialities of Dedifferentiated Cells

The results obtained from this series of experiments is summed up in the following tables. Table I shows the numbers and stages of embryos which resulted from the transplantation of clumps of blastema cell nuclei and from the gastrula endoderm cell nuclei used as controls. Percentage fertility refers to the percentage of normally fertilized eggs which developed to normal gastrulae stages in the particular batch of eggs used. As can be seen, this factor is variable and it is essential that it is determined for each experimental series. Table II shows an analysis of the results obtained with epidermal cell nuclei; the fully differentiated cell type, and gastrula endoderm cell nuclei transplanted as controls.

It is obvious that a low figure is obtained during the early stages of development from both the test embryos and the gastrula endoderm cell nuclei controls if only the number arrested at a particular stage is counted. In the case of the blastema and epidermal cell nuclei, this is due to the fact that very few host eggs showed any signs of development anyway, but with the gastrula nuclei this low figure is due to the fact that most embryos survived the earlier stages and went on to develop further. In table III, which compares all the final results, the figure appearing in each column represents not only the numbers of embryos arrested at a particular stage but also those which survived that stage and developed further.

Blastema cell nuclei caused a total of 5.4% of the host eggs to cleave normally and this is extremely close to the percentage achieved with the epidermal cell nuclei. Dead blastema cell nuclei, by contrast, were unable to support any cleavage at all. When, however, these two percentages are compared with the number of gastrula endoderm nuclei which resulted in normal cleavage (57.3%), then a very large difference is found. The same correlation and discrepancy is found when a comparative analysis of the percentage total cleavage is made (table III). Blastema nuclei transplants gave rise to 4.3% of embryos which formed blastulae compared to the 5.1% of embryos forming blastulae with epidermal cells. However, with the gastrula endoderm nuclei, 57.3% of the host eggs developed as far as blastula stages. The same pattern is repeated in the percentage of embryos which formed gastrulae; 1.1% resulting from the blastema cell nuclei and 0.65% from the epidermal cell nuclei. The equivalent figure obtained with gastrula endoderm nuclei was 44%. None of the blastema or epidermal nuclei resulted in development beyond very early gastrula stages (stage $10\frac{1}{4}$) [NIEUWKOOP and FABER, 1956]. That is to say, there was a definite con-

Table I. Results from the nuclear transplantation of clumps of blastema cell nuclei and single gastrula endoderm nuclei[1]

% Fertility	Blastema cell nuclei (clump)				Gastrula endoderm cell nuclei (single)				
	Total transplants	Normal cleavage	Arrested blastulae	Arrested gastrulae	Total transplants	Arrested blastulae	Arrested gastrulae	Arrested neurulae	Normal larvae
70	26	1	1	–	8	1	2	–	2
80	31	3	1	1	8	1	–	–	3
80	33	1	1	–	7	–	1	–	3
60	30	2	1	–	8	–	1	–	2
90	35	1	–	1	12	2	2	2	4
70	31	2	2	–	12	3	1	–	2
	186	10	6	2	55	7	7	2	16
	100%	5.4%	3.2%	1.1%	100%	12.7%	12.7%	3.6%	29.1%

1 Figures showing survival indicate the numbers which were arrested at a particular stage. The percentage fertility refers to the quality of the batch of eggs used as hosts in a particular series.

Table II. Results from the nuclear transplantation of clumps of epidermal cell nuclei and control, single, gastrula cell nuclei[1]

% Fertility	Epidermal cell nuclei (clump)				Gastrula endoderm cell nuclei (single)				
	Total transplants	Normal cleavage	Arrested blastulae	Arrested gastrulae	Total transplants	Arrested blastulae	Arrested gastrulae	Arrested neurulae	Normal larvae
80	32	–	–	–	8	1	–	1	4
70	34	3	2	–	10	–	2	–	3
80	30	1	1	–	8	–	–	–	4
60	27	2	1	1	12	3	–	2	2
70	32	3	3	–	9	1	2	1	1
	155	9	7	1	47	5	4	4	14
	100%	5.8%	4.5%	0.65%	100%	10.6%	8.5%	8.5%	29.8%

1 Conventions as in table I.

Table III. Comparison of the results obtained from the transplantation of blastema cell nuclei and epidermal cell nuclei ('clump' method) and the single gastrula endoderm cell nuclei[1]

	%cleaving normally	%forming blastulae	%forming gastrulae	%forming neurulae	%giving feeding stages
Blastema cell nuclei (clump)	5.4%	4.3%	1.1%	–	–
Epidermal cell nuclei (clump)	5.8%	5.15%	0.65%	–	–
Gastrula endoderm cell nuclei (single)	57.3%	57.3%	44.5%	35.3%	29.1%

1 The percentages have been claculated not only from the numbers of embryos arrested at a particular stage of development but also those that went on to develop further.

centration of pigment in the region of the presumptive blastopore and the dorsal lip started to form but never became crescentic or extended ventrally. In the case of the gastrula endoderm nuclei, 35.3% reached neurula stages and 29.1% survived to feeding larval stages.

D. The Suspension of Gene Function or a Return to a Pleuripotent State?

Surveying the results in table III as a whole, it is immediately obvious that the figures from the blastema cell nuclei show a high degree of similarity to the results obtained with the epidermal cell nuclei (the control differentiated type). In fact, the slight differences between them are certainly not significant when consideration is given to the degree of accuracy that can be expected with this technique. On the other hand, the differences in the results from the blastema nuclei and the gastrula endoderm nuclei (the control undifferentiated cell type) are extremely marked. The conclusion may be drawn that, as the results from the blastema cell nuclei resemble so closely those obtained with the differentiated cell type and differ in such a marked way from those of the undifferentiated cell type, the blastema cell nuclei more closely resemble the differentiated than the undifferentiated cell type. These findings may be interpreted as meaning that the blastema cell genome is still differentiated despite the fact that the cytoplasm has lost its specialized morphological characteristics or specific proteins and, indeed, has also lost most of the synthetic machinery that would be required to produce them, though whether this is cause or effect is not very easy to decide.

A multiplicity of factors could conceivably cause a suspension of specific protein synthesizing systems in the dedifferentiated cell and lead to the loss of morphological complexity. It should be remembered that both the circumstances that lead to dedifferentiation, i.e., tissue culture and epimorphic regeneration, place the cells in physiologically abnormal conditions and it might well be anticipated that there would be a temporary suspension of some relatively unessential processes.

Dedifferentiation is in any case a rather slow process taking some 3–4 days in *Xenopus* for the myofibrils to disappear. HOLTZER [1961] found that up to the 7th to 8th days following amputation there were still traces of myosin present in the regenerating salamander tail and that, although it was absent for a period of 5–7 days, it reappeared by the 12th to 13th day. Its presence for 3–4 days is not surprising in view of the reports of the long lived mRNA which is sometimes found in vertebrates; for example, haemoglobin synthesis in mammalian erythrocytes [e.g., KRUH and BORSOOK, 1956; CAMERON and PRESCOTT, 1963]. It is not known at what time after amputation the genes responsible for the production of specialized protein cease to be functional or whether indeed the genes do cease to produce mRNA. If the 'switching off' of the genes is almost instantaneous or extremely rapid, the presence of characteristic proteins several days later could be due to the presence of long lived mRNA. In any event, the cells do not remain in the dedifferentiated state for very long and characteristic proteins soon reappear in the cytoplasm.

One may regard the blastema as a collection of totipotent cells with the entire range of the genomic spectrum available for transcription, which redifferentiate according to the site that they happen to occupy solely under the influence of the stump tissues. Alternatively, the blastema may be regarded as a collection of apparently similar, yet in fact, genomically differentiated cells in which there has been a suspension of regions of the genome which normally produce the specialized proteins of the cell. When considering the loss of specialized proteins from the cytoplasm of dedifferentiating cells, due consideration should be given to the fact that the cells of the blastema undergo a period of rapid mitosis and that this in itself could be related to a loss of rough endoplasmic reticulum [e.g., PALADE, 1955] and a reduced capacity of the cell to perform protein synthesis. This would not mean that the genes were in any way 'switched off' during dedifferentiation but rather, that the machinery for translating the mRNA into protein was deficient.

Interesting work, which may shed some light on the way in which tissue-specific cells of the blastema could rearrange themselves into their tissue types, has been done by MOSCONA [1962, 1963] and HUMPHREYS [1963] who have shown that if cells from different species of sponge are mixed randomly, the cells are able to reorganize themselves into their specific types. HOLTFRETER [1939] was the first to illustrate a similar phenomenon with amphibian embryonic cells and since that time a great deal of evidence has substantiated the concept of tissue specificity [e.g., TOWNES and HOLTFRETER, 1955]. This could explain the ability of the dedifferentiated cells to reorder themselves into groups of similar tissue type after mitosis is complete; an ability which they seem to retain even if the young blastema is separated from the influences of the stump [STOCUM, 1968].

Concerning the reversibility of differentiation, the experiments of GURDON [1962] have shown that it is possible to obtain fertile adult *Xenopus* from a small percentage of differentiated intestinal epithelial cells. This experiment has been criticized because of the close association between the intestinal cells and the primordial germ cells but the intestinal cells which GURDON used as a source of differentiated nuclei for nuclear transplantation were sufficiently differentiated to exhibit a brush border which would make them easily distinguishable and unlikely to be confused with any more 'embryonic' cells in the region. In his latest series of experiments on the transplantation of cultured cells [GURDON and LASKEY, 1970] he has removed all cells of endodermal origin and, consequently, all primordial germ cells.

There appears to be an interesting species difference between *Xenopus* and *Rana* as shown by similar experiments performed by SMITH [1965]. He transplanted somatic endoderm cells from the anterior quarter of the intestine of *Rana* at a stage considerably earlier than those used by GURDON, and found that he was unable to obtain any normal tadpoles and only 1 out of a total of 382 transfers succeeded in completing gastrulation. In the same series of experiments, SMITH transplanted primordial germ cells from a slightly later stage (st. 25 [SHUMWAY, 1940]) and found that 40% would develop into normal tadpoles. He interprets these results as indicating that there is a marked difference in the *'developmental capacity'* (SMITH's italics) of the two types of cell nuclei transplanted and thinks that the intestinal cells become restricted in terms of their ability to direct normal development.

In the experiments of BURGESS [1967] neither the cells of the epidermis nor those of the blastema were ever noted as giving rise to development past early gastrula stages. This could be a criticism of the method of transplanting nuclei in groups instead of singly. However, the experiments of KING and MCKINNELL [1960] with renal adenocarcinoma cells of *Rana* produced 2 abnormal embryos which survived to neurula stages and 9 which completed gastrulation, out of a total of 142 transplants (this is the total number of transplants performed and not the total number of transplants which went on to cleave normally). These figures are higher than some experiments carried out with single cell nuclei and do at least show that nuclei transplanted in this way are capable of development past gastrulation. There is a great deal of evidence which shows that development up to late blastula stages does not depend on the functioning of the genes but on the information which is laid down in the cytoplasm during oogenesis [BELITSINA *et al.*, 1964; SPIRIN and NEMER, 1965; KRIGSGABER and NEYFAKH, 1972]. The experiments of NEIFAKH [1960, 1964] have shown that heavy irradiations with x-rays or γ-rays at any time between fertilization and the formation of the blastula, will result in the failure of gastrulation. GROSS and COUSINEAU [1963, 1964] have found that development is normal in sea urchin eggs allowed to develop in actinomycin D until blastula stages at almost the same rate as in normal controls. There is also the evidence of BRACHET *et al.* [1962] that in amphibian lethal crosses, development will proceed until they are arrested as blastula stages. The fact that all the transplants from the blastema nuclei transplants failed to complete gastrulation may indicate that the genome is incapable of directing complete and normal development.

E. Concluding Remarks

It could be argued from the particular to the general, that all cellular differentiation is an irreversible procedure. However, to draw such a sweeping conclusion from the highly particular instance of blastema formation during the regeneration of the hind limb of *Xenopus* is not justified. Furthermore, the experiments of GURDON have indicated that in at least the case of differentiated intestinal epithelium, it would seem likely that differentiation is reversible. The experiments of GURDON and LASKEY [1970] on cultured cells have yielded results not unlike those obtained with blastema cells. Out of a total of 3546 transfers, 124 (3%) gave complete cleavage, 107 (3%) were arrested as blastulae, 13 (0.4%) formed abnormal gastrulae, 3 gave near normal tadpoles and 1 normal feeding tadpole was obtained. Having regard to the degree of accuracy that can reasonably be expected with this method, it would appear that the cultured cells used by GURDON and LASKEY were unable to support normal development.

The whole scheme of gene regulation and function and cellular differentiation and dedifferentiation is obviously not as simple as it was thought to be even in the simplest of organisms. BURGESS *et al.* [1969] have discovered a factor in *E. coli* which dictates the regions on the DNA template that RNA polymerase can recognize; similar work by BAUTZ *et al.* [1969] and DUNN and BAUTZ [1969] has confirmed these findings. OPPENHEIM [1970] has found an anti-repressor which appears to be another factor in some aspects of the gene regulatory story, and FJELLSTEDT [1970] has found 'super suppressors' or nonsense suppressors in the Saccharomyces (yeast), yet all these systems may be assumed to be far less complex than those found in the Metazoa. In the last few years BALTIMORE [1970] and TEMIN and MIZUTANI [1970] have shown that RNA tumour viruses contain an enzyme which uses viral RNA as a template for the synthesis of DNA. This enzyme has been called 'reverse transcriptase' and its existence was first postulated by Temin in 1964 who put forward the theory to explain the induction of cancer by RNA viruses. RNA dependent DNA polymerase has since been reported in a large number of RNA-containing tumour viruses [MIZUTANI *et al.*, 1970 and SPIEGELMAN *et al.* 1970] and also in non-tumour viruses [LIN and THORMAR, 1970], and SCOLNICK *et al.* [1971] have now demonstrated the presence of reverse transcriptase in normal (i.e. non-cancerous) human cells in tissue culture. These cells showed no evidence of viral infection and this indicates that RNA dependent DNA polymerase may play a role in normal cells.

A great deal of information has accumulated indicating the function of RNA in priming DNA synthesis [VERMA, 1971; CHANG and BOLLOM, 1972; LEIS and HURWITZ, 1972 and WELLS *et al.* 1972]. KELLER [1972] has shown that the replication of *E. coli* DNA by a polymerase from human tumour KB cells is prevented by the presence of ribonuclease H, an enzyme which destroys the RNA part of DNA-RNA hybrids, and LARK [1972] has found that rifampicin will inhibit the replication of *E. coli* DNA if it is administered ten minutes before replication is due to commence. These results may be interpreted as meaning that the absence of RNA, either because it has been destroyed or because its synthesis has been inhibited, prevents the synthesis of DNA.

The experiments of BRUTLAG *et al.* [1971] have shown that whereas rifampicin prevents the formation of double stranded phage M13 DNA from the single stranded state, chloramphenicol does not. Since the action of chloramphenicol is to inhibit protein synthesis it appears that it is the formation of RNA itself and not that of protein intermediaries which is required for the synthesis of DNA. This conclusion is supported by WICKNER *et al.* [1972] who have found that the formation of phage M13 DNA requires RNA synthesis in an *in vitro* situation. It has now been proposed by SUGINO *et al.* [1972] that the role of RNA in DNA synthesis involves the transcription of the RNA which then acts as a primer for DNA synthesis.

It was suggested in 1968 by OKAZAKI *et al.* that the synthesis of DNA is a discontinuous process with regions or segments of the DNA being formed intermittently and later joining together. This idea appeared to conflict with the concept of RNA primed DNA synthesis until OKAZAKI's group [SUGINO, 1972] proposed that each individual DNA segment is primed by RNA and showed that the segments are heavier than if they were composed exclusively of DNA. Furthermore it is only on extraction with ribonuclease that the segments show the density characteristic of single stranded DNA. SUGINO *et al.* [1972] have also shown that in *E. coli*, RNA is covalently linked to nascent short DNA fragments.

Evidence of this nature leads to the speculation that differentiation may be controlled by the distribution of maternal RNA in the oocyte. During cleavage the distribution pattern of RNA would be reflected in different cells and transcription by means of reverse transcriptase could then result in DNA replicas. These could prime different regions of the genome causing differential gene function and hence morphological differentiation. Cells having undergone the process of dedifferentiation as in the case of regeneration blastema may owe their apparent stability of expression and restriction of genetic potentialities to such a mechanism.

It is now twelve years since JACOB and MONOD [1961]propounded their brilliant theory of gene control in bacteria yet in spite of many excellent attempts, an understanding of gene control in Metazoa still remains a fascinating and tantalizing problem.

References

AGARWAL, K. L.; BÜCHI, H.; CARUTHERS, M. H.; GUPTA, N.; KHORANA, H. G.; KLEPPE, K.; KUMAR, A.; OHTSUKA, E.; RAJBHANDARY, U. L.; VAN DE SANDE, J. H.; SGARAMELLA, V.; WEBER, H. and YAMADA, T.: Nature *227:* 27 (1970).

ALLFREY, V. G.; LITTAU, V. C. and MIRSKY, A. E.: Proc. nat. Acad. Sci., U.S. *49:* 414 (1963).

ANTON, H.J.; In V. KIORTSIS and H.A.L. TRAMPUSCH Regeneration in animals, pp. 377–395 (North-Holland Publ. Co., Amsterdam 1965).

BALTIMORE, D.: Nature *226:* 1209 (1970).

BARKER, K. L. and WARREN, J. C.: Proc. nat. Acad. Sci. U.S. *56:* 1298 (1966).

BARR, G. C. and BUTLER, J. A. V.: Nature *199:* 1170 (1963).

BAUTZ, E. K.; BAUTZ, F. A. and DUNN, J. J.: Nature *223:* 1022 (1969).

BAZILL, G. W. and PHILPOT, J. St. L.: Biochim. biophys. Acta *76:* 223 (1963).

BELITSINA, N. V.; AITKHOZHIN, M. A.; GAVRILOVA, L. P. and SPIRIN, A. S.: Biochemistry (USSR) *29:* 315 (1964). (English transl.)

BISCHLER, V.: Rev. suisse Zool. *33:* 431 (1926).

BISCHLER, V. and GUYENOT, E.: C.R. Soc. Biol. *94:* 968 (1926).

BLOCH, D. P.: Proc. nat. Acad. Sci., U.S. *48:* 324 (1962a).

BLOCH, D. P.: J. Histochem. Cytochem. *10:* 137 (1962b).

BODEMER, C. W. and EVERETT, N. B.: Develop. Biol. *1:* 327 (1959).

BOIVIN, A.; VENDRELY, R. and VENDRELY, C.: C.R. Acad. Sci. *226:* 1061 (1948).

BONNER, J. and HUANG, R. C.: J. molec. Biol. *6:* 169 (1963).

BONNER, J. and HUANG, R. C.: In Histones, their role in the transfer of genetic information Ciba Found. Study no. 24, pp. 18–41 (Churchill, London 1966).

BONNER, J.; DAHMUS, M. E.; FAMBROUCH, D.; HUANG, R. C.; MARUSHIGE, K. and YUAN, Y. H.: Science *159:* 47 (1968).

BOUBLIK, M., BRADBURY, E.M., CRANE-ROBINSON, C. and RATTLE, H.W.E. Nature New Biol. *229:* 149 (1971).

BRACHET, J.; BIELIAVSKY, N. and TENCER, R.: Bull. Acad. roy. Méd. Belg. *48:* 255 (1962).

BREUER, C. B. and FLORINI, J. R.: Biochemistry *5:* 3857 (1966).

BRIGGS, R. and KING, T. J.: Proc. nat. Acad. Sci., U.S. *38:* 455 (1952).

BRIGGS, R. and KING, T. J.: J. exp. Zool. *122:* 485 (1953).

BRIGGS, R. and KING, T. J.: J. Morph. *100:* 269 (1957).

BRUTLAG, D., SCHEKMAN, R. and KORNBERG, A.: Proc. nat. Acad. Sci., U.S. *68:* 2826 (1971).

BURGESS, A. M. C.: J. Embryol. exp. Morph. *18:* 27 (1967).

BURGESS, R. R.; TRAVERS, A. A.; DUNN, J. J. and BAUTZ, E. K. F.: Nature *221:* 43 (1969).

BURSTEIN, C.; COHN, M.; KEPES, A. and MONOD, J.: Biochim. biophys. Acta *95:* 634 (1965).

BUSCH, H. and DAVIS, J. R.: Cancer Res. *19:* 1241 (1958).

BUSCH, H.; STEELE, W. J.; HNILICA, L. S.; TAYLOR, C. W. and MAVIOGLU, H.: J. cell. comp. Physiol. *62*/Suppl. 1: 95 (1963).

BUTLER, E. G.: J. exp. Zool. *65:* 271 (1933).

BUTLER, E. G.: Anat. Rec. *62:* 295 (1935).

CAHN, R. L. and CAHN, M. B.: Proc. nat. Acad. Sci., U.S. *55:* 106 (1966).

CAMERON, I. L. and PRESCOTT, D. M.: Exp. Cell Res. *30:* 609 (1963).

CARLSON, B. M.: J. Morph. *122:* 249 (1967).

CHALKLEY, D. T.: J. Morph. *94:* 21 (1954).

CHANG, L.M.S. and BOLLUM, F.J.: Biochem. Biophys. Res. Commun. *46:* 1354 (1972).
CLARK, R.J. and FELSENFELD, G.: Nature New Biol. *229:* 101 (1971).
COON, H. G.: Proc. nat. Acad. Sci., U.S. *55:* 66 (1966).
CRUFT, H. J.; MAURITZEN, C. M. and STEDMAN, E.: Phil. Trans. Roy. Soc. Lond. B *241:* 93 (1957).
CRUFT, H. J.; MAURITZEN, C. M. and STEDMAN, E.: Proc. Roy. Soc. Lond. B *149:* 36 (1958).
CRUFT, H. J. and LEAVER, J. L.: Nature *192:* 556 (1961).
DE HAAN, R. L.: J. exp. Zool. *133:* 73 (1956).
DENT, J. N.: J. Morph. *100:* 61 (1962).
DOUNCE, A. L. and HILGARTNER, C. A.: Exp. Cell Res. *36:* 288 (1964).
DREW, R. M. and PAINTER, R. B.: Radiat. Res. *11:* 535 (1959).
DUNN, J. J. and BAUTZ, E. K.: Biochem. biophys. Res. Commun. *36:* 925 (1969).
EBISUZAKI, K.: J. molec. Biol. *7:* 379 (1963).
EGGERT, R. C.: J. exp. Zool. *161:* 369 (1966).
EKHOLM, R. and SJÖSTRAND, F. S.: J. Ultrastruc. Res. *1:* 178 (1957).
ELSDALE, T. R.; GURDON, J. B. and FISCHBERG, M.: J. Embryol. exp. Morph. *8:* 437 (1960).
ENGLESBERG, E.; SQUIRES, C. and MERONK, F., jr.: Proc. nat. Acad. Sci., U.S. *62:* 1100 (1969).
FJELLSTEDT, T. A.: J. Bact. *101:* 108 (1970).
FORSYTH, J. W.: J. Morph. *79:* 287 (1946).
FRANKHAUSER, G. and HUMPHREY, R. R.: Proc. nat. Acad. Sci., U.S. *29:* 344 (1943).
GEORGIEV, G. P.; ANANIEVA, L. N. and KOSLOV, J. V.: J. molec. Biol. *22:* 365 (1966).
GILBERT, W. and MÜLLER-HILL, B.: Proc. nat. Acad. Sci., U.S. *56:* 1891 (1966).
GILBERT, W. and MÜLLER-HILL, B.: Proc. nat. Acad. Sci., U.S. *58:* 2415 (1967).
GODLEWSKI, E.: Roux' Arch. Entw. Mech. *114:* 108 (1928).
GOSS, R. J.: J. Embryol. exp. Morph. *6:* 638 (1958).
GROSS, P. R. and COUSINEAU, G. H.: Biochem. biophys. Res. Commun. *10:* 321 (1963).
GROSS, P. R. and COUSINEAU, G. H.: Exp. Cell Res. *33:* 368 (1964).
GURDON, J. B.: Develop. Biol. *4:* 256 (1962).
GURDON, J. B. and LASKEY, R. A.: J. Embryol. exp. Morph. *24:* 227 (1970).
HAY, E. D.: Amer. J. Anat. *91:* 447 (1952).
HAY, E. D.: J. biophys. biochem. Cytol. *4:* 583 (1958).
HAY, E. D.: Develop. Biol. *1:* 555 (1959).
HAY, E. D.: In D. RUDNICK Regeneration, p. 177 (Ronald Press, New York 1962).
HAY, E. D. and FISCHMAN, D. A.: Anat. Rec. *136:* 208 (1960).
HAY, E. D. and FISCHMAN, D. A.: Develop. Biol. *3:* 26 (1961).
HERTWIG, G.: Roux' Arch. Entw. Mech. *111:* 292 (1927).
HINDLEY, J.: 6th Int. Congr. Biochemistry, New York 1964, abstracts, p. 61.
HNILICA, L. S. and BUSCH, H.: J. biol. Chem. *238:* 918 (1963).
HNILICA, L. S. and KAPPLER, H. A.: Science *150:* 1470 (1965).
HNILICA, L. S.; JOHNS, E. W. and BUTLER, J. A. V.: Biochem. J. *82:* 123 (1962).
HOLTFRETER, J.: Arch. exp. Zellforsch. *23:* 169 (1939).
HOLTZER, H.: In D. RUDNICK Synthesis of molecular and cellular structure, pp. 35–87 (Ronald Press, New York 1961).
HOLTZER, H.; ABBOTT, J.; LASH, J. and HOLTZER, S.: Proc. nat. Acad. Sci., U.S. *46:* 1533 (1960).
HORN, E. C. and WARD, C. L.: Proc. nat. Acad. Sci., U.S. *43:* 776 (1957).

HUANG, R. C. and BONNER, J.: Proc. nat. Acad. Sci., U.S. *48:* 1216 (1962).
HUANG, R. C.; BONNER, J. and MURRAY, K.: J. molec. Biol. *8:* 54 (1964).
HUMPHREYS, T.: Develop. Biol. *8:* 27 (1963).
IMAMOTO, F.: Nature *220:* 31 (1968).
IMAMOTO, F.: Molec. gen. Genet. *106:* 123 (1970).
ITO, J. and IMAMOTO, F.: Nature *220:* 441 (1968).
IWAI, K.: In J. BONNER and P. Ts'o The nucleohistones, pp. 59–65 (Holden-Day, San Francisco 1964).
IZAWA, M.; ALLFREY, V. G. and MIRSKY, A. E.: Proc. nat. Acad. Sci., U.S. *50:* 811 (1963).
JACOB, F. and MONOD, J.: J. molec. Biol. *3:* 318 (1961).
JOHRI, M. M. and VARNER, J. E.: Proc. nat. Acad. Sci., U.S. *59:* 269 (1968).
KAEMPFER, R. D. R. and MAGASANIK, B.: J. molec. Biol. *27:* 475 (1967).
KAISER, A. D. and JACOB, F.: Virology *4:* 509 (1957).
KARCZMAR, A. G. and BERG, G. G.: J. exp. Zool. *117:* 139 (1951).
KELLER, W.: Proc. nat. Acad. Sci., U.S. *69:* 1560 (1972).
KIDSON, C.: Biochem. biophys. Res. Commun. *21:* 283 (1965).
KIDSON, C. and KIRBY, K. S.: Nature *203:* 599 (1964).
KIM, K-H. and COHEN, P. P.: Proc. nat. Acad. Sci., U.S. *55:* 1251 (1966).
KING, T.J. and MCKINNELL, R.G.: In 14th Annual Symposium on Fundamental Cancer Research at the M.D. Anderson Hospital and Tumor Institute at Houston, Texas. Cell physiology of neoplasia, pp. 591–615 (University of Texas Press, Houston 1960).
KONIGSBERG, I.R.: Science *140:* 1273 (1963).
KOSSEL, A.: The protamines and histones (Longmans, Green, London 1921).
KRIGSGABER, M.R. and NEYFAKH, A.A.: J. Embryol. exp. Morph. *28:* 491 (1972).
KRUH, J. and BORSOOK, H.: J. biol. Chem. *220:* 905 (1956).
LARK, K.G.: J. Molec. Biol. *64:* 47 (1972).
LAUFER, H.: J. Embryol. exp. Morph. *7:* 431 (1959).
LEIS, J.P. and HURWITZ, J.: J. Virol. *9:* 130 (1972).
LEVANDER, G.: Nature *155:* 148 (1945).
LEVESON, J. E. and PEACOCKE, A. R.: Biochim. biophys. Acta *123:* 329 (1966).
LEVESON, J. E. and PEACOCKE, A. R.: Biochim. biophys. Acta *149:* 311 (1967).
LIN, F.H. and THORMAR, H.: J. Virol. *6:* 702 (1970).
LITTAU, V.C.; BURDICK, C.J.; ALLFREY, V.G. and MIRSKY, A.E.: Proc. nat. Acad. Sci., U.S. *54:* 1204 (1965).
LOPASHOV, G.V. and SOLOGUB, A.A.: J. Embryol. exp. Morph. *28:* 521 (1972).
MANNER, H. W.: J. exp. Zool. *122:* 229 (1953).
MARKERT, C. L. and URSPRUNG, H.: Develop. Biol. *7:* 560 (1963).
MARUSHIGE, K. and BONNER, J.: J. molec. Biol. *15:* 160 (1966).
MEANS, A. R. and HAMILTON, T. H.: Proc. nat. Acad. Sci., U.S. *56:* 1594 (1966).
MIRSKY, A.E.: In L.C. DUNN Genetics in the twentieth century, pp. 127–153 (Macmillan, New York 1951).
MIRSKY, A.E. and RIS, H.: Nature *163:* 666 (1949).
MIZUTANI, S., BOETTIGER, D. and TEMIN, H.M.: Nature *228:* 424 (1970).
MORGAN, T.H.: Embryology and genetics (Columbia University Press, New York 1934).
MORRILL, C. V.: J. exp. Zool. *25:* 107 (1918).

MOSCONA, A. A.: J. cell. comp. Physiol. *60*/Suppl. 1: 65 (1962).
MOSCONA, A. A.: Proc. nat. Acad. Sci., U.S. *49:* 742 (1963).
NEYFAKH, A. A.: Dokl. Akad. Nauk. SSSR *132:* 1458 (1960).
NEYFAKH. A. A.: Nature *201:* 880 (1964).
NIEUWKOOP, P. D. and FABER, J.: Normal table of *Xenopus laevis* (Daudin) (North-Holland Publ. Co., Amsterdam 1956).
NILSSON, O.: Exp. Cell Res. *14:* 341 (1958).
OKAZAKI, R., OKAZAKI, T., SAKABE, K., SUGIMOTO, K. and SUGINO, A.: Proc. nat. Acad. Sci., U.S. *59:* 598 (1968).
OPPENHEIM, A. B.: Nature *226:* 31 (1970).
O'STEEN, W. K. and WALKER, B. E.: Anat. Rec. *139:* 547 (1961).
PALADE, G. E.: J. biophys. biochem. Cytol. *1:* 59 (1955).
PALADE, G. E. and PORTER, K. R.: J. exp. Med. *100:* 641 (1954).
PATEL, G. and WANG, T. Y.: Biochim. biophys. Acta *95:* 314 (1965).
PATRICK, J. and BRIGGS, R.: Experientia *20:* 431 (1964).
PAUL, J. and GILMOUR, R. S.: Nature *210:* 992 (1966a).
PAUL, J. and GILMOUR, R. S.: J. molec. Biol. *16:* 242 (1966b).
PAUL, J. and GILMOUR, R. S.: J. molec. Biol. *34:* 305 (1968).
PHILLIPS, D. P. M.: Progr. Biophys. biophys. Chem. *12:* 211 (1961).
PIRROTTA, V. and PTASHNE, M.: Nature *222:* 541 (1969).
PORTER, K. R.: J. Histochem. Cytochem. *2:* 346 (1954).
PORTER, K. R. and MACHADO, R. D.: J. biophys. biochem. Cytol. *7:* 167 (1960).
PRADA, N.: Tumori *32:* 151 (1946).
PTASHNE, M.: Proc. nat. Acad. Sci., U.S. *57:* 306 (1967a).
PTASHNE, M.: Nature *214:* 232 (1967b).
RIDDIFORD, L. M.: J. exp. Zool. *144:* 25 (1960).
RIGGS, A. D. and BOURGEOIS, S.: J. molec. Biol. *34:* 361 (1968).
RIGGS, A. D.; BOURGEOIS, S.; NEWBY, R. F. and COHN, M.: J. molec. Biol. *34:* 365 (1968).
ROSE, S. M.: J. exp. Zool. *108:* 337 (1948).
ROYCHOUDHURY, R.; DATTA, A. and SENS, S. P.: Biochim. biophys. Acta *107:* 346 (1965).
SALPETER, M. M. and SINGER, M.: Develop. Biol. *2:* 516 (1960).
SAUER, M. E. and WALKER, B. E.: Radiat. Res. *14:* 633 (1961).
SCHOTTE, O. E. and HARLAND, M.: J. Morph. *73:* 329 (1943).
SCOLNICK, E.M., AARONSON, S.A., TODARO, G.J. and PARKS, W.P.: Nature *229:* 318 (1971).
SETTERFIELD, G.; NEELIN, J. M.; NEELIN, E. M. and BAYLEY, S. T.: J. molec. Biol. *2:* 416 (1960).
SHUMWAY, W.: Anat. Rec. *78:* 139 (1940).
SMITH, L. D.: Proc. nat. Acad. Sci., U.S. *54:* 101 (1965).
SPIEGELMAN, S., BURNY, A., DAS, M. R., KEYDAR, J., SCHLOM, J., TRAVNICEK, M. and WATSON, K.: Nature *228:* 430 (1970).
SPIRIN, A. S. and NEMER, M.: Science *150:* 214 (1965).
STEDMAN, E. and STEDMAN, E.: Nature *152:* 556 (1943).
STEDMAN, E. and STEDMAN, E.: Nature *166:* 780 (1950).
STEEN, T. P.: J. exp. Zool. *167:* 49 (1967).
STEEN, T. P. and THORNTON, C. S.: J. exp. Zool. *154:* 207 (1963).
STOCUM, D. L.: Develop. Biol. *18:* 441 (1968).
STONE, L. S.: Anat. Rec. *106:* 89 (1950).

STONE, L. S.: J. exp. Zool. *113:* 181 (1952).
STONE, L. S.: Amer. J. Ophthal. *36:* 31 (1953).
STONE, L. S.: J. exp. Zool. *129:* 505 (1955).
SUGINO, A., HIROSE, S. and OKAZAKI, R.: Proc. nat. Acad. Sci., U.S. *69:* 1863 (1972).
SWIFT, H.: Exp. Cell Res., Suppl. *9:* 54 (1963).
SWIFT, H.: In J. BONNER and P. TS'O, The nucleohistones, pp. 169–183 (Holden-Day, San Francisco 1964).
TEMIN, H. N. and MIZUTANI, S.: Nature *226:* 1211 (1970).
THORNTON, C. S.: J. Morph. *62:* 17 (1938a).
THORNTON, C. S.: J. Morph. *62:* 219 (1938b).
TOWNES, P. L. and HOLTFRETER, J.: J. exp. Zool. *128:* 53 (1955).
TRAMPUSCH, H. A. L. and HARREBOMEE, A. E.: In V. KIORTSIS and H. A. L. TRAMPUSCH Regeneration in animals and related problems, pp. 341 (North-Holland Publ. Co., Amsterdam 1965).
URSPRUNG, H. and MARKERT, C. L.: Develop. Biol. *8:* 309 (1963).
VERMA, I. M., MEUTH, N. L., BROMFELD, E., MANLY, K. F. and BALTIMORE, D.: Nature New Biol. *233:* 131 (1971).
WALLACE, H.: J. Morph. *112:* 261 (1963).
WANG, T. Y.: Proc. nat. Acad. Sci., U.S. *54:* 800 (1965).
WANG, T. Y.: J. biol. Chem. *242:* 1220 (1967).
WEISS, P.: Roux' Arch. Entw. Mech. *104:* 359 (1925).
WEISS, P.: Principles of development (Holt New York 1939).
WELS, R. D., FLÜGEL, R. M., LARSON, J. E., SCHENDEL, P. F. and SWEET, R. W.: Biochemistry *11:* 621 (1972).
WICKNER, W., BRUTLAG, D., SCHEKMAN, R. and KORNBERG, A.: Proc. nat. Acad. Sci., U.S. *69:* 965 (1972).
XEROS, N.: Nature *194:* 682 (1962).
ZUBAY, G. and WATSON, M.R.: J. biophys. biochem. Cytol *5:* 51 (1958).

Author's address: Dr. ANN M. C. BURGESS, The London Hospital Medical College, Turner Street, *London E. 1.* (England)

Neoplasia and Cell Differentiation, pp. 153–188
(Karger, Basel 1974)

The Effect of Chemicals with Gene-Inhibiting Activity on Regeneration

ALEXANDER WOLSKY

Departement of Radiology, New York University Medical Center, New York, N.Y.

Contents

Introduction

Regeneration processes have often been compared to neoplastic growth and it was pointed out repeatedly that there is a fundamental similarity between these two categories of phenomena. Both in regeneration and in neoplastic development seemingly undifferentiated ('embryonic') cells appear to be multiplying in a localized region of a fully developed organism. It is, therefore, quite legitimate that in a book on neoplasia and cell differentiation regeneration phenomena should be discussed in detail. The usefulness of regenerating organisms, especially vertebrates, as test objects for testing carcinostatic substances was also pointed out [BIEBER and HITCHINGS, 1959; LEHMANN, 1961; WOLSKY and VAN DOI, 1965, and others]. Although there is a long way from results obtained on a regenerating salamander or a frog larva to their application in clinical cancer chemotherapy, the relationship and close similarity is quite clear and can yield useful results in the future.

It is therefore surprising to find that this approach to cancer research was not used on a larger scale and especially that hundreds of 'gene-inhibitory' antibiotics used in cancer chemotherapy were not tested more often or more thoroughly on regenerating systems. This chapter will review the relatively few cases in which the obvious advantages of this approach were utilized and will point out the relationship between chemical composition of the substances and their biological activity wherever possible.

Because of the neoplasia-oriented nature of this review, the regeneration processes which will be discussed will be almost exclusively the ones occurring in lower vertebrates, in particular amphibians, which have a spectacular capacity to rebuild lost organs, such as limbs, tail, or the lens of the eye. Compared with them, other vertebrates are relatively poor regenerators, although the constant, physiological replacement of certain tissues (intestinal epithelium, blood cells), or the sometimes extensive tissue repair after damage (liver, muscle) in higher vertebrates, including mammals, deserves attention and occasional reference will be made to the effect of antibiotics on such phenomena too.

The process of regeneration in amphibians and the factors controlling it are described in detail in another chapter [CARLSON, pp. 60–105]. Here it suffices to say that the replacement of lost parts in these organisms is achieved, in rather sharp contrast to various invertebrate groups, almost exclusively by dedifferentiation and rearrangement of various old tissues in the stump. The regeneration blastema, as the rudiment of the new organ is called, is made up from cells which have lost their various specialized structures and returned to an undifferentiated, 'juvenile' condition. At least they all look alike and look like true embryonic cells but how far they have regained their original 'pluripotent' status is still a matter of dispute. In recent years an excellent review [THORNTON, 1968] and a number of fine books [e.g., HAY, 1966; GOSS, 1969; ROSE 1971] have been published on regeneration and those who wish to gain more background information on the subject are referred to these sources. The biochemical aspects of regeneration are treated in detail by FLICKINGER [1967] and SCHMIDT [1968].

As this chapter deals with 'gene-inhibitory' chemicals, our first task must be to give a definition of the gene. Even with a superficial knowledge of genetic concepts one is bound to realize that the term is today almost like a dirty word not to be used in better circles. Euphemisms, such as cistron, or genetic material are substituted for it, to indicate that there is more to it than meets the eye. The spectacular revolution of ideas, expressed almost symbolically by the double helix, and such metaphors as 'transcription', 'translation', or 'genetic code', make the old concept of the gene as a corpuscular unit, a 'bead on a string', indeed obsolete. Today, the DNA → RNA → Protein trinity is the 'central dogma' of the faith of all molecular biologists. It is true that in this trinity DNA has the primacy (though this article of faith has been badly shattered recently by the demonstration by BALTIMORE [1970] and TEMIN and MIZUTANI [1970] of the existence of RNA-directed DNA synthesis in certain instances). A segment of DNA containing a sufficient number of nucleotide triplets in a 'meaningful' sequence comes nearest to the traditional gene concept. But it is so intimately linked to the 'information', 'messenger', 'template' and similar ideas (it is not without significance that these concepts can be best described by metaphors) that it cannot be logically separated from these. Nor can it be separated from the polypeptide chain assembled on a 'messenger' RNA 'template' in ribosomes. This is the 'primary gene product', the somewhat nebulous and not clearly defined 'genabhängiger Wirkstoff' of the early German pioneers of physiological genetics [cf. GOLDSCHMIDT, 1955] and clearly also a part of the genetic mechanism, inseparable from its nucleic acid partners in the 'trinity'.

The 'trinity' is also a logical successor to the 'one gene–one enzyme' concept of BEADLE [cf. e.g., BEADLE, 1945, 1948], which has been severely criticized by several authors [e.g., GOLDSCHMIDT, 1955; WADDINGTON, 1962] but had a basic truth in it. After all, before the chemical nature of the genetic material was clarified, many theoreticians were inclined to suspect proteins rather than nucleic acids as the material basis of heredity [cf. e.g., FREY-WYSSLING, 1938, p. 186 ff.] and there was a tendency in some early speculations not only to correlate, but to identify genes with enzymes [e.g., HAGEDOORN, 1911; GOLDSCHMIDT, 1916].

As today the 'gene', whatever that word means, has to be conceived as a functional rather than structural unit [PICKEN, 1960], it is obvious that the various aspects of its function must be viewed as a whole. Therefore, when discussing 'gene-inhibiting' substances, we shall in the following consider those antimetabolites which interfere with the function of the chromosomal nucleic acids in general, irrespective of the particular aspect which is affected, whether it is the replication of DNA strands, or the transcription of the DNA 'information' into RNA 'message'. or the translation of this 'message' into amino acid sequence in the polypeptide chain of the 'primary gene product'.

As mentioned above, very few gene-inhibiting substances were so far applied to regenerating systems with the aim of analyzing their effect on the regeneration process. Even smaller is the number if we consider only those which affect directly the primary component of the genetic mechanism, DNA. There is a group of substances which affect DNA in a manner which is strikingly similar to the effect of x-rays, γ-rays and, to a lesser extent and in a somewhat different manner, short-wave ultraviolet radiation. These radiation effects manifest themselves in abnormal or completely suppressed mitosis but also in the production of invisible changes in the genetic material, i.e., mutations. Chemical substances which have the same effects, i.e., are antimitotic and mutagenic, are often called 'radiomimetic'. But this is a relatively small group of chemicals and there are many others which interfere with the genetic material in a somewhat different manner but still quite effectively. We propose to deal here not only with the regeneration-inhibiting effect of the 'radiomimetic' substances in the strict sense but also with all other chemicals which affect, in one way or another, the genetic mechanism and through it the regeneration process. A fairly reasonable classification of these chemicals is the one adopted by BALIS [1968] and with some modifications and simplifications this will be followed here. It divides the substances according to their established or suspected mode of action into 5 categories:

1) substances interfering with purine and pyrimidine synthesis and interconversion;
2) substances affecting replication and transcription;
3) nucleic acid analogs acting by incorporation;
4) alkylating agents;
5) inhibitors of protein synthesis

A number of these substances satisfy the criteria of the definition of an 'antibiotic' in the strict sense, which requires that the substance not only interferes with vital processes (especially those of pathogenic microorganisms) but also be produced by microorganisms. But there is an even greater number of substances which have very similar effects but are either the products of higher organisms, animals or plants (e.g., colchicine) or are synthetic products not encountered in nature. Perhaps the expression 'antimetabolite' would be a more suitable collective term to include all the substances which will be surveyed here but only if we use it in its widest sense, to include under biochemical processes affected not only those of continuous metabolism but also those which form the basis of morphological and morphogenetic phenomena.

It must be kept in mind, however, that the mechanism by which a certain substance exerts its effect on processes of regeneration is seldom known in all its details. In fact, many of the 'gene-inhibitory' antimetabolites are pleiotropic in their effect, i. e., affect regeneration – which is itself a complex process – in more than one way. This pleiotropy manifests itself sometimes even on the biochemical level. Even in the case of actinomycin, where chemical composition and mode of action were so admirably correlated in a number of fine studies (see below), there is still a good deal of uncertainty in detail and the main effect upon the DNA-directed RNA synthesis cannot explain all observed phenomena [cf. e.g., REVEL *et al.*, 1964; HONIG and RABINOWITZ, 1965]. Similarly, colchicine, whose main effect (or perhaps more correctly, its most conspicuous effect) – the disintegration of the mitotic spindle in cell division – is well known, affects metabolic processes in other ways too and its full range of biological activity is still shrouded in a cloud of uncertainties.

Because of these circumstances the classification as outlined above, is somewhat artificial and arbitrary. In many instances a certain substance could be classified under two or even more headings. Nevertheless, the classification based on the mode of action is still the best way to bring order to the subject and wherever difficulties arise due to the fact of pleiotropy they will be pointed out and clarified as far as possible.

Substances Interfering with Purine and Pyrimidine Synthesis and Interconversion

Purine and pyrimidine synthesis and interconversion have, understandably, received a great deal of attention from biochemists, parallel with the gradual unravelling of the paramount role which they play in life processes as essential components of nucleic acids. The foundation for the purely biochemical studies was laid by several research groups in the decade following World War II. Naturally, no detailed account will be given here of these highly complex biochemical processes, the more so because an excellent recent survey of the subject exists in BALIS', 'Antagonists and Nucleic Acids' [1968], to which frequent reference will be made here and in the following sections. It suffices to point out in this connection that studies concentrating on purine synthesis in chicken liver have clarified the sources from which the components of the purine ring are obtained and also the sequence of events, step by step, by which the ring is actually constructed on the ribosephosphate complex of the nucleotide. The simultaneous demonstration that exogenous purines can be also incorporated directly into nucleic acids was a further important step in understanding how these master molecules of life are formed [BALIS, 1968, chap. 1]. Curiously, pyrimidine synthesis seems to follow different pathways as in this case the ring is formed as an aglycone and then attached 'ready made' to the phosphoribosyl component [BALIS, 1968, chap. 3].

Naturally, these studies have helped to identify not only the factors which are involved in the normal processes of synthesis and interconversion but also a number of substances which are specifically inhibiting them or interfering with their normal course. From the great number of these antagonists several were identified on the basis of their antimitotic or tissue growth inhibiting properties and the study of regeneration under their influence was repeatedly one of the tools in the analysis of their mode of action. In certain instances the lead to more precise biochemical studies was actually given by observations concerning the visible morphological effects on cell division, tissue growth and regeneration. This was particularly the case with the most spectacular of the 'antimitotic', or 'karyoclastic', or 'mitostatic' substances–colchicine. Since the discovery by DUSTIN and co-workers that this alkaloid stops mitosis in a peculiar manner at metaphase ('c-mitosis'), an enormous number of investigations was undertaken in various directions to analyze the phenomenon [cf. EIGSTI and DUSTIN, 1955]. It is worth noting that in more recent studies colchicine was replaced by colcemide (demecolcin), a substance which is very similar in its chemical composition to colchicine, can be obtained from the same plant sources and is about 20 times more effective in producing the same symptoms as colchicine [EIGSTI and DUSTIN, 1955]. Other substances, having the same or similar effect as colchicine, but neither chemically related, nor offering any clues as to their mode of action on the chemical level, such as podophyllin [CORNMAN and CORNMAN, 1951] or vinblastine [HODES *et al.*, 1960] have also been investigated.

The cytological basis of the mitostatic effect of these substances is the disintegration of the mitotic spindle, which prevents the separation and pole-ward movement of chromatids. It is also known, however, for quite some time that several of these antimitotic agents, especially colchicine, have other effects besides the disintegration of the mitotic spindle. ÖSTERGREN [1944] noted for the first time the well-known chromosome contracting effect of colchicine, while BEAMS and EVANS [1940] noted a decrease of cytoplasmic viscosity, a rather vague expression. In more recent years, purely biochemical research has added a good deal to the clarification of the possible mode of action of colchicine and other functionally (though not chemically) related compounds. These studies clearly indicate interference with the processes of nucleotide synthesis even though the exact mechanism is far from being clarified. It was reported as early as 1949 that colchicine interferes with enzymes which degrade deoxyribonucleotides. Later both NAD and AMP were found in increased, while ATP, GTP and UTP in decreased amount in mammalian liver after colchicine treatment. Furthermore, colchicine as well as vinblastine reduce the incorporation of uridine into transfer RNA but not into ribosomal RNA [cf. BALIS, 1968, chap. 2]. Colchicine also inhibits xanthine oxidase in rat liver [AFFONSO *et al.*, 1961]. These and similar studies clearly indicate that the various natural antimitotic alkaloids come under the heading of 'gene inhibitors', even though their exact mode of action on the 'gene' remains obscure.

Both colchicine and vinblastine have been applied to regenerating systems. Apart from experiments with Velban (vinblastine sulfate) on regenerating mammalian liver [LUICKX and VAN LANKER, 1966] and on tissues, such as spleen and bone marrow, involved in physiological regeneration [VAN LANKER *et al.*, 1966], the standard materials for vertebrate regeneration, i.e., the amphibian limb and tail, were used to test the effect of these substances. An early experiment [WOLSKY, 1941], reported in a preliminary form, showed that colchicine in concentration of 1:20,000 completely inhibits tail regeneration in the adult newt, *Molge cristata (Triton cristatus),* when specimens are placed in the solution immediately after amputation and are left there (with periodic replacement of the solution). Other experiments have shown that colchicine in this concentration is not toxic and does not interfere with the respiratory metabolism of amphibian embryos [WOLSKY, 1940]. Two weeks after amputation the stump was still covered with a thin epidermis having flattened, often pycnotic nuclei and there was practically no regeneration blastema. Some enlarged blood vessels, surrounded by fibrous connective tissue were observed near the amputation surface. However, when colchicine-treated specimens are washed and returned to water after 2-weeks treatment, their regeneration capacity returns and a delayed blastema formation takes place.

THORNTON [1943] investigated the effect of colchicine on regenerating forelimbs of larval *Ambystoma* salamanders, using extremely high concentrations (1:1,000; 1:500). He noted not only complete absence of regeneration but also an excessive dedifferentiation, followed by regression and resorption of the stump which has lead to a complete disappearance of the limb up to the shoulder region. This excessive regression was observed only when the colchicine treatment started immediately after amputation.

When it started 9 days later, the regeneration was suppressed but no regression of the stump was observed. Thornton correlates these findings with the results obtained by Butler and Puckett [1940] after x-irradiating amputated limbs, and similar results of Butler and Schotté [1941] and others, with amputated limbs, in which the stump was denervated. In these cases a similar excessive regression was noted. Butler has suggested that perhaps dedifferentiation and blastema formation are antagonistic processes and any treatment which inhibits blastema formation will necessarily lead to excessive dedifferentiation and, eventually, to disintegration and resorption of the 'useless' dedifferentiated cells.

Thornton's experiments had their counterpart in the extensive studies of Lehmann and his co-workers in Switzerland [Lüscher, 1946a, b; Lehmann, 1947; cf. also Lehmann, 1961]. Using regenerating tails of larvae of the South African clawed toad, *Xenopus laevis* (which later became a standard test-object in a number of related studies), they have obtained essentially similar results as Thornton, although using colchicine in greater dilution and applying it to the amputated tail in different ways. Particularly interesting is the finding that a tail in the process of regeneration, if wounded and exposed for only one hour to colchicine (1:2,000) will be completely resorbed. These authors also confirmed the observation, made on adult *Molge cristata,* that after removal of colchicine the tails regain their regeneration capacity and start to replace the lost organ. They point out that this is in sharp contrast to the effect of x-rays, which destroy the regeneration capacity permanently. The resumption of regenerative processes after temporary inhibition was observed especially after a skilful local treatment, in which larvae, immobilized with MS 222 (Sandoz) were placed on a glass slide and only the amputated surface was exposed to colchicine by covering it for $\frac{1}{2}$ h with filter paper soaked with the colchicine solution [Lüscher, 1946a, b]. Regression of stump tissues was noted by Lüscher when the regenerating tail tissues of the larvae were exposed to colchicine for a longer period. The first 3 days produce no visible effect but then a great number of arrested mitoses appear and regression sets in on the 5th day after amputation.

From these various studies it is evident that colchicine has a strong inhibitory effect on amphibian regeneration from the beginning. This effect is rather uniform on widely differing material which includes both urodeles and anurans, larval as well as adult stages and different organs, such as limbs and tails. It is fairly obvious that this effect is achieved primarily by the prevention of mitotic cell divisions. The thick epidermal cap, which is so characteristic for other types of regeneration inhibition (see next section), is missing here, which is quite significant. But as a good deal of the regeneration processes, especially blastema formation, is not mitosis dependent, it must be assumed that other colchicine effects, on the level of purine and pyrimidine metabolism, are also involved. Especially the striking regression phenomena, reported both by LÜSCHER and THORNTON, indicate such influences.

Vinblastine (VLB, Velban), which is considered colchicine-like in its mode of action [cf. NEUSS *et al.,* 1964; SENTEIN, 1964], has apparently a different effect on amphibian regeneration, according to FRANCOEUR [1968]. This author notes that in his experiments (tail regeneration of *Rana pipiens* larvae) vinblastine has a selective inhibitory effect on mesodermal elements, such as notochord and muscle. But the observations on mammalian liver regeneration under the influence of vinblastine, mentioned above [LUICKX and VAN LANKER, 1966] indicate that there is also a direct influence on nucleic acid synthesis. In a second paper FRANCOEUR and WILBER [1968] report a peculiar effect of vinblastine on blood islands in regenerating *Rana pipiens* tails, producing abnormally enlarged and proliferating blood vessels. The authors correlate this phenomenon with the known carcinostatic properties of the substance, claiming that the richer blood supply is detrimental to malignant growth [cf. GARB, 1968].

Another type of inhibition of the synthesis of nucleic acid components is produced by folic acid antagonists, such as aminopterin. It is well established that several enzyme systems, involved in one group of reactions, the addition of one carbon component to the purine ring, need folic acid as a co-enzyme. The various folic acid antagonists, which are structurally more or less similar to folic acid, can inhibit this reaction. For normal function, folic acid has to be reduced to tetrahydrofolic acid (THF) and this reaction is catalyzed by folic reductase. The antagonists appear to prevent this reaction by inactivating folic reductase [cf. BALIS, 1968, Chap. 4]. It is interesting to note that colchicine was also reported to depress folic acid activity in whole blood [LUKETIC *et al.,* 1965] although there is no chemical similarity between it and the antifolates.

The role of folic acid antagonists on cell multiplication and tissue growth was first noticed in studies on malignant growth. The observation that aminopterin can produce temporary regression in cases of acute leukemia in children [FARBER *et al.,* 1948] was a significant first step in cancer chemotherapy, followed by many similar ones. The effect of aminopterin and amethopterin on amphibian regeneration was the subject of a thorough study by GEBHARDT and FABER [1966] and both the methods applied and the results obtained were quite different from those of other experiments reviewed here. The authors studied forelimb regeneration in *Ambystoma mexicanum* and the antifolates were administered in a single dose (100 mg/kg body weight) orally. Forelimbs were amputated either at the level of the upper arm (operation A) or in the distal region of the lower arm, near to the wrist (operation B). The treatment with folic antagonists did not produce a quantitatively significant inhibition though the authors mention a general growth retardation and, occasionally, a reduced length of the lower arm. But the main result was qualitative: a frequent occurrence of either polydactyly or oligodactyly in a phase-specific and region-specific fashion. Polydactyly, which was of the postaxial type, occurred in 1/3 of the cases but only in the type A operations (amputation at upper arm level) and only when the antifolate treatment was applied within 10 days after amputation. At this time the regeneration blastema was in the cone stage. Oligodactyly was produced both in the A and B type operations in later stages, even if the antifolate was administered as late as the time when the first digital rudiment appeared. The maximum incidence of oligodactyly (100%!) was produced when the aminopterin dose was given 14 days after amputation, at the time when the regeneration blastema was in the paddle stage. The authors note that in the type B operations (amputation at distal lower arm level) the sensitive phase for inducing oligodactyly was shorter and the optimal response occurred also earlier. As mentioned, no polydactyly was produced in forelimbs amputated at this level.

The effects of aminopterin and methotrexate could be counteracted by administering to the treated animals citrovorum factor (5-formyl, 5,6,7,8 tetrahydrofolic acid) which clearly indicates that the antifolates acted in the manner as in other systems. It is particularly interesting that a third folic acid analog, 10-methyl pteroylglutamic acid, had no adverse effect on limb regeneration. This substance has a methyl group substitution in position 10 but no substituted amino group in the 4 position as aminopterin and methotrexate.

The data indicate that the interference produced by the antifolates is not so much regeneration inhibiting as rather teratogenic. GEBHARDT and FABER compared their data with several teratogenic studies on embryonic development and found certain similarities, notably the fact that polydactyly was induced by teratogens in earlier, oligodactyly in later embryonic stages. Why the folic acid antagonists have this specific effect – especially polydactyly – on regeneration while other inhibitors of purine syntesis do not show it, is an interesting question which awaits further investigation. It should be mentioned here, however, that BIEBER and HITCHINGS [1959] in their extensive studies with various inhibitors (see section after next) have also used aminopterin on tail regeneration of *Rana pipiens* larvae and, although they found a strong inhibitory effect, do not mention any particular malformations produced by it. Of course, the tail with its rather simple and uniform structure is not so suitable for a detailed teratogenetic analysis as the limb, which has a great variety of gross structural as well as histological characteristics.

The school of LEHMANN [HAHN and LEHMANN, 1960], which used tail regeneration of *Xenopus* larvae repeatedly as a sensitive quantitative test to evaluate the effect of various 'morphostatic' substances, has tested among others, one of the lesser-known analogs of pteridine, quinoxalin (3576 CIBA). In 1:100,000 concentration it reduced regeneration by 36/ (i.e., the average length of regenerates treated with the substance was at the end of the experiment 36% shorter than that of the controls). Whether quinoxalin, which is the analog of only one part of the folic acid molecule, exerts its effect by the same mechanism as the true folic acid antagonists, is not known and somewhat doubtful.

Besides the folic acid antagonists there is another characteristic group of inhibitors of purine and pyrimidine metabolism (synthesis as well as catabolism), which can be best described as analogs. They resemble closely in their molecular structure the biological purines and pyrimidines and can, therefore, interact with the enzyme systems of these latter ('substrate competition'?). Not all analogs have this effect or have it to the same extent, the most powerful ones being the fluoro-analogs, such as fluoro-orotic acid (FO), fluorouracil (FU), this latter both as aglycone and attached to deoxyribose (FUDR). Others, as for example iodo-deoxyuracil (IUDR) not only block purine synthesis, notably the synthesis of thymine (from orotic or formic acid), but also become incorporated into DNA, which is an entirely different kind of 'gene-inhibition' and will be reviewed in a later section.

Fluorouracil was applied to the study of tail regeneration in '4 to 6 cm tadpoles' (collected in ponds of Connecticut, U.S.A., presumably bullfrog, *Rana catesbeyana*) by DUMONT and SOHN [1963]. The substance was administered in repeated doses of 0.5 mg intraperitoneally and has, in the authors' words, completely suppressed blastema formation.

One of the substances, which belongs to this group because of its effect, even though it cannot be called an analog on the basis of its chemical structure, is hydroxyurea. It was shown to be an effective inhibitor of certain tumors. It was successfully applied also to regenerating mammalian liver, where it was shown to inhibit DNA but not RNA synthesis [SCHWARZ *et al.,* 1965]. It is of interest that in cultured mammalian cells (Chinese hamster) it affects the S phase of the mitotic cycle (DNA synthesis) and lethally damages cells which are in this phase at the time of exposure. Cells in the G_1 phase survive but cannot enter the S, while cells in the G_2 phase will even complete mitosis (M) and will be stopped only at the end of the next G_1 [SINCLAIR, 1965].

Antimetabolites Interfering with Replication and Transcription

While the antimetabolites discussed so far can be considered as closely associated with the synthesis of 'genes' and – with a certain stretch of the term – could be termed radiomimetic, the substances to be discussed here are more involved with the proper functioning of the genetic material. In particular, they are interfering with the genes' primary function, which is transfer and propagation of information and instruction from the static nuclear genes (DNA) to the more mobile RNA components, i.e., 'transcription' in the metaphor language of the 'basic dogma'.

The substance which comes immediately to mind in this connection is actinomycin, especially actinomycin D, which in the last decade rapidly became a household tool for all studies connected with the primary gene function of transcription. This relatively small molecule, made of a phenoxazone chromophore and two lactone rings, selectively inhibits RNA synthesis on DNA templates, while affecting DNA to a lesser degree [REICH *et al.,* 1961, 1962]. Details of the mode of action of actinomycin on the molecular level, which was studied by several workers [cf. e.g., HAMILTON *et al.,* 1963, REICH, 1964] will not be discussed here. But it should be mentioned that the model, according to which actinomycin attaches directly to guanine components of the DNA double helix and is located in its minor grove, creating thereby a steric inhibition for RNA polymerase activity, is still generally accepted as an explanation. In higher concentrations actinomycin interferes with DNA polymerase too, which must be due to some secondary effects on the DNA helix, probably preventing strand separation. This is reflected in the higher melting temperature of DNA, which was treated with more concentrated actinomycin [HASELKORN, 1964].

Actinomycin was applied repeatedly to regenerating amphibian systems. WOLSKY and VAN DOI [1965] used both 5-day-old *Rana pipiens* larvae and 2-month-old *Ambystoma punctatum* larvae to investigate the effect of actinomycin D on tail regeneration [see also VAN DOI and WOLSKY, 1964]. CARLSON [1965, 1966, 1967a] studied limb regeneration in adult *Triturus viridescens* and later, in a detailed analysis of the role of various tissue components in the actinomycin inhibition, in the black axolotl *Siredon mexicanum* [CARLSON, 1967b, 1969]. Although the studies of CARLSON and of WOLSKY and VAN DOI were carried out independently from each other, and on different material with slightly different methods, the results were strikingly similar. Not only a strong overall inhibition of regeneration was registered in all cases but also a greatly reduced blastema formation and an abnormal thickening of the apical epidermis. The significance of this latter phenomenon is not quite clear but cannot be overlooked.

There are many indications that the epithelial cap plays an active role in the removal of tissue debris and in the initiation of the dedifferentiation processes which precede blastema formation, presumably by producing proteolytic enzymes [cf. e.g., POLEZHAEW and FAWORINA, 1935; IDE-ROZAS, 1936; SINGER and SALPETER, 1961; SCHMIDT and WEARY, 1963; HAY, 1965, 1966, chap. 2]. If this is the case, then actinomycin obviously obstructs this activity by inhibiting RNA, and subsequent proteolytic enzyme synthesis. The hyperplasia of the epithelial cap might be then an expression of a regulatory effort to compensate for the paucity or lack of proteolytic enzymes with the accumulation of more epidermal cells in the apical cap. The latest investigations of CARLSON on the regeneration of axolotl limbs whose epidermis was treated with actinomycin while the rest was untreated [CARLSON, 1967b, 1969] seems to point in this direction: the treatment delayed the formation of the epidermal cap but also the dedifferentiation and blastema formation.

Actinomycin has quantitatively different effects on blastema formation and growth, on the one hand, and on mitosis, on the other. Low concentrations (1 μg/ml) mildly inhibit the former but allow mitosis to proceed in the regenerating system [WOLSKY and VAN DOI, 1965]. These authors were led to postulate that the proteins for mitosis may be synthesized on mRNA templates, which are qualitatively different from the templates of the proteins of blastema formation (proteolytic enzymes, others). The mRNA for mitotic proteins might be the long.lived type while the mRNA templates of the proteins needed for regeneration (blastema formation) seem to have a shorter life span. However, these differences may not exist for all cell types as the work of ROTHSTEIN *et al.* [1966] on the effect of actinomycin upon bullfrog lens indicates. Here actinomycin promptly inhibits mitosis.

Some problems of regeneration on the cyto- and histochemical level were attacked with the help of actinomycin by YAMADA [1966] and YAMADA and ROESEL [1966]. These authors, using such techniques as fluorescence microscopy and autoradiography demonstrated directly the inhibition of protein synthesis by actinomycin treatment in the case of Wolffian lens regeneration in *Triturus viridescens.* There was prompt inhibition when actinomycin was applied before stage VI of lens regeneration, i.e., before fiber differentiation actually starts in the outgrowth of the iris margin. However, in later stages the same treatment could not prevent lens fiber differentiation. This indicates that the messenger RNA, needed for the synthesis of fiber proteins, is of the long-lived type. This is in good agreement with the findings of SCOTT and BELL [1965] on mRNA utilization during lens development in the chick embryo.

These various studies with actinomycin clearly indicate the unusually great heuristic value of this antibiotic in regeneration research. One can safely predict that future research will show further uses of actinomycin in this field. For example, the question of the penetration of this substance (and other similar antibiotics) into aquatic organisms, such as an amphibian larva, when it is added to the water surrounding the organisms, is still unsolved. The use of radioactive (^{3}H) actinomycin could be of help here as it was recently in clarifying the same question concerning echinoderm eggs and embryos [THALER *et al.*, 1969; VILLEE and GROSS, 1969; GREENHOUSE, 1970, 1971].

Phleomycin, an antibiotic lesser known and lesser used than actinomycin, resembles the latter insofar as it attaches itself to the DNA molecule (at thymidine components), a fact demonstrated by several authors with different methods [cf. e.g., PIETSCH and GARRETT, 1968]. But in contrast to actinomycin, its effect is on DNA synthesis rather than DNA-directed RNA synthesis. It leaves this latter process unaffected which gives phleomycin a particular place in molecular biology. Its use makes it possible to study the time relationship between gene replication and primary gene function. A study of this type with the help of phleomycin was undertaken by PIETSCH and co-workers who used as test object regenerating rat muscles. Their investigations are the more valuable as they compare results with earlier ones on the effect of actinomycin D and colchicine on the same material [PIETSCH, 1961, 1964; PIETSCH and MACALISTER, 1965].

Muscle injury, according to these and earlier studies [cf. LASCH *et al.*,

1957] elicits within 48 hours a rapid proliferation of mononucleated cells in the wound region. Two days later (i.e., 4 days after injury) newly developing muscle fibers appear in great number in the regeneration area. Proliferation of mononucleated cells can be arrested with colchicine in the period when waves of mitotic cell division occur in the coagulum, i.e., between 48 and 72 hours after injury. Actinomycin is effective only in a short period, just before the appearance of differentiating muscle fibers, around 72 hours. Before or after this time actinomycin has no effect. Phleomycin, on the other hand, exerts its effect when administered before the period of mitoses, about 24 to 36 hours after wounding the muscle. From these data it can be concluded that the mRNA templates needed for muscle differentiation are formed and discharged shortly before the visible signs of actual differentiation appear. The genetic material (DNA) itself, which directs the mRNA synthesis, is formed by replication $2\frac{1}{2}$ to 3 days earlier.

A substance which chemically hardly fits in the same category as actinomycin but seems to have similar effects, and was widely used in regeneration research, is β-mercaptoethanol. It has a number of seemingly unrelated biological effects, attributed primarily to its mercaptol (SH) component. It inhibits mitosis by interrupting spindle formation [MAZIA, 1958; MAZIA and ZIMMERMAN, 1958] and by the same mechanism produces endomitosis and polyploidy [JACKSON, 1963]. (In these respects mercaptoethanol resembles colchicine rather than actinomycin.) SRINIVASAN *et al.* [1964] have shown that β-mercaptoethanol decreases considerably the incorporation of radioactive (^{3}H) cytidine in both nuclear and nucleolar RNA, even at low concentrations (10^{-4} M). The same authors found a relative inhibition of the transfer of the radioactive label from nucleus to cytoplasm. These effects strongly resemble the effect of actinomycin, but the mechanism by which mercaptoethanol achieves them is still unknown even though the SH group is obviously suspected (upsetting the normal SS $\rightleftharpoons$ SH equilibrium?).

The antimitotic and RNA-inhibiting properties of mercaptoethanol can easily explain its strong inhibitory effects on developmental processes, which several workers observed. This is true both of embryonic development [RAPKINE and BRACHET, 1951; BRACHET, 1958; BRACHET and DELANGE-CORNIL, 1959; WOLSKY and WOLSKY, 1968], where there are indications that the morphogenetic response of competent anlagen to embryonic inductions is stronger affected than the inductor itself [BRACHET and DELANGE-CORNIL, 1959; WOLSKY and WOLSKY, 1968], and of regeneration

[HAHN and LEHMANN, 1960b, cf. also HAHN, 1959; HAHN and LEHMANN, 1958, 1960a; HAHN *et al.*, 1961]. The regeneration experiments were made with the standard test material of the LEHMANN school, the regenerating tail tip of *Xenopus laevis* larvae of about 25 to 35 mm length and HAHN and LEHMANN point out that their material is far more sensitive to mercaptoethanol than other objects used in developmental-biological experiments, such as echinoderm eggs or amphibian embryos. MAZIA used mercaptoethanol on sea urchin eggs in 0.075 M concentration; WOLSKY and WOLSKY on amphibian embryos (for short 'pulse' treatment) in 0.05 M; BRACHET on the same material (for prolonged treatment) in 0.0033 M, while *Xenopus* tail regenerates were treated with only 0.0004 to 0.00005 M! However, when HAHN and LEHMANN applied the substance to the regenerating *Xenopus* tail, their primary aim was not only to test its morphostatic (i.e., regeneration-inhibiting) effect in itself but also to analyze its mode of action. In particular, they compared its effect to that of another SH-carrying substance, 5,7-dimercapto-thiazolo [5,4-d] pyrimidine (E 96). Qualitatively, the inhibition was the same with both of these SH-substituted antimetabolites (though E 96 had to be used in higher concentrations and did not produce more than 60% inhibition) but the detailed analysis, made with simultaneous and sequential combination showed that their mode of action is quite different and indeed antagonistic. The authors postulate that the SH group of β-mercaptoethanol acts indirectly, in the early phases of regeneration, as an activator of proteolytic enzymes of the cathepsin type [cf. also HAHN, 1959; HAHN *et al.*, 1961; JENSEN *et al.*, 1956]. The strong morphostatic effect is secondary, via these enzymes, which upset the balance of protein metabolism in the regenerating tail tissue and lead to a severe loss of protein. E 96, on the other hand, seems to act as a purine inhibitor. The mercaptoethanol studies were part of a more elaborate project of LEHMANN and co-workers, undertaken to test the possible synergism (or antagonism) of a number of morphostatic substances in order to clarify the way in which they interfere with processes of regeneration. These studies will be discussed further below.

Here, however, brief mention should be made of a number of studies which indicate that foreign RNA in itself can interfere with the normal RNA metabolism in developing systems, including regenerating amphibian tissues. POLEZHAEV, for many years a pioneer in the study of inducing and stimulating regeneration in non-regenerating (anuran) amphibians, such as adult frogs, and in regenerating (urodele) forms, in which the regenerative capacity was suppressed by x-irradiation, has used in recent studies liver

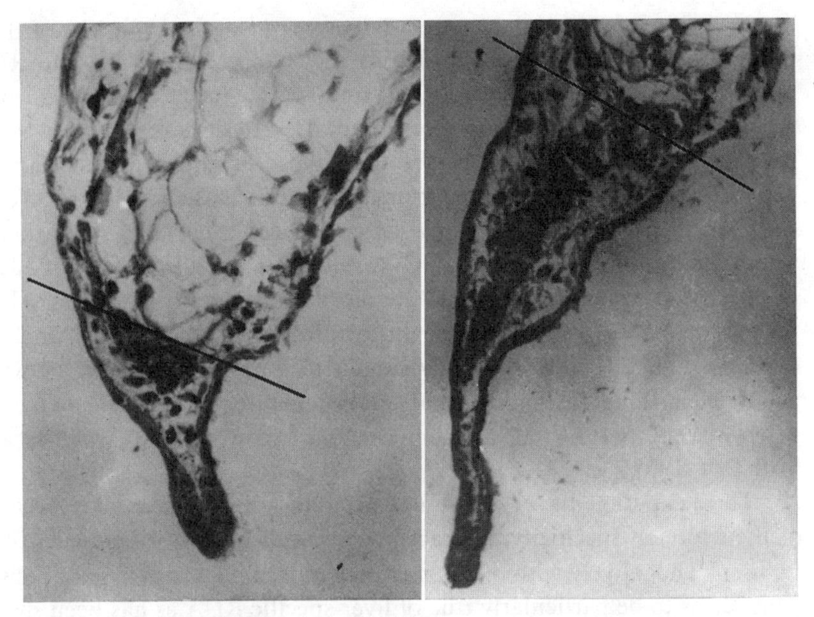

Fig. 1. Longitudinal sections through the tail tips of young (12-day-old) *Rana pipiens* larvae in the process of regeneration, 2 days after amputation. Left: normal regeneration; right: regenerate of a specimen, placed immediately after amputation in a 0.12 mg/ml solution of an RNA preparation from adult liver of the same species. Transverse lines indicate the level of amputation. Note compact column of strongly stained, closely packed juvenile notochord cells in continuation of the transsected notochord in control, only a small accumulation of similar cells at the tip of the notochord of the RNA treated regenerate. Enlargement in both photographs ×130. (Original.)

RNA preparations, particularly in Mexican salamanders, receiving massive doses of total body irradiation. The RNA preparations, injected peritoneally, restored the regeneration capacity to a certain extent. This work is summarized in two papers in Russian, with brief English summaries [POLEZHAEV, 1966a, b], and also in a recent English translation of POLEZHAEV's book on the subject [POLEZHAEV, 1972].

Similar results were reported on a smaller scale by SMITH and CRAWFORD [1969] when adult *Rana pipiens* frogs with amputated limbs were treated with intraperitoneal injections of RNA preparations from liver tissue of the same species. Instead of ordinary wound healing, a certain amount of dedifferentiation and blastema formation were observed.

Other workers have found that the addition of alien RNA has inhibitory effects on the process of regeneration [WOLSKY *et al.,* 1966; WOLSKY and

WOLSKY, 1967, 1968]. Adult newts *(Triturus viridescens),* with half of their tail amputated and injected in 3 to 5 day intervals with a crude RNA preparation from adult liver of the same species (each time 0.5 ml of a 0.1 mg/ml RNA in Ringer solution) had after 15 days regenerates which were on the average 21% shorter than those of the controls. (Similar experiments with RNA prepared from tail tissue increased regeneration over the controls by 26%.) *Rana pipiens* larvae, placed after tail amputation in somewhat stronger (0.12 mg/ml) RNA solutions, made from liver of adults of the same species, regenerated 33% shorter tails than the controls. These regenerates were compared also histologically to the controls and the differences (fig. 1) were strikingly similar to those found earlier between actinomycin-inhibited and normal tail regenerates of *Ambystoma* larvae when weak (0.001 mg/ml) actinomycin solutions were used [WOLSKY and VAN DOI, 1965].

These experiments seem to indicate that the alien nucleic acids, because of their foreign specificity, interact in some still unknown manner ('competition'?) negatively with the tissues' own inborn nucleic acid metabolism. This seems to be particularly true of liver-specific RNA as has been shown in a number of studies on tumor growth [NIU, 1960; DE CARVALHO and RAND, 1961; AKSENOVA *et al.,* 1962]. These various authors have found some striking inhibitions of tumor growth after incubating transplantable tumor cell suspensions with liver RNA. AKSENOVA *et al.* have, for example, observed no tumor growth when cells of a transplantable strain (Malyugin) were incubated for 8 hours with liver RNA of 0.1 mg/ml concentration.

Naturally, these data do not actually contradict POLEZHAEV's positive results with liver RNA because the situation is obviously quite different when the investigated tissues' own nucleic acid metabolism is disturbed, as in POLEZHAEV's experiments. It is quite conceivable that after the severe damage, which the x-irradiation must have inflicted upon the regenerating tissues of POLEZHAEV's axolotls, any addition of exogenous RNA could have only ameliorated the situation.

Similarly, the results of SMITH and CRAWFORD can be explained by assuming that the exogenous RNA from liver tissue has provided a certain morphogenetic stimulus in a case where such stimulus is normally nonexistent.

But when the nucleic acid mechanism is intact, as in the frog larvae of the experiment by WOLSKY and WOLSKY *et al.,* interference from exogenous RNA seems to be detrimental, especially if it comes from such highly specific tissue as the liver.

Nucleic Acid Analogs Acting by Incorporation

In a previous section it was pointed out that certain purine and pyrimidine analogs interfere with the synthesis and catabolism of the normal, biologically active purines and pyrimidines by blocking the activity of enzymes which the normal processes need. These analogs acted thus similarly to folic acid antagonists and were themselves antagonists of purine and pyrimidine metabolism. There are, however, other analogs which act not as antagonists of synthesis or interconversion but rather by being 'mistaken' for the normal purines or pyrimidines and by being incorporated into nucleic acid polymers in place of the normal constituents. They thus 'contaminate', or 'adulterate', the nucleic acids with useless substitutes. Such 'contamination' by analog incorporation can, understandably, destroy or, at least, seriously impair the biological activity of a nucleic acid, especially DNA, which gives these analogs, acting by incorporation, their great significance in nucleic acid research.

Of course, it is not always known for sure whether a certain analog acts by its virtue as an inhibitor of an enzyme reaction connected with biological purines and pyrimidines, or by its capacity of 'fooling' the polymerizing enzyme system and making it to incorporate the worthless substitute into its product. But in most cases the mechanism has been clarified (mostly by tracing the path of labeled components) and it has been established that such well-known analogs as 8-azaguanine, 2-thiouracil, 5-bromouracil, iododeoxyuridine, 6-mercaptopurine, 5-azacytidine and many others are, indeed, incorporated into DNA or RNA or both, even though a few questions remain open [cf. BALIS, 1968, chap. 6].

From the great number of these 'incorporating' analogs, which as mitotic inhibitors play a great role in cancer chemotherapy [cf. BIESELE, 1958], only relatively few have been applied to the study of regeneration. Some of the widely used ones, like 6-mercaptopurine, azathymine, thioguanine and others, have been tested on amphibian material by BIEBER *et al.* [1954, 1961], BIEBER and HITCHINGS [1959], YANKOW [1965] and CHERAYIL [unpublished].

BIEBER *et al.* [1954] have shown that 6-mercaptopurine inhibits considerably the regenerative growth of amputated tails of *Rana pipiens* larvae and increasing concentrations (from 1 to 199 mg%) increase the effect in a linear manner from 10% ('blastema index', based on the formula *Length of treated blastema/Length of control blastema* = 0.90) to 80% (blastema index 0.20). Thioguanine, on the other hand, has no effect in low concentrations (0.1 to 1.0 mg%) but produces a rapidly increasing inhibition in concentrations higher than 1.0 mg%. Adenine, adenylic acid and guanylic acid can reverse the inhibition of both analogs to a certain extent. Two other purine analogs, 8-azaguanine and 2,6-diaminopurine, had no effect on tail regeneration during the period of observation, which was 7 days.

BIEBER and HITCHINGS [1959] compared, in the framework of the American Cancer Society's survey [cf. GELLHORN and HIRSCHBERG, 1953] the regeneration inhibiting capacity of 26 carcinostatic substances, using again tail regeneration of *Rana pipiens* larvae as test material. The strongest effect was obtained with 6-mercaptopurine. (Curiously, both colchicine and chloramphenicol, which other authors used successfully to inhibit regeneration, were ineffective in these tests.) A more detailed comparison was made between mercaptopurine, benzimidazole (another purine inhibitor) and aminopterin. All three had strong inhibitory effects on regeneration of *Rana pipiens* tails but there were indications that these substances (especially aminopterin in contrast to the two purine inhibitors) acted in different phases of the regeneration process (see section on synergism). The authors stress especially the differences between the regression (dedifferentiation) and blastema formation phases.

YANKOW [1965], using lower concentrations of 6-mercaptopurine and thioguanine (3.33×10^{-4} M), as well as two further analogs, 5-bromodeoxycytidine (BCDR) and 5-iodo-deoxyuridine (IUDR), obtained only slight inhibitions in larval tail regeneration of *Rana pipiens* but considerable inhibition in the tail regeneration of adult newts *(Triturus viridescens)*. In this latter case the control regenerates have grown to 2.3 mm in 24 days, while the regenerates treated with the analog solutions were only 1.1 to 1.5 mm long.

Both BIEBER and co-workers and YANKOW applied the treatment to their experimental specimens *in toto,* i.e., submerging them in the solutions for the duration of the experiments. On the other hand, CHERAYIL [1963, unpublished] applied another analog, azathymine, to adult *Triturus viridescens* after tail amputation in the form of intraperitoneal injections (0.5 ml of a 4 mg/ml solution every days) and obtained considerable inhibition in 4 weeks.

None of these authors give much information about the histological findings on the inhibited or retarded regenerates. The little information that is available does not give clues as to the mechanism of action, except perhaps YANKOW's remark that the apical epidermal cap was 'missing or poorly organized in all cases', with 'numerous intercellular spaces'. This seems to indicate at least that the mode of action of the analogs is different from that of actinomycin, which—as pointed out above—produces thikkened epidermal caps. BIEBER *et al.* [1954] note that blastema formation, which 'involves dedifferentiation (morphological and biochemical) of existing cells, preceding cell multiplication', is not affected by 8-azaguanine and 2,6 diaminopurine, but the subsequent cell proliferation is slightly affected. This too, points up the rather sharp difference between these analogs and actinomycin in their histological effect.

Alkylating Agents

The expression 'alkylating agent' is used in biology in a somewhat different sense than in chemistry. A number of biologically active substances belong here which can react with a wide variety of functionally important groups in several hormones, enzymes and nucleic acids. Chemically, this effect is achieved by replacing hydrogen with an alkyl group or an alkyl derivative. From our point of view the most important biological alkylators are those which interfere with protein and nucleic acid synthesis. In this category we encounter again the phenomenon of pleiotropy, a number of substances which are not only alkylating agents but also act as inhibitors of purine synthesis, like the diazo compounds azaserine and 6-diazo-5-oxo-L-norleucine (DON).

Interest in alkylating agents goes back to the infamous mustard gas (bis chloroethyl sulfide, 'H') and nitrogen mustard (bis chloroethyl amine, 'HN_2') of World War I. Their mechanism of alkylation is thoroughly investigated [cf. BALIS, 1968, chap. 7] and quite often the notion of biological alkylation is associated exclusively with the action of mustards. It was not before the end of World War II that purely biological studies were undertaken with these substances and it turned out that they are both powerful antimitotic and mutagenic agents. Their antimitotic and mutagenic action closely resembles the effect of x-rays and the expression 'radiomimetic substances' is often applied to alkylating agents. (The wider application of the term to all substances which interfere in one way or another with gene activity is not quite justified but often done.) This similarity indicates that their effects are primarily exerted on nucleic acids and processes dependent on nucleic acids, including protein synthesis. But their actual mode of action is still not clear. In fact, a number of data indicate that while nitrogen mustard inhibits mitosis, it does not prevent DNA, RNA or protein synthesis. This was shown by CASPERSSON *et al.* [1963] with ultraviolet cytophotometry and by LEVIS *et al.* [1964] with autoradiography of radioactive thymidine incorporation. Nitrogen mustard is radiomimetic even in this cytochemical behavior as CASPERSSON *et al.* [1963] found that moderate doses of X-rays have the same effect, i.e., continuous nucleic acid and protein synthesis in the absence of mitosis.

In one of the few applications of an alkylating agent to the study of regeneration this characteristic dissociation of mitotic activity from nucleic acid synthesis was clearly demonstrated by BASSLEER *et al.* [1963] and by MATAGNE-DHOOSCHE [1964]. Using myleran (dimethanesulfonoxybutane) on regenerating tails of the Mexican salamander, *Ambystoma mexicanum* (immersion of whole body after amputation in various concentrations, ranging from 2.5×10^{-6} to 1.1×10^{-3} M) these authors noted a marked inhibition of the regeneration. In the experiments of MATAGNE-DHOOSCHE [1964] the average length of the control regenerates was, after 25 days, 2.0 mm, while that of specimens treated with 5×10^{-5} M myleran was 1.5 mm, and that of the individuals exposed to a 4.4×10^{-4} M solution was only 0.5 mm. At the same time there was a marked difference between the epidermal cells of the control and myleran-treated regenerates, analyzed by MATAGNE-DHOOSCHE. The control cases showed two distinct nuclear size categories in the epidermis, one 5.4, the other 10.8 arbitrary units (each unit equivalent to a sphere of 7.8 μm diameter and 247.5 μm^3 volume). The myleran-treated regenerates contained, in addition, a third category of epidermal nuclei which measured about 20 units and also some odd sizes. The percentage of nuclei falling in the different categories was also different. The myleran-treated regenerates had more epidermal nuclei in the higher categories and less in the lower (5.4 unit) category. At the same time there was a clear indication of mitotic inhibition. The mitotic index was 1 to 2 per 1,000 cells in the myleran-treated specimens, 3 to 14 per 1,000 in the controls. All this seems to agree well with the chemical findings showing uninhibited nucleic acid synthesis in mitotically inhibited nuclei.

An earlier study [KARCZMAR, 1948] on the effect of the more commonly used alkylating agent, nitrogen mustard (methyl-bis β-chloroethyl amine), upon regeneration in *Ambystoma punctatum* was not carried to cytological details but also reported definite inhibition of regeneration.

YANKOW [1965] used uracil mustard, both alone and in combination with analogs (see above) on regenerating tails of adult *Triturus viridescens*. In 3.33×10^{-4} M concentration the mustard inhibited regeneration moderately. While the control tails have grown in 24 days, on the average, by 2.3 mm, the mustard-treated tails were only 1.5 mm, i.e., the regeneration was inhibited by 35%. The results of the combination of uracil mustard with thioguanine and other analogs, as well as the whole question of synergistic and antagonistic effects of gene inhibiting substances on regeneration will be discussed in a later section.

Inhibitors of Protein Synthesis

On the basis of the definition of a gene, as adopted here, i.e., identifying it with the whole DNA → RNA → Protein trinity, it is obvious that agents which inhibit protein synthesis belong to the subjects of this review. There are a number of chemicals, mostly natural antibiotics produced by various species and strains of *Streptomyces*, which belong here because they interfere with the third step of the genetic mechanism. Among the better known ones are puromycin, streptomycin, chloromycetin (chloramphenicol) and several others. Their mode of action is not always the same and often probably pleiotropic. Thus, for example, puromycin, which inhibits protein synthesis probably by occupying an active site on the ribosome [NATHANS and NEIDLE, 1963] and thereby blocking further amino acid addition because it lacks a carboxyl group, also acts as an inhibitor of nucleic acid synthesis [cf. BALIS, 1968, chap. 8]. Similarly, streptomycin and, to a lesser degree, dihydrostreptomycin precipitate nucleic acids [MOSKOWITZ, 1963], but also inactivate ribosomes in 'sensitive' systems by forming complexes with them and preventing transfer of amino acids from the amino acyl-tRNA to protein [ERDÖS and ULLMANN, 1959, 1960]. Streptomycin and some related antibiotics (kanamycin, neomycin) also have the rather unique effect on the genetic mechanism that they produce a 'misreading' of the genetic code. Presumably by distorting the ribosomes, they make a certain triplet sequence ('code word') on the mRNA appear to the approaching amino acyl-tRNA molecules as if it were another 'code word'. Consequently, a wrong amino acid will be inserted in the polypeptide chain at that point [SPOTTS and STANIER, 1961; GORINI and KATAJA, 1964; ANDERSON *et al.*, 1965].

Several of the better known protein synthesis-inhibiting antibiotics were applied to the study of their effect on amphibian regeneration. LIVERSAGE and co-workers studied the effect of both puromycin and chloramphenicol on limb regeneration in adult *Triturus (Diemictylus) viridescens* [BURNET and LIVERSAGE, 1964; LIVERSAGE and COLLEY, 1965]. Puromycin, injected daily in the peritoneum in doses of 0.05 ml of a 0.01 mg% solution, produced strong inhibition of the regeneration blastema. In 91% of the cases the regenerates remained in a permanent blastema stage or formed small cones. (The results were evaluated 40 days after amputation.) In 9% of the treated specimens – altogether 87 were treated – the regenerates reached the palette stage but had 'particularly abnormal histogenesis'. The PAS reaction was negative. While the stump tissues were relatively normal the blastema was composed of abnormally small cells.

Chloramphenicol was given either in intramuscular injections in the vicinity of the regenerate or intraperitoneally. The daily doses varied from 0.5 to 5.0 mg. (This latter seems to be abnormally high.) Chloramphenicol markedly delayed regeneration but did not stop it entirely and regeneration blastemas were formed. As with puromycin, the majority of the treated specimens regenerated only a small cone in 40 days; the best cases were in the palette stage. After the injections were stopped there was some increased regeneration but the growth was highly abnormal, very short and narrow, with or without shortened, abnormally shaped fingers. The histological studies supported the external findings. The dedifferentiation of bones was greatly delayed and up to the 25th post-amputation day few cells accumulated in the blastema. The later growth, up to the 40th day, was also retarded and abnormal.

Somewhat similar results were reported recently by PROCACCINI and DOYLE [1970] concerning the influence of streptomycin on regeneration, even though these authors used different material and different methods of treatment. Young larvae, 5 or 10 days old, of *Rana clamitans* were used and their tail regeneration investigated. After tail amputation the larvae were placed immediately in streptomycin solutions in concentrations ranging from 0.9 to 1.3 mg/ml. The authors emphasize the toxicity of the substance for regenerating larvae and point out that a solution strong enough to inhibit regeneration completely is also the LD_{50} dose. The same concentration, however, does not kill intact, non-regenerating larvae of the same age which indicates that the regeneration process in itself alters the whole physiology of the organism.

Lower concentrations of streptomycin, between 0.9 and 1.1 mg/ml had some enhancing effect on wound healing and produced an enlarged epidermal cap, in the same manner as actinomycin. Higher doses, however,

(1.2, 1.3 mg/ml) inhibited the epidermal proliferation. The dedifferentiation process was not inhibited by the antibiotic but the redifferentiation (histogenesis) was abnormal, especially in the higher concentrations, which resulted in disorganization of the muscle segments producing regenerates of abnormal shape. In a recent publication the same authors report on the effect of streptomycin upon limb regeneration in adult *Triturus (Diemictylus)* salamanders [PROCACCINI and DOYLE, 1972]. While confirming the earlier finding that streptomycin does not inhibit the formation of the epidermal cap over the regeneration blastema, except in high concentrations, they find, after applying 40 μg/g streptomycin (intraperitoneally, in one single dose just prior to amputation) marked alterations in the blastemal elements, leading to frequent external malformations (oligodactyly). The authors suggest that 'one of the important effects of streptomycin is its binding irreversibly to the ribosomal surface, and causing translational errors resulting in the production of inoperative protein'.

These data show the severe consequences of protein deprivation on regeneration. Obviously, those proteins which are needed for growth and differentiation are affected most but some enzymes involved in dedifferentiation must have been also in short supply, as the reports on delayed dedifferentiation indicate. The thickening of the epidermal cap under the influence of lower concentrations of streptomycin is worth further investigation.

Some preliminary experiments in the author's laboratory with kanamycin in much lower concentrations (10 mg% and 25 mg%) showed no effect on tail regeneration of early *Rana pipiens* larvae or more advanced *Ambystoma punctatum* larvae. The results are, however, not quite conclusive yet.

Finally, a substance should be mentioned here, because it is an analog of an amino acid, but otherwise has probably no particular connection with protein synthesis. It is N-dichloroacetyl-DL-serine which was shown by LEVI *et al.* [1960] to be carcinostatic in certain cases, is morphostatic on sea urchin eggs [WOLSKY and WOLSKY, 1960] and was applied to regenerating forelimbs of the newt *Triturus (Diemictylus) viridescens,* by GOMES [1964]. As N-dichloroacetyl-DL-serine is chemically related to azaserine (O-diazoacetyl-L-serine), one of the most versatile pleiotropic antimetabolites [cf. BALIS, 1968], it is impossible to say what its mode of action could be. In the experiments of GOMES it mildly inhibited regenerative growth but had little or no noticeable effect on dedifferentiation and blastema formation.

Synergism of Gene-Inhibiting Substances

One of the fundamental questions in the study of regeneration and its inhibition by chemicals is whether the inhibition can be made more effective when several 'morphostatic' substances are applied together to the same process and, thus, several aspects of gene function are interfered with simultaneously. Understandably, not only developmental biologists but also oncologists are interested in this problem, since the combination of drugs in cancer chemotherapy is becoming more widely used [cf. e.g., GOLDIN and MANTEL, 1957]. Nevertheless, morphostatic synergism can be better studied on the well analyzed and fairly well (though not completely) understood processes of regeneration than on the far more enigmatic malignant proliferations. LEHMANN and his school [for a summary of their work cf. LEHMANN, 1961, 1962] must be given credit for concentrating on this aspect while at the same time recognizing the bearing of their studies on cancer research and the close parallelism between inhibition of regeneration by chemicals and cancer chemotherapy.

The theoretical basis of morphostatic synergism is the thesis, often stressed by LEHMANN [cf. e.g., LEHMANN, 1945] that development is a 'kombinative Einheitsleistung', i.e., a single achievement (literally a unit performance), brought about by the combined effect of several factors, usually acting in temporal sequence or spatial-causal relation. These factors have a definite chemical (metabolic) basis and the old idea of CHILD [cf. e.g., CHILD, 1941] that quantitative metabolic differences, forming a definite pattern (usually along an axis, 'axial gradients') produce qualitative morphological differences, is revived in LEHMANN's concept of a metabolically determined morphogenetic potential as the basis of developmental processes.

Two sectors of metabolism seem to contribute mostly to the creation of a certain morphogenetic potential, protein and nucleotide (purine and pyrimidine) metabolism. LEHMANN and co-workers stress the important role of cathepsin-like proteolytic enzymes in the former and point out that a number of studies have shown the existence of a characteristic cathepsin pattern in the *Xenopus* tail which undergoes phase-specific changes during regeneration [DEUCHAR *et al.*, 1957; LEHMANN, 1957a]. The same was shown for catalase activity [HAHN, 1959]. A higher level of metabolic (especially anabolic) activities clearly increases the morphogenetic potential. In the case of regeneration the replacement of lost parts will proceed slower or faster according to the rate of metabolism.

All this means that developmental processes can be influenced by chemical factors which interfere with metabolism (notably protein and nucleotide metabolism) and they can attack metabolism at several points, either simultaneously or consecutively.

In regeneration processes the sequential nature of events is quite obvious and distinct phases are well defined and recognized. The great role of the dedifferentiation phase makes vertebrate regeneration particularly suitable to study the different phases and their susceptibility to various inhibitors. LEHMANN's contention that cathepsin-like proteolytic enzymes play a distinct role in the morphostatic mechanism seems to implicate dedifferentiation. An excessive dedifferentiation and stump resorption can have disastrous effects long before the constructive phases can manifest themselves. THORNTON's and LÜSCHER's early observation of colchicine-induced resorption of stumps, resembling the effect of stump denervation, would deserve further analysis. But, of course, the excessive activity of proteolytic enzymes and a disturbance of the equilibrium between anabolism and catabolism of proteins can just as well interfere with the later phases of blastema formation and differentiation.

In LEHMANN'S elaborate terminology 'morphostatic' (or 'histostatic') sometimes means an action or capacity to prevent cell growth and differentiation, as contrasted to 'antimitotic' which means arresting cell division. LEHMANN [1961] points out that certain substances are strong morphostatic inhibitors but not antimitotic. Others, however, are both morphostatic and antimitotic. In this distinction two different strategies of gene inhibiting substances are already implied. But there are other distinctions. LEHMANN contrasts, for example, substances which influence protein turnover (especially the cathepsin activators) to those which affect purine metabolism. These latter may or may not inhibit mitosis.

Put to experimental test, LEHMANN's concepts turned out to be correct. A number of experiments have demonstrated the existence of synergism of certain morphostatic substances and the lack of synergism in others. Particularly interesting is the study of the different effects of the two SH-substituted substances, β-mercaptoethanol and 5,7-dimercapto-thiazolo(5,4-d)pyrimidine (E 96), and their interaction with each other and with other morphostatic substances [VON HAHN and LEHMANN, 1960]. As mentioned in an earlier section, both of these substances are 'morphostatic' when tested on *Xenopus* larval tail regeneration (mercaptoethanol definitely stronger than E 96). But in combination they are antagonistic. For example, mercaptoethanol (1:64,000) alone in a certain series produced 54% inhibition (regenerating tails' average length 46% of that of controls), E 96 alone (1:32,000) produced 33% inhibition. Combined with each other (each in the same concentration) the inhibition was 15%.

Both substances were combined with colchicine, 1,2-dihydroxy-3-methyl-(7)6-ethoxy-quinoxalon (3576), pyrazolo pyrimidine (18994), diethoxy-2,6-bis ethylimino benzoquinone (17121) and 4-amino-6-methyl-heptanone-(3)-hydrochloride (E 9) respectively [formulae in HAHN and LEHMANN, 1960]. The results were quite different. For example, with colchicine mercaptoethanol is indifferent, E 96 synergistic; with 3567 mercaptoethanol is synergistic and E 96 is indifferent; with 18994 mercaptoethanol is synergistic, E 96 slightly antagonistic. (With E 9 both are synergistic, with 17121 both indifferent.)

It is regrettable that with several of these substances the exact mode of action is still not known. Quinoxalone's morphostatic effect can be alleviated by adenine, folic acid and aminopterin. This fact, taken together with the substance's molecular structure would indicate some kind of interference with purine synthesis. Structural similarity indicates the same for 18994. This is the probable reason that they are not synergistic with E 96 which is itself a purine analog. On the other hand, both 3576 and 18994 are synergistic with mercaptoethanol which would seem to indicate that the latter is involved in another step of gene activity (RNA synthesis ?). HAHN and LEHMANN have also studied the phase specificity of the two SH-substituted chemicals, mercaptoethanol and E 96, and found that the former acts in a later phase of regeneration, while E 96 acts in an earlier phase. These various observations allow a number of conjectures and educated guesses about the possible mode of interaction, but the exact mechanism of synergism (and antagonism) of these and other substances will have to wait until further data will be available.

On a more modest scale, YANKOW [1965] investigated the possible synergistic effect of other morphostatic antimetabolites, widely used in experimental and clinical cancer chemotherapy. His approach and interpretation was somewhat different from that of HAHN and LEHMANN as he simply diluted one solution with an equal amount of another (each originally in 3.33×10^{-4} M concentration), to see if the mixture had the same inhibitory effect as the components alone in double strength. This was the case with combinations of uracil mustard (UM), 6-mercaptopurine (MP) and thioguanine (TG) in adult *Triturus* tail regeneration. As mentioned in a previous section, all three antimetabolites alone reduce regeneration of *Triturus* tails considerably, by almost 50%, uracil mustard somewhat less (length of control regenerates after 24 days 2.3 mm, UM 1.5 mm, MP 1.2 mm, TG 1.3 mm). The result of combination UM/MP was 1.2 mm, UM/TG 1.5 mm, MP/TG 1.1 mm. This means that in none of the 3 combinations was the effect below the single effect of the weaker component in double strength and in one case, MP/TG, it was actually somewhat above the single effects of both components.

Unfortunately, there are no data about the effect of the 3 antimetabolites alone in half strength (i.e., on 1.66×10^{-4} M concentration). Therefore, it is not sure that the above data actually represent synergism. In the experiments of HAHN and LEHMANN [1960], substance 18994, for example, had the same inhibitory effect (89%) both in 1:32,000 and in 1:64,000 concentration, but substance 3576 caused 36% inhibition in 1:100,000 concentration, only 16% in 1:200,000 strength. But if we assume that reduction of the concentration of the antimetabolites to half strength in YANKOW's experiments would have reduced their inhibitory effect considerably if they were applied alone, then we can say that the combinations in the manner as they were carried out do indicate a certain synergism.

The extensive studies of BIEBER and HITCHINGS [1959], referred to above, brought out some pertinent points about possible synergism of regeneration inhibitors, although it was not the aim of these authors to investigate this problem. In studying separately the effect of several inhibitors upon *Rana pipiens* larval tail regeneration they noted a striking qualitative difference between the effect of 6-mercaptopurine (and another antipurine, benzimidazole), on the one hand, and aminopterin, on the other, even though the quantitative results were about the same. The former show their effect 48 hours after application (especially if the treatment begins soon after amputation), the latter's effect is not manifested for 5 to 6 days. Also, the antipurines seem to have a strong effect on dedifferentiation while aminopterin does not seem to affect this process. The authors actually express the view that the purine antimetabolites act in an early phase of the regeneration process while aminopterin acts later. This is in good agreement with the view of LEHMANN and his co-workers concerning the phase specificity of regeneration inhibiting substances.

Phase specificity of regeneration inhibitors was also stressed by PIETSCH and co-workers [cf. PIETSCH and CORBETT, 1968] when comparing mammalian muscle regeneration under the influence of actinomycin, colchicine and phleomycin respectively.

All these data indicate that the investigation of the problem of synergism would need some further background study on the chemistry of the biological activity of the various combinations and until this is available, the studies on synergism of morphostatic substances can be considered only as a hopeful beginning.

Conclusions

The studies surveyed in this chapter can be summed up very briefly in the following points:

1. One fact emerges clearly, which students of regeneration have realized long ago [cf. CARLSON, in this volume] – regeneration is a complex process with several phases and contributing elements and the term 'combinative unit-performance', coined for developmental processes in general [LEHMANN, 1945], is nowhere more appropriate than in the characterization of 'secondary development', i.e., regeneration.

2. This being so it must be also clear that antibiotics (in the widest sense), which inhibit genes (also in the widest sense) have an almost infinite variety of attack points and their 'strategy' and mode of action differs widely from case to case.

3. The unravelling of the different mechanisms by which antibiotics inhibit regeneration is a major task in which many research workers were and are actively engaged. Despite their efforts and achievements, the picture which their results present is still very incomplete. Further studies will be well worth the effort as there are many similarities between regeneration and neoplastic growth and results obtained with inhibitors of regeneration might be applicable to cancer research.

References

AFFONSO, O. R.; MITIDIERI, R. and VILLELA, G. G.: Effect of colchicine on rat liver xanthine oxidase. Nature *192:* 666–667 (1961).

AKSENOVA, N. N.; BRESLER, V. M.; VOROBYEV, V. I. and OLENOV, J. M.: Influence of ribonucleic acids from the liver on implantation and growth of transplantable tumors. Nature *196:* 443–444 (1962).

ANDERSON, W. F.; GORINI, L. and BRECKENRIDGE, L.: Role of ribosomes in streptomycin activated suppression. Proc. nat. Acad. Sci., U.S. *54:* 1076–1083 (1956).

BALIS, M. E.: Antagonists and nucleic acids (North-Holland Publ. Co., Amsterdam 1968).

BALTIMORE, D.: RNA-dependent DNA polymerase in virions of RNA tumor viruses. Nature *226:* 1209–1211 (1970).

BASSLEER, R.; COLLIGNON, P. et MATAGNE-DHOOSCHE, F.: Effets cytologiques du Myleran

sur le muscle strié de l'embryon de poulet *in ovo* et de la queue de l'axolotl en régénération. Arch. Biol., Liège *74:* 79 (1963).
BEADLE, G. W.: Biochemical genetics. Chem. Rev. *37:* 15–96 (1945).
BEADLE, G. W.: Physiological aspects of genetics. Ann. Rev. Physiol. *10:* 17–46 (1948).
BEAMS, H. W. and EVANS, T. C.: Some effects of colchicine upon the first cleavage of *Arbacia punctulata.* Biol. Bull. *79:* 188–189 (1940).
BIEBER, S.; BIEBER, R. and HITCHINGS, G. H.: Activities of 6-mercaptopurine and related compounds on embryonic and regenerating tissues of *Rana pipiens.* Ann. N.Y. Acad. Sci. *60:* 207–211 (1954).
BIEBER, S. and HITCHINGS, G. H.: Effects of growth inhibitors on amphibian tail blastema. Cancer Res. *19:* 112–115 (1959).
BIESELE, J. J.: Mitotic poisons and the cancer problem (Elsevier, Amsterdam 1958).
BRACHET, J.: Effects of β-mercaptoethanol on morphogenesis in amphibian eggs. Nature *181:* 1736–1737 (1958).
BRACHET, J. et DELANGE-CORNIL, M.: Recherches sur le rôle des groupes sulfhydriles dans la morphogenèse. Devel. Biol. *1:* 79–100 (1959).
BUTLER, E. G. and PUCKETT, W. O.: Studies on cellular interaction during limb regeneration in *Ambystoma.* J. exp. Zool. *84:* 223–239 (1940).
BUTLER, E. G. and SCHOTTÉ, O. E.: Histological alterations in denervated nonregenerating limbs in urodele larvae. J. exp. Zool. *88:* 307–341 (1941).
CARLSON, B. M.: Inhibition of limb regeneration in the adult newt, *Triturus viridescens* treated with actinomycin D. Doklady Akad. Nauk SSSR *171:* 229–232 (1966) (In Russian with summary in English).
CARLSON, B. M.: The effect of actinomycin D upon epidermal-mesodermal interactions in limb regeneration. Amer. Zool. *7:* 702 (1967a).
CARLSON, B. M.: The histology of inhibition of limb regeneration in the newt *Triturus* by actinomycin D. J. Morph. *122:* 249–264 (1967b).
CARLSON, B. M.: Inhibition of limb regeneration in the axolotl after treatment of the skin with actinomycin D. Anat. Rec. *163:* 389–402 (1969).
CASPERSSON, T.; FARBER, S.; FOLEY, G. E. and KILLANDER, D.: Cytochemical investigations on the nucleolus-ribosome system. Effects of actinomycin D and nitrogen mustard. Exp. Cell Res. *32:* 529–552 (1963).
CHERAYIL, B. X.: A quantitative and qualitative study of the effect of 6-azathymine on the tail regeneration of *Triturus viridescens;* Diss., Fordham University (in manuscript) (1963).
CHILD, C. M.: Patterns and problems of development (Chicago Univ. Press, Chicago 1941).
CORNMAN, I. and CORNMAN, M. E.: The action of podophyllin and its fractions on marine eggs. Ann. N.Y. Acad. Sci. *51:* 1443–1481 (1951).
DEUCHAR, E. M.; WEBER, R. and LEHMANN, F. E.: Differential changes of cathepsin activity in regenerating tail of *Xenopus* larvae related to protein breakdown and total nitrogen. Helv. physiol. pharmacol. Acta *15:* 212–229 (1957).
DE CARVALHO, S. and RAND, H. J.: Comparative effects of liver and tumor ribonucleic acids on the normal liver and the Novikoff hepatoma cells of the rat. Nature *189:* 815–817 (1961).
DUMONT, A. E. and SOHN, N.: Effect of 5-fluorouracil on regeneration in tadpoles. Nature *199:* 617 (1963).
EIGSTI, O. J. and DUSTIN, P., jr.: Colchicine, in Agriculture, medicine, biology and chemistry, pp. 1–470 (Iowa State Coll. Press, Ames 1955).

Farber, S.; Diamond, L. K.; Mercer, R. D.; Sylvester, R. R., jr. and Wolff, J. A.: Temporary remissions in acute leukemia in children produced by folic acid antagonist, 4-aminopteroylglutamic acid (aminopterin). New Engl. J. Med. *238:* 787–793 (1948).

Flickinger, R. A.: Biochemical aspects of regeneration; in R. Weber The biochemistry of animal development, vol. 2 (Academic Press, New York 1967).

Francoeur, R. T.: General and selective inhibition of amphibian regeneration by vinblastine and actinomycin. Oncology *22:* 218–226 (1968).

Francoeur, R. T. and Wilber, C. G.: Amphibian regeneration and the teratogenic effects of vinblastine. Oncology *22:* 302–311 (1968).

Frey-Wyssling, A.: Submikroskopische Morphologie des Protoplasmas und seiner Derivate (Bornträger, Berlin 1938).

Garb, S.: Neglected approaches to cancer. Sat. Review *51:* 54–58 (1968).

Gebhardt, D. O. E. and Faber, J.: The influence of aminopterin on limb regeneration in *Ambystoma mexicanum*. J. Embryol. exp. Morph. *16:* 143–150 (1966).

Gellhorn, A. and Hirschberg, E.: Investigations of diverse systems for cancer chemotherapy screening. Cancer Res., Suppl. *3* (1953).

Goldin, A. and Mantel, N.: The employment of combinations of drugs in chemotherapy of neoplasia: a review. Cancer Res. *17:* 635–654 (1957).

Goldschmidt, R.: Genetic factors and enzyme action. Science *43:* 98–100 (1916).

Goldschmidt, R.: Theoretical genetics (Univ. of California Press, Berkley 1955).

Gomes, L.: Effect of N-dichloroacetyl-dl-serine on regeneration in the forelimb of *Diemictylus viridescens*. Proc. Soc. exp. Biol. Med. *115:* 204–206 (1964).

Gorini, L. and Kataja, E.: Phenotypic repair by streptomycin of defective genotypes in *E. coli*. Proc. nat. Acad. Sci., U.S. *51:* 487–493 (1964).

Goss, J. R.: Regeneration (Saunders, Philadelphia 1970).

Greenhouse, G.: Permeability of sea urchin embryos to actinomycin D. Amer. Zool. *10:* 524 (1970).

Greenhouse, G.; Hynes, R. O. and Gross, P. R.: Sea urchin embryos are permeable to actinomycin. Science *171:* 686–689 (1971).

Hagedoorn, A. L.: Autocatalytic substances, the determinants of inheritable characters. Vortr. u. Aufsätze üb. Entwicklungsmech., vol. 12 (Engelmann, Leipzig 1911).

Hahn, H. P. von: Der Einfluss morphostatischer Stoffe auf Proteasen regenerierender Gewebe. Oncologia *12:* 120–129 (1959).

Hahn, H. P. von: Catalase activity in the regenerating tail tip of *Xenopus* larvae and the effect of 3-amino-1,2,4 triazole. Experientia *15:* 379–380 (1959).

Hahn, H. P. von and Lehmann, F. E.: Die Veränderung der Kathepsinaktivität im regenerierenden Schwanz der Xenopuslarve unter dem Einfluss morphostatischer Hemmstoffe. Helv. physiol. pharmacol. Acta *16:* 107–126 (1958).

Hahn, H. P. von and Lehmann, F. E.: Beeinflussung von Kathepsinaktivität und Regenerationsleistung durch morphostatische Hemmstoffkombinationen. Helv. physiol. pharmacol. Acta *18:* 198–218 (1960a).

Hahn, H. P. von and Lehmann, F. E.: Verschiedenartige synergistische Effekte zweier SH-substituierter Morphostatika (β-Mercaptoethanol und 5,7-Dimercapto-thiazolo [5,4-d] pyrimidin). Rev. suisse Zool. *67:* 353–370 (1960b).

Hahn, H. P. von; Niehus, B.; Scholl, A. and Lehmann, F. E.: Growth inhibition by chemicals which have dissimilar effects on cathepsin and acid phosphatase activity in tail regenerates of *Xenopus* larvae. Naturwissenschaften *48:* 386–387 (1961).

HAMILTON, L. D.; FULLER, W. and REICH, E.: X-ray diffraction and molecular model building studies of the interaction of actinomycin with nucleic acids. Nature *198:* 538–540 (1963).

HASELKORN, R.: Actinomycin D as a probe for nucleic acid secondary structure. Science *143:* 682–684 (1964).

HAY, E.: Limb development and regeneration; in R. L. DE HAAN and H. URSPRUNG Organogenesis, pp. 315–336 (Holt, Reinhardt and Winston, New York 1965).

HAY, E.: Regeneration (Holt, Reinhardt and Winston, New York, 1966).

HODES, M. E.; ROHN, R. J. and BOND, W. H.: Vinblastine. I. Preliminary clinical studies. Cancer Res. *20:* 1041–1049 (1960).

HONIG, C. R. and RABINOWITZ, M.: Actinomycin D: inhibition of protein synthesis unrelated to effect on template RNA synthesis. Science *149:* 1504–1506 (1965).

IDE-ROZAS, A.: Die cytologischen Verhältnisse bei der Regeneration von Kaulquappenextremitäten. Roux' Arch. Entw. Mech. *135:* 552–608 (1936).

JACKSON, M.: Polyploidy and endoreduplication in human leukocyte cultures treated with beta-mercaptoethanol. Exp. Cell Res. *31:* 194–198 (1963).

JENSEN, P. K.; LEHMANN, F. E. and WEBER, R.: Catheptic activity of regenerating tail of *Xenopus* larvae and its reaction to histostatic substances. Helv. physiol. pharmacol. Acta *14:* 188–201 (1956).

KARCZMAR, A. G.: The effect of menthyl-bis (β-chloroethyl) amine (nitrogen mustard), hydroquinone and thyroxine on regeneration. Anat. Rec. *101:* 712 (1948).

LASH, J. V.; HOLTZER, N. and SWIFT, H.: Regeneration of mature skeletal muscle. Anat. Rec. *128:* 679 (1957).

LEHMANN, F. E.: Einführung in die physiologische Embryologie (Birkhäuser, Basel 1945).

LEHMANN, F. E.: Chemische Beeinflussung der Zellteilung. Experientia *3:* 223–232 (1947).

LEHMANN, F. E.: Synergistische und antagonistische Hemmstoffkombinationen bei der Schwanzregeneration der Xenopuslarve. Helv. physiol. pharmacol. Acta *15:* 431–443 (1957a).

LEHMANN, F. E.: Die Schwanzregeneration der Xenopuslarve unter dem Einfluss phasenspezifischer Hemmstoffe. Rev. suisse Zool. *64:* 533–546 (1957b).

LEHMANN, F. E.: Chemisch gehemmtes Wachstum von Regeneraten und Tumoren und die Dynamik gewebseigener Proteasen. Verh. Naturforsch. Ges. Basel *70:* 45–80 (1959).

LEHMANN, F. E.: Action of morphostatic substances and the role of proteases in regenerating tissues and in tumor cells. Adv. Morphogen. *1:* 153–187 (1961).

LEHMANN, F. E.: Zellbiologische und biochemische Probleme der Morphogenese. 13. Coll. Ges. Physiol. Chem., Mosbach/Baden, pp. 1–20 (Springer, Berlin 1962).

LEHMANN, F. E. und DETTELBACH, H. R.: Histostatische Wirkungen von Aminoketonen auf die Schwanzregeneration der Xenopuslarve. Rev. suisse Zool. *59:* 253–259 (1952).

LEHMANN, F. E. und SCHOLL, A.: Morphostatische Effekte der Liponsäure und des Nicotinsäureamid auf die regenerierende Schwanzspitze von Xenopuslarven. Naturwissenschaften *49:* 187–188 (1962).

LEVI, I.; BLONDAL, H. and LOZINSKY, E.: Serine derivative with antitumor activity. Science *131:* 666 (1960).

LEVIS, A. G.; MARIN, G. and DANIELLI, G. A.: Differential inhibition of RNA and DNA synthesis by nitrogen mustard in cultured mammalian cells. Caryologia *17:* 427–431 (1964).

LUICKX, A. and VAN LANKER, J. L.: Metabolic effects of vinblastine on deoxyribonucleic acid and ribonucleic acid synthesis in regenerating liver. Lab. Invest. *15:* 92–100 (1966).

LUKETIC, G. C.; SANTINI, R. and BUTTERWORTH, C. E.: Depression of whole blood folate activity by colchicine. Proc. Soc. exp. Biol. Med. *120:* 13–16 (1965).

LÜSCHER, M.: Die Wirkung des Colchicins auf die an der Regeneration beteiligten Gewebe im Schwanz der *Xenopus*-Larve. Rev. suisse Zool. *53:* 683–734 (1946a).

LÜSCHER, M.: Die Hemmung der Regeneration durch Colchicin beim Schwanz der Xenopuslarve und ihre entwicklungsphysiologische Wirkungsweise. Helv. physiol. pharmacol. Acta *4:* 465–494 (1946b).

MATAGNE-DHOOSCHE, F.: Action cellulaire du Myleran chez l'axolotl au cours de la regénération. Arch. Biol., Liège *75:* 93–106 (1964).

MAZIA, D.: SH-compounds in mitosis. I. The action of mercaptoethanol on the eggs of the sand dollar *Dendraster excentricus.* Exp. Cell Res. *14:* 484–494 (1958).

MAZIA, D. and ZIMMERMANN, A. M.: SH-compounds in mitosis. II. The effect of mercaptoethanol on the structure of the mitotic apparatus in sea urchin eggs. Exp. Cell Res. *15:* 138–153 (1958).

MOSKOWITZ, M.: Differences in precipitability of nucleic acids with streptomycin and dihydrostreptomycin. Nature *200:* 335–337 (1963).

NEUSS, N.; JOHNSON, I. S.; ARMSTRONG, J. G. and JANSEN, C. J.: The Vinca alkaloids. Adv. Chemother. *1:* 123–142 (1964).

NIU, M. C.: Effects of ribonucleic acid on mouse ascites cells. Science *131:* 1321 (1960).

ÖSTERGREN, G.: Colchicine mitosis, chromosome contraction, narcosis and protein chain folding. Hereditas *30:* 429–467 (1944).

PICKEN, L.: The organisation of cells and other organisms (Clarendon Press, Oxford 1960).

PIETSCH, P.: The effect of colchicine on regeneration of mouse skeletal muscle. Anat. Rec. *139:* 167 (1961).

PIETSCH, P.: Time-dependent inhibition of myogenesis by actinomycin D. Nature *203:* 1177 (1964).

PIETSCH, P. and CORBETT, C.: Competitive effects of phleomycin and mercuric chloride *in vivo.* Nature *219:* 933–934 (1968).

PIETSCH, P. and GARRETT, H.: Primary site of reaction in the *in vitro* complex of phleomycin in DNA. Nature *219:* 488–489 (1968).

PIETSCH, P. and MACALISTER, S. B.: Replication and the activation of muscle differentiation. Nature *208:* 1170–1172 (1965).

POLEZHAEV, L. V.: Restoration of regenerative capacity suppressed by x-irradiation. Izvestia Akad. Nauk SSSR, Ser. Biol. *1966 (1)* 37–58 (1966a) (In Russian with summary in English).

POLEZHAEV, L. V.: Mechanism of recovery of the regenerative capacity suppressed by x-irradiation. Izvestia Akad. Nauk. SSSR, Ser. Biol. *1966 (2):* 254–165 (1966b) (In Russian with summary in English).

POLEZHAEV, L. V.: Loss and restoration of regenerative capacity in tissues and organs of animals (Harvard University Press, Cambridge 1972).

POLEZHAEV, L. V. und FAWORINA, W. M.: Über die Rolle des Epithels in den anfänglichen Entwicklungsstadien einer Regenerationsanlage der Extremität beim Axolotl. Roux' Arch. Entw. Mech. *133:* 701–727 (1935).

PROCACCINI, D. J. and DOYLE, C. M.: Streptomycin induced teratogenesis in developing and regenerating amphibians. Oncology *24:* 378–387 (1970).

PROCACCINI, D. J. and DOYLE, C. M.: The inhibition of limb regeneration in adult *Diemictylus viridescens* treated with streptomycin. Oncology *26:* 393–404 (1972).

RAPKINE, L. et BRACHET, J.: Recherches sur le rôle des groupes sulfhydriles dans la morphogenèse. I. Action des inhibiteurs des groupes SH sur l'œuf entier et sur les explantats dorsaux et ventraux chez les Amphibiens. Implantation de proteins sulfhydrilées. Bull. Soc. Chim. biol. *33:* 427–438 (1951).

REICH, E.: Actinomycin: correlation of structure and function of its complexes with purines and DNA. Science *143:* 684–689 (1964).

REVEL, M.; HIATT, H. I. and REVEL, J. P.: Actinomycin D: an effect on rat liver homogenates unrelated to its action on RNA synthesis. Science *146:* 1311–1313 (1964).

ROSE, S. M.: Regeneration: key to understanding normal and abnormal growth and development (Appleton-Century-Croft, New York 1971).

ROTHSTEIN, H.; FORTIN, J. and SONNEBORN, D.: Inhibition of DNA synthesis and cell division by actinomycin D. Experientia *22:* 294–295 (1966).

SCHMIDT, A. J.: Cellular biology of vertebrate regeneration and repair (Univ. of Chicago Press, Chicago 1968).

SCHMIDT, A. J. and WEARY, M.: Localization of acid phosphatase in the regenerating forelimb of the adult newt, *Triturus.* J. exp. Zool. *136:* 301–328 (1963).

SCHWARZ, S.: SCHWARZ, M.: GARAFOLO, M.; STERNBERG, S. S. and PHILIPS, F. S.: Hydroxy urea: inhibition of deoxyribonucleic acid syntesis in regenerating liver of rats. Cancer Res. *25:* 1867–1887 (1965).

SCOTT, R. B. and BELL, E.: Messenger RNA utilization during development of the chick embryo lens. Science *147:* 405–407 (1965).

SENTEIN, P.: L'action de la vincaleukoblastine sur la mitose chez *Triturus helveticus* Raz. Chromosoma *15:* 416–456 (1964).

SINCLAIR, W. K.: Hydroxyurea; differential lethal effects on cultured mammalian cells during the cell cycle. Science *150:* 1729 (1965).

SINGER, M. and SALPETER, M. M.: Regeneration in vertebrates: the role of wound epithelium; in M. X. ZARROW Growth in licing systems, pp. 277–311 (Basic Books, New York 1961).

SMITH, ST. D. and CRAWFORD, G. L.: Initiation of regeneration in adult *Rana pipiens* limbs by injection of homologous liver nuclear RNP. Oncology *23:* 299 (1969).

SPOTTS, C. R. and STANIER, R. Y.: Mechanism of streptomycin action on bacteria: a unitary hypothesis. Nature *192:* 633–637 (1961).

SUEOKA, N. and KARO-SUEOKA, T.: Transfer RNA and cell differentiation; in J. N. DAVIDSON and W. E. COHEN Progress in nucleic acid research and molecular biology, vol. 10, pp. 45–87 (1970).

TEMIN, H. M. and MIZUTANI, S.: RNA-dependent DNA polymerase in virions of Rous sarcoma virus. Nature *226:* 1211–1213 (1970).

THALER, M. M.; COX, M. C. L. and VILLEE, C. A.: Actinomycin D: uptake by sea urchin eggs and embryos. Science *164:* 832–834 (1969).

THORNTON, C. S.: The effect of colchicine on limb regeneration in larval *Ambystoma* J. exp. Zool. *92:* 281–295 (1943).

THORNTON, C. S.: Amphibian limb regeneration. Adv. Morphogenes. *7:* 205–249 (1968).

VAN DOI, N. and WOLSKY, A.: The effect of actinomycin D on the regeneration of *Ambystoma* larvae. Amer. Zool. *4:* 432–433 (1964).

VAN LANKER, J. L.: FLANGOS, A. L. and ALLEN, J.: Metabolic effects of vinblastine. I. The effects of vinblastine on nucleic acid synthesis in spleen and bone marrow. Lab. Invest. *15:* 1291 (1966).

VILLEE, C. A. and GROSS, P. R.: Uptake of actinomycin by sea urchin eggs and embryos. Science *166:* 402–403 (1969).

WADDINGTON, C. H.: New patterns in genetics and development (Columbia Univ. Press, New York 1962).

WOLSKY, A.: Untersuchungen über die Wirkung des Colchicins bei Amphibien. I. Wirkung auf den Sauerstoffverbrauch der Keime. Arb. Ung. Biol. Forschungsinst. (Tihany) *12:* 352–356 (1940).

WOLSKY, A.: 'Matter' and 'form' in experimental morphology. Szent Istvan Akad. Ertes. *26:* 248–261 (1941) (In Hungarian with summary in English).

WOLSKY, A. and VAN DOI, N.: The effect of actinomycin on regeneration processes in amphibians. Trans. N.Y. Acad. Sci. Ser. II *27:* 882–893 (1965).

WOLSKY, A. and WOLSKY, M. DE I.: Induction and competence in morphogenesis: amphibian lens development from non-lens ectoderm, influenced by mercaptoethanol. Oncology *22:* 290–301 (1968).

WOLSKY, A. and WOLSKY, M. DE I.: Nucleic acid control of regeneration in amphibian larvae. Ann. Embryol. Morphogénét., Suppl. *1:* 248–249 (1969).

WOLSKY, M. DE I.; MONTEIL, J. and WOLSKY, A.: Tail regeneration in the newt *Triturus (Diemictylus) viridescens* under the influence of RNA preparations from adult tissues of the same species. Amer. Zool. *6:* 614 (1966).

WOLSKY, M. DE I. and WOLSKY, A.: The effect of a carcinostatic antimetabolite on the development of sea urchin eggs. Anat. Rec. *138:* 391 (1960).

WOLSKY, M. DE I. and WOLSKY, A.: Inhibition of tail regeneration in larvae of the leopard frog, *Rana pipiens,* by RNA preparations from the adult liver of the same species. Amer. Zool. *7:* 174 (1967).

YAMADA, T.: Control of tissue specificity: the pattern of cellular synthetic activities in tissue transformation. Amer. Zool. *6:* 21–31 (1966).

YAMADA, T. and ROESEL, M. E.: Effects of actinomycin D on lens regenerating systems. J. Embryol. exp. Morph. *12:* 713–725 (1964).

YANKOW, M.: Effect of growth inhibitors and their combinations on the tail regeneration of amphibia. Trans. N.Y. Acad. Sci. Ser. II *27:* 878–881 (1965).

Author's address: Prof. Dr. ALEXANDER WOLSKY, Department of Radiology, New York University Medical Center, 550 First Avenue, *New York, NY 10016* (USA)

Neoplasia and Cell Differentiation, pp. 189–233
(Karger, Basel 1974)

Cell Surfaces in Neoplasia

GERALD C. EASTY

Chester Beatty Research Institute, Institute of Cancer Research, Royal Cancer Hospital, London

Contents

I. Introduction

The plasma membrane-surface complex is intimately involved in the initial stages, at least, of any interaction between cells and their environment. In addition to transport and associated surface metabolic functions possessed by all cells, the cell surface plays a very important role in cell—cell and cell—substrate interactions involved in the establishment of normal morphogenetic patterns during embryonic development, and in the regulation and maintenance of normal cellular functions in adult tissues. During the last twenty years evidence has accumulated that the abnormal growth and behavior of neoplastic cells is associated with alterations in the surfaces of these cells relative to their normal counterparts. Inevitably, early attempts to define the differences between the surface of normal and tumor cells exclusively in terms of a single parameter have failed. It has become apparent that there may be a multiplicity of changes in the surfaces of cells which have undergone neoplastic transformation. Certain *aspects* of the behavior of malignant cells associated with the surface have emerged, however, which if not common to all malignant cells have been detected in many of the systems investigated. Thus, in malignant cells in general, there appears to be a degree of loss of contact control of movement and proliferation and deficiencies in contact mediated communication. Also, reduced cellular adhesion of malignant cells has long been implicated in tumor invasion and metastasis. The recent discovery of tumor specific antigens on the cell surface has provided a basis, yet to be realized, for a truly specific therapy of cancer. We may then ultimately come full circle and utilize the aberrant surface properties responsible for many of the unfortunate consequences of malignant transformation for its successful treatment.

II. Composition of the Plasma Membrane-Surface Complex

Our limited knowledge of the chemical composition and molecular architecture of the structures which separate the interior of cells from their environments makes a precise definition of the plasma membrane complex very difficult. There are many properties and functions which have been ascribed to the peripheral structures of cells. Cell movements accompanied by alterations in the shapes of cells may involve the participation of structures located within the cytoplasm just beneath the surface. Further out there is a region of low dielectric constant and high electrical resistance which is almost certainly responsible for many of the permeability properties of the cell, and exterior to this lies a region of high dielectric constant and low resistance which is probably intimately involved in cell contact, adhesion and recognition processes. Any of the many functions connected with the surface may be associated primarily with a specific constituent of the plasma membrane complex, but reactions involving one constituent may lead to alterations in others.

A. Problems Associated with Isolation

The problems involved in the chemical analysis of the plasma membrane complex are formidable. Two main approaches have been used. The first involves physical disruption of the cells followed by attempts to separate the plasma membrane from all other cell constituents including membranes of intracellular organelles. The second involves the isolation of components of the membrane complex without destroying the viability of the cells. Both approaches, but particularly the first, yield material which can be grossly contaminated with constituents from the cell interior. CURTIS [1967], WEISS [1967a] and STECK and WALLACH [1970] have critically reviewed many of the problems associated with these procedures. Under optimum conditions involving, for example, the enzymic release of surface components from a clone of cells accompanied by minimum loss of cell viability, the unavoidable presence of a small percentage of dead cells and the release of cell constituents by the viable cells can result in the contamination of the surface component with an excess of non-surface material. Even before the cells are disrupted there is the possibility that their surfaces may have adsorbed substances from their environment which strictly should not be considered as components of the cell surface. The Lewis blood group substances in man appear to be reversibly adsorbed from the plasma [SNEATH and SNEATH, 1959], and the Forssman antigens may be acquired by certain cells if they are cultured in serum containing this antigen [COOMBS *et al.*, 1961]. In addition, antigenic substances may cease to be detectable on the surfaces of cells as a result of culturing the cells under certain conditions. CHESSIN *et al.* [1965] found that cells from primary human amnion cultures were strongly H antigen positive but that after 30 days in culture the blood group activity was greatly diminished. As cells grown in tissue culture for long periods are increasingly used for the isolation and analysis of plasma membranes and the analysis of surface-associated functions, the dangers of loss or acquisition of antigens, or of other substances not so readily detected, cannot be overlooked.

Most attempts to prepare intact plasma membrane complexes have been carried out under conditions which were far from ideal, often involving the use of tissues containing many different types of cells and probably accompanied by loss of soluble membrane constituents, adsorption of contaminants, rearrangements of membrane structure, and all the complex and largely unidentified changes produced by the action of enzymes released from the interior of the cells. For these reasons most chemical analyses of plasma membrane complexes have been restricted to two types of cell, the mammalian erythrocyte which contains no intracellular structures, and the Schwann cell which, when it has wrapped itself around an axon forms a structure (myelin) which is composed almost entirely of the plasma membrane complex. With other cell types the usual procedure for isolating the membranes has involved homogenization or disruption of the cells, followed by separation of the various membrane constituents by centrifugation techniques of varying complexity. Recently, a most promising technique, originally introduced by WARREN *et al.* [1966], has been used with considerable success by HICKS and KETTERER [1970] and BARLAND and SCHROEDER [1970]. WARREN's technique involves strengthening the plasma membrane by addition of fluorescein mercuric acetate and zinc ions which are applied in hypotonic solution. The hypotonicity of the medium causes the cells to swell, lifting the plasma membrane from the underlying cell cytoplasm which remains largely unswollen. BARLAND and SCHROEDER [1970] have adapted this method to give a rapid and simple procedure for the isolation of plasma membranes from cells grown on glass. After swelling, the culture is agitated and membranous vesicles, reported to be remarkably free of other cell organelles, are released into the medium from which they are concentrated by centrifugation. As the authors point out, this procedure of isolating membranes from monolayer cultured cells reduces the possible loss of surface constituents which can occur when cultured cells are first released from the culture vessel surface by treatment with proteolytic enzymes.

B. Chemical Composition

The gross chemical composition of preparations of isolated membranes varies with the isolation procedures employed. Mammalian erythrocyte membranes contain 20-40% lipid, 80–60% protein and about 5% carbohydrate [MADDY, 1966].

1. Protein

Removal of lipid from erythrocyte ghosts by extraction with alcohol-ether leaves a residue containing denatured protein given the name stromatin by JORPES [1932], but later shown to be a suspension of myelin forms by FURCHGOTT [1940]. Other lipid extracting procedures have yielded rather ill-defined components containing protein, carbohydrate and complexed lipid in varying proportions [MOSCOVITCH and CALVIN, 1952]. Erythrocyte ghost protein has been isolated in a soluble form by MADDY [1964], who found the material to be heterogeneous, of high molecular weight (150,000–700,000), and containing sialic acid. Further evidence of the heterogeneity of human red cell membrane proteins has been provided by ROSENBERG and GUIDOTTI [1968, 1969] and POULIK and LAUF [1969]. LENARD [1970] solubilized these red cell membrane proteins, which were free of hemoglobin, and separated them into 14 different molecular weight classes which could not be interconverted, suggesting that they constituted different polypeptide chains rather than aggregates of identical smaller units. Four major components had molecular weights varying from 86,000 to 255,000 and together accounted for 60–65% of the total membrane protein. Most of the associated carbohydrate was found in one protein fraction. Studies employing the techniques of infrared spectroscopy, optical rotatory dispersion and circular dichroism, on preparations of myelin [MADDY and MALCOLM, 1965], erythrocyte membranes [LENARD and SINGER, 1966] and plasma membranes of Ehrlich ascites carcinoma cells [WALLACH and ZAHLER, 1966; WALLACH and GORDON, 1968] have also provided evidence for the occurrence of protein and information concerning its conformation in plasma membrane complexes. Considerable uncertainty still surrounds the details of the conformation of membrane proteins, but in a critical discussion of the data available CHAPMAN and WALLACH [1968] concluded that there is evidence for a substantial proportion of the protein being in the α-helix form, some of which is located in the interior of the membrane, stabilized by hydrophobic interactions, while the ionic side chains of the proteins are on the exterior surface of the membrane.

Other protein constituents of the plasma membrane complex include enzymes, such as ATP'ases, which have been used as markers for identifying isolated cell surface membranes [EMMELOT *et al.*, 1964]. GRÖSCHEL-STEWART *et al.* [1970] have reported the demonstration of the presence of actomyosinlike protein on the surfaces of cultured chick embryo muscle and liver cells using immunofluorescent techniques. JERNE [1967] and MITCHISON [1967] have suggested that lymphocytes and other immunologically competent cells may possess preformed antibodies as an integral part of their cell surface, and that these γ-globulin receptors may be involved in the regulation of proliferation and γ-globulin synthesis by the cells following contact with the appropriate antigen. Direct evidence for the existence of γ-globulin molecules on the surfaces of normal lymphoid cells of mice has recently been obtained by PARASKEVAS *et al.* [1970] using a reverse immune adherence technique. Immunologically active glycoproteins or glycopeptides have been isolated from intact red cells [KLENK and UHLENBRUCK, 1960] or their stroma [SPRINGER *et al.*, 1966], but most of the immunological specificity resides in the carbohydrate portions of the glycoproteins [MORGAN and WATKINS, 1959].

2. Lipid

The most detailed and accurate analyses of the lipids of plasma membranes have been carried out on erythrocytes. As most, if not all, of the lipids present in mammalian erythrocytes are associated with the plasma membrane complex, either the isolated ghosts or whole red cells provide suitable material for the study of these plasma membrane lipids. The lipids of these membranes and those of other cells consist of about 25–30% cholesterol (the only neutral lipid) [NELSON, 1967], 55–65% phospholipids, while the remainder consists of either glycolipids or unspecified lipids. Comparisons of the lipid composition of erythrocytes and myelin with that of mitochondria and other intracellular organelles indicate that glycolipids occur exclusively in plasma membranes [FLEISCHER *et al.*, 1967], although it does not follow that the plasma membranes of all cell types contain glycolipid. ROUSER *et al.* [1968] have reviewed in considerable detail the lipid composition of membranes of animal cells and the reader is recommended to consult their article for detailed information. Cholesterol is reasonably constant in quantity from species to species, but there are large species differences in the proportions of different glycolipids, phospholipids and fatty acid constituents. The lipid composition of the plasma membranes of erythrocytes can be influenced by alterations in the animal's diet, specifically in the fatty acids of the phospholipids [DE GIER and VAN DEENEN, 1964]. Exchange of phospholipids between mammalian erythrocytes and serum lipoproteins has been observed by SAKAGAMI *et al.* [1965], and between cultured chick embryo fibroblasts and serum proteins present in the culture medium [PETERSON and RUBIN, 1969]. PETERSON and RUBIN suggested that some of the phospholipids could be released from the cell in combination with cellular lipoproteins (mol.wt. 5×10^6) which were themselves constituents of the plasma membrane. Release of phospholipid-lipoprotein complexes could result from the sloughing of small spherical blebs of membranes frequently observed in culture medium which has been in contact with cells, particularly those subjected to shearing forces and other mild insults. Exchange of erythrocyte membrane cholesterol with plasma cholesterol occurs readily [ASHWORTH and GREEN, 1964], while cholesterol of myelin appears to be metabolically, and probably structurally, very stable. These differences may be related to the differences in environment and function of these two types of membrane, the tightly wrapped myelin functioning as an electrical insulator while the whole erythrocyte surface is in contact with blood plasma.

3. Carbohydrate

A carbohydrate-rich layer has been detected histochemically at the surfaces of many types of mammalian cells by light microscopy [GASIC and GASIC, 1962a] and electron microscopy [GASIC and BERWICK, 1963; RAMBOURG and LEBLOND, 1967]. These carbohydrates are present as glycoprotein and glycolipid in the plasma membrane complexes. In addition to evidence derived from chemical analysis of isolated membranes for the occurrence of plasma membrane carbohydrates [MADDY, 1964; SPRINGER *et al.*, 1966; WOODIN and WIENECKE, 1966], the carbohydrates are responsible for most of the antigenicity associated with the cell surface, as demonstrated by the removal of many surface antigens by treatment of intact cells with proteolytic enzymes [MORTON, 1962] accompanied by the release of sialoglycopeptides possessing blood group activity [COOK *et al.*, 1960]. All cells so far investigated have a net negative surface charge at physiological pH, and studies of the electrophoretic mobilities of intact cells following treatment with specific enzymes, particularly neuraminidase, have indicated that this surface charge is for many cells mainly due to the presence of sialic acid at the surface [COOK *et al.*, 1961; WALLACH and EYLAR, 1961; FORRESTER *et al.*, 1962].

4. Nucleic Acid

There have been a number of reports of the presence of RNA associated with preparations of isolated plasma membrane complexes of certain types of cells [TAKEUCHI and TERAYAMA, 1965; WALLACH and ZAHLER, 1966] in quantities varying from 0.5% to 4% dry weight. Undoubtedly, some of these preparations have been contaminated with soluble or ribosomal RNA originally located within the cytoplasm, but the observations of WEISS and MAYHEW [1966, 1967, 1969] and WEISS and SINKS [1970] of the decrease of net electronegative charge at the surfaces of intact cells following treatment with RN'ase provides the most convincing evidence for the existence of RNA at the surface of cells. Intact RNA in solution is rapidly degraded by nucleases released from many types of cells in culture and by nucleases present in blood plasma. Presumably, the RNA on the surfaces of certain types of cells is appreciably resistant to these types of RN'ases. The function of RNA at the cell surface is at present unknown but WEISS [1968] has suggested that it may play a part in cell recognition processes and in the transfer of information between cells.

III. Structure

A. Bimolecular Lipid Layer Model

For more than 30 years the plasma membrane was considered to be a continuous bimolecular lipid layer in which the polar groups of membrane phospholipids faced outwards [DANIELLI and DAVSON, 1934–35]. To account for the low interfacial tension observed at the outer surfaces of cells, DANIELLI and DAVSON proposed that a layer of protein molecules was absorbed on the outer surface in contact with the polar groups. As this simple continuous bimolecular lipid leaflet model appeared to be too impermeable to water and small ions by comparison with the known properties of cell membranes it was modified by STEIN and DANIELLI [1956] to contain polar pores lined with protein extending through the thickness of the lipid layer. On the basis of the similarity in appearance of electron micrographs of plasma and other membranes from many types of cells ROBERTSON [1957] proposed a model for both the plasma membrane and the membranes of intracellular organelles which was essentially that of DANIELLI and DAVSON. The membranes, particularly after permanganate fixation, appeared as a three-layered unit consisting of two electron-dense lines 20 Å wide separated by a less dense region 35 Å wide.

B. Globular Micelle Model

In recent years, however, many observations have been made which indicate that many functions of biological membranes cannot satisfactorily be accounted for on the basis of the bimolecular lipid leaflet alone. Electron microscopic observations of macromolecular lipid complexes led LUCY and GLAUERT [1964] to propose a model for biological membranes in which a varying proportion of the membrane phospholipids are arranged to form globular micelles (40 Å diameter) with the polar groups facing the exterior. It was suggested that these micelles were arranged in penta- or hexagonal array with 4-Å pores between the micelles, the structure being stabilized by a layer of protein or glycoprotein on either side of the plane of micelles. Units of globular protein possessing enzymic or hormonal activity could be interspersed between the lipid micelles. LUCY [1968] has given a critical account of models proposed for cell membranes and emphasizes that the globular micelles of lipid might be in dynamic equilibrium with the bimolecular lipid leaflet which probably constitutes the permeability barrier of the plasma membrane, with the globular micelles occurring as local modifications related to specific functions. Evidence supporting the occurrence of globular micelles in regions of the plasma membranes has come mainly from electron microscopy and x-ray diffraction studies. SJÖSTRAND and ELFVIN [1964] observed globular components in membranes having diameters of about 50 Å, separated by stained septa 10–20 Å wide, and proposed a model for membrane structure very similar to that of LUCY. BLAISIE *et al.* [1965], on the basis of data obtained from electron microscopic and x-ray diffraction studies, reported the presence of arrays of 40 Å diameter particles in the surface membranes from the retina of the frog. Globular units in hexagonal arrays have been observed in tight junctions of epithelial cells [REVEL and KARNOVSKY, 1967]. The luminal membrane of the superficial cells lining the rat urinary bladder has been shown to contain a substructure of hexagonally arranged subunits visible in negatively stained preparations [HICKS and KETTERER, 1969]. Subsequently, WARREN and HICKS [1970] achieved further resolution of these arrays by the linear integration technique of MARKHAM *et al.* [1964] and showed that they consisted of 12 subunits arranged in a stellate configuration. The fluidity of biological membranes has been recently demonstrated by FRYE and EDIDIN [1970] who followed the mixing of mouse and human surface antigens after fusion of mouse and human cells by Sendai virus. The antigens of the two species were distinguished by antibodies labelled with either red or green fluorochromes, and from the speed of formation of mosaic surface staining and the inability of metabolic inhibitors to delay the mingling of antigens it seemed highly probable that the membrane of the heterokaryon possessed fluid properties at physiological temperatures.

The use of electron microscopy for the elucidation of the molecular structure of biological membranes is complicated by the number and complexity of the processes to which the membranes are subjected before an image is obtained. The uncertainty concerning the precise sites of reaction of the fixatives and stains, the possible modifications of structure and phase changes which may be produced by their reactions with membrane constituents, combined with the necessity to dehydrate and embed the specimen before sectioning, make it difficult to conclude with certainty that the fine structural details of membranes revealed in the electron microscope always correspond with those which existed in the membranes of the functioning cell. Negative staining techniques using phosphotungstate [BRENNER and HORN, 1959] have been used for examining isolated membranes [CUNNINGHAM and CRANE, 1964; WHITTAKER, 1966] and do not necessarily involve the use of fixatives, embedding or sectioning. The validity of the image produced by negative staining of lipid containing materials has been questioned, however, by FINEAN and RUMSBY [1963] because it involves dehydration of lipoproteins which can result in structural reorganization as shown by x-ray diffraction studies. BANGHAM and HORNE [1964] have also criticized the application of the negative staining technique to lipid-containing material on the basis of the development of considerable electrostatic stress at the lipid-water interface during drying in the presence of the ions of negative stains, which could result in the formation of micelles from bimolecular leaflets [HAYDON and TAYLOR, 1963].

The introduction of the freeze-etching technique [MOOR *et al.*, 1961] promises to provide many of the answers to the problems associated with the artifacts, known or suspected, connected with the elucidation of membrane structure by electron microscopic examination of sections or negatively stained preparations. In the freeze-etching technique the living specimens are snap frozen under conditions not incompatible with the preservation of cell viability, fractured in a vacuum, a thin layer of ice sublimed from the fractured surface at $-100\,^{\circ}C$ and the fractured surface shadowed with heavy metal and replicated with carbon. The shadowed carbon replica is then examined in the electron microscope. The technique has been used mainly for examining the structures of plant cells [MOOR and MÜHLETHALER, 1963]. There has been considerable controversy concerning the actual location of the plane of fracture revealed when membranes are examined. BRANTON [1966], from an examination of freeze-etched root tip cells, concluded that the true outer surface of the membrane is rarely seen in these preparations. He considered that freeze-etching split the membranes revealing structure inside the membrane, and concluded that the globular particles frequently observed on many freeze-etched membrane faces represented micellar configurations, while the smooth region between the globules represented the inner surface of a continuous lipid leaflet. MOOR [1966], on the contrary, was of the opinion that the surface of the membrane is mainly exposed by this technique, and concluded that the observations supported the unit membrane concept of the continuous bimolecular lipid leaflet. The question seems to have been resolved in favour of BRANTON, at least as far as the freeze-etching of erythrocytes is concerned. TILLACK and MARCHESI [1970] labelled the outer surface of human erythrocyte ghosts with fibrous actin which can be recognized in freeze-etched preparations. Such preparations showed that actin was absent from the fracture plane of the membrane containing the scattered 85 Å particles which characterize the fracture plane. Similarly, DA SILVA and BRANTON [1970] linked ferritin molecules covalently to both the inner and outer surfaces of erythrocyte ghosts before freeze-etching, and were unable to detect ferritin in the fracture plane containing 85 Å particles. Their work provides convincing evidence for the existence of a natural cleavage plane within the membrane, probably formed by the juncture of methyl end groups of the lipid bilayer, and containing scattered globular units whose chemical nature remains to be determined.

Globular and threadlike structures have been detected on the surfaces of cells in regions of close contact, and lipid appears to be a constituent of at least one of these structures. In an extremely elegant electron microscope analysis of the intercellular junctions of mouse liver GOODENOUGH and REVEL [1970] demonstrated that the zonula occludens formed by the fusion of apposing cells consists of a meshwork of branching and anastomosing threadlike contacts which seal the lumen of the bile canaliculus from the liver intercellular space. The gap junction which is characterized in section by a 20-Å gap between the apposed membranes consists of a polygonal lattice of subunits with a 90–100-Å center-to-center spacing, with minute pits on the other 'face' of the membrane which appear to lie between the units of the polygonal lattice. Extractions of membrane preparations with aqueous acetone revealed that 60% acetone removes the polygonal lattice of the gap junction, and analysis of the extract revealed a phospholipid composition characteristic of the whole membrane. Acetone extraction had no effect on the structure of the zonula occludens which, however, lost its capacity of excluding substances such as lanthanum.

In conclusion, the proportion of globular units present in the plasma membrane complex appears to vary with the cell type, and their distribution in the membrane of a single type of cell may be further related to regions of the membrane possessing specialized permeability and other functions. Thus, the insulating properties of the myelin sheath may depend on the membrane possessing an almost exclusively bimolecular lipid leaflet structure, while the greater plasma membrane permeabilities necessary for the functioning of other types of cell may be related to the presence of globular micelle units or protein molecules.

IV. Chemical Composition of the Plasma Membrane Complexes of Normal and Malignant Cells

Attempts to find characteristic differences have been severely limited by the difficulties, previously referred to, of obtaining adequate quantities of membrane preparations from comparable normal and malignant cells in a sufficiently pure form. Most analyses have been made on whole tissues or cells, and even if it is assumed, for example, that the analysis of lipids reflects the lipid composition of the membranes, it must be remembered that any differences in lipid composition may be related to alterations in the proportions of intracellular membranes rather than the plasma membranes. For example, the endoplasmic reticulum is scanty in many tumor cells [MERCER, 1963] and decreases in cells of tissues undergoing neoplastic transformation following exposure to carcinogens [PORTER and BRUNI, 1959].

A. Lipids

Most analyses of lipids of normal and malignant tissues have not revealed any consistent or striking differences in the proportions of the various lipid components [NOVIKOFF, 1957; VEERKAMP *et al.*, 1961], but any differences in lipid composition relating specifically to the plasma membrane could easily have been totally obscured by similarities in the composition of the bulk of intracellular lipids. Perhaps the most significant work on lipids associated with the membranes of normal and malignant cells has originated from the work of RAPPORT and his colleagues. RAPPORT *et al.* [1959] isolated a substance, which they named Cytolipin H, in much larger quantities from tumors than from normal tissues. This material, which was capable of combining with antitumor antibodies, was later identified as a simple glycosphingolipid—lactosyl ceramide. If FLEISCHER *et al.* [1967] are correct in their assertion that glycolipids are found exclusively in the plasma membranes of cells, then these differences in glycolipid content may reflect significant differences in the lipid composition of the plasma membrane complexes of normal and malignant cells. The association of Cytolipin H with surface antigenic activity lends strong support to this suggestion. On the basis of this and later work RAPPORT [1969] suggested that malignant transformation of normal cells involves alterations in the glycosphingolipid content of their plasma membranes such that the number of complex molecules is reduced while the number of simpler molecules, such as Cytolipin H, is increased. Support for this suggestion has come from HAKOMORI and MURAKAMI [1968] who analyzed the glycosphingolipids extracted from three types of cultured cells: normal hamster fibroblasts (BHK 21-C13) which do not readily give rise to tumors following inoculation into hamsters, and two other derived lines, one of which had undergone spontaneous transformation *in vitro* and gave tumors three weeks after inoculation, and one (Py) which had been transformed with polyoma virus and gave tumors in hamsters within one week, using standardized inoculation procedures. They found that the 'normal' cells contained 4–5 times more monosialo Cytolipin H/mg protein than the Py cells and 10 times less Cytolipin H. The glycolipid composition of the 'spontaneously' transformed cells was intermediate between the 'normal' and more malignant Py cells. There was also a fairly good correlation between the lack of parallel orientation of the cells on monolayer culture and the content of Cytolipin H.

Subsequent work by HAKOMORI *et al.* [1968] on the glycolipid content of cultured hamster, mouse and human fibroblasts and their virus transformed derivatives gave a rather less clear picture. Although the quantities per mg protein of the hematosides, N-acetylhematoside and N-glycolylhematoside were significantly decreased in the virus transformed cells, the quantity of Cytolipin H (lactosyl ceramide) and glucosyl ceramide increased only slightly for virus transformed hamster cells and actually decreased in virus transformed human fibroblasts. The story becomes further confused by the observations of MORA *et al.* [1969] of the ganglioside content of lines of 'normal', virus- and spontaneously transformed mouse cells. In contrast with the earlier observations of HAKOMORI and MURAKAMI [1968] they did not find a marked or consistent decrease in the monosialo Cytolipin H content in virus transformed cells, and, indeed, for some cell lines the content of this substance was lower in the normal cells than in the malignant cells. They did find, however, significant reductions in the mono- and disialoceramide tetrahexosides present in the virally transformed cells, but not in spontaneously transformed cells. At present, there seems to be reasonably consistent evidence for a decrease in the content of some complex glycolipids of virus transformed cells compared with their normal counterparts. The evidence for such a decrease in *in vitro* spontaneously transformed cells is less strong, and for naturally occurring human tumors what evidence there is at present indicates the converse, since SEIFERT and UHLENBRUCK [1965] found an increase in mono- and disialo-Cytolipin H in brain tumors compared with normal brain tissue.

B. Carbohydrates

A number of independent workers who have compared a variety of different carbohydrate-containing substances present in the surface membrane complexes of normal and transformed cells have found that many of the substances which on the basis of preliminary observations appeared to be specific for malignant cells can, in fact, be detected on the surfaces of normal (general neonatal or embryonic) cells when the normal cells have been subjected to mild proteolytic treatment. For example, BURGER and GOLDBERG [1967] identified an antigenic determinant containing N-acetylglucosamine specific for the surfaces of virus transformed cells, but BURGER [1969] later demonstrated the presence of this determinant on the surfaces of normal cells after treatment with trypsin. Similarly, HAKOMORI *et al.* [1968] found that treatment of their normal cells with trypsin exposed the hematosides characteristic of the virus transformed cell surfaces, while HÄYRY and DEFENDI [1970] detected antigens, previously considered to be specific for virus transformed malignant cells, on the surfaces of both normal cell lines and primary cultures of embryonic mouse and hamster fibroblasts following treatment of the normal cells with trypsin or chymotrypsin. Thus, the malignant transformation of normal cells, particularly when viruses are involved, may result in the repression of synthesis of trypsin-labile surface materials in the virus transformed cell, but, obviously, other processes may be responsible.

Gross differences in the carbohydrates of membrane components of 3T3 mouse cells and SV40 transformed cells have been reported [WU *et al.*, 1969; ONODERA and SHEININ, 1970]. ONODERA and SHEININ compared the composition of glucosamine-containing components released from the surfaces of 3T3 and virus transformed cells by brief treatment with trypsin which did little damage to the cells. Analysis of the incompletely digested surface materials revealed the presence of at least 5 fractions from the 3T3 cells while only 3 fractions were obtained from virus transformed cells. WU *et al.* [1969] prepared membrane fractions of 3T3 and SV40 transformed cells and compared the compositions of membrane glycoproteins and glycolipids. The virus transformed cell membranes contained a much lower sialic acid and galactosamine content and a higher content of glucosamine than their normal counterparts. Similar, but not as extensive differences were detected between the normal 3T3 and spontaneously transformed cells derived from the 3T3 line.

WALLACH [1969] has presented a most stimulating hypothesis derived from the existence of mutants of neurospora possessing different membrane functions and properties, apparently derived from single amino acid replacements in membrane structural proteins [WOODWARD and MUNKRES, 1966]. By analogy he suggests that neoplastic transformation involves the introduction of inappropriate protein components into plasma membranes, either replacing or additional to normal components, and these new components would be responsible, ultimately, for many or all of the changes in social behavior, metabolism and growth control which have been associated with malignancy. There is at present very little information available on the protein content or amino acid composition of plasma membranes of comparable normal and malignant cells.

C. Sialic Acid and Other Ionogenic Groups

The sialic acid content of the plasma membrane complex of normal and malignant cells has excited considerable interest for many years, but no simple conclusion can be reached as yet from consideration of the available data. BARKER *et al.* [1959] and KAWASAKI [1958] found that the sialic acid content of some human tumors was higher than that of comparable normal tissues, and the histochemical techniques employed by DEFENDI and GASIC [1963] indicated that much of the sialic acid may be associated with the cell surface. But others have found decreased amounts of total sialic acid [OHTA *et al.*, 1968] or membrane associated sialic acid [WU *et al.*, 1969] in cell lines transformed with oncogenic viruses compared with their 'normal' precursors. Most of the other evidence concerning chemical and configurational differences between the plasma membrane complexes of normal and malignant cells has been derived from studies of the surface charges of the cells, particularly following treatment with reagents which may specifically remove or modify certain ionogenic groups. Another approach which has often been combined with studies of the surface charges of cells involves the isolation, identification and measurement of materials released from surfaces following enzyme treatment of intact cells. LANGLEY and AMBROSE [1964, 1967] reported the isolation of a sialoglycopeptide from Ehrlich ascites cells after trypsin treatment, while WALBORG *et al.* [1969] isolated what was probably a family of sialoglycopeptides from Novikoff ascites hepatoma cells by treatment with papain. WALBORG *et al.* [1969] imply that their conditions of treatment were such that contamination of the fractions with materials released from damaged cells was not significant. As similar extraction procedures were not carried out on comparable normal cells in either of these studies, their relevance to the process of malignant transformation is not known.

Critical reviews of the results of investigations of the surface ionogenic groups of normal and malignant cells and of the difficulties involved in their interpretation have been written by CURTIS [1967], WEISS [1967a] and COOK [1968]. It would seem highly imprudent in the light of present knowledge to conclude that the surfaces of all or even a majority of malignant cells can be distinguished from their normal counterparts by a higher negative charge density [AMBROSE *et al.*, 1956] or by a greater contribution of ionized sialic acid residues to the ζ-potentials [FUHRMANN, 1965]. Other workers have not found an increased electrophoretic mobility [VASSAR, 1963] or increased neuraminidase-susceptible sialic acid [WEISS and SINKS, 1970] to be a general characteristic of the surfaces of malignant cells. Although FORRESTER [1965] found that one clone of polyoma virus transformed hamster cells had a higher electrophoretic mobility than the untransformed cells, another clone of transformed cells had a mean mobility indistinguishable from that of the normal cells. COOK and JACOBSON [1968] have even observed that the electrophoretic mobilities of normal lymphoid cells from AK mice were significantly greater than those of comparable lymphoblastic leukemia cells, contrary to the type of result obtained by RUHENSTROTH-BAUER [1965] who obtained a higher range of mobilities of malignant human lymphocytes compared with their normal counterparts. Similarly, WEISS and SINKS [1970] found no consistent differences between normal and malignant human lymphoid cells. Considering the difficulties involved in the identification of the different classes of lymphoid cells and the influence of the immune status of the host on their electrophoretic mobilities, such a spread of results is hardly surprising. Results similar to those of WEISS and SINKS were obtained by PATINKIN *et al.* [1970a] from a comparison of the electrophoretic mobilities of normal lymphoid cells and macrophages with those of leukemic cells of a number of mouse strains. No significant differences with respect to pH-mobility relationships, affinity for calcium or susceptibility to neuraminidase were observed. They also confirmed the observation of MAYHEW and WEISS [1968] that the highly invasive macrophage had the lowest mobility of all the cells tested and a relatively small amount of neuraminidase-susceptible sialic acid present on its surface. PATINKIN *et al.* [1970b] also compared primary chick embryo cultures with equivalent Rous sarcoma virus transformed cells and could detect no significant differences in the surface charge, calcium binding, or reduction in mobility following neuraminidase treatment. As these experiments did not involve the use of virus transformed cells derived from clones of 'normal' cells used by FORRESTER *et al.* [1964], it is not surprising that PATINKIN *et al.* [1970b] found considerable variation in mobility between different experiments, and that the spread of mobilities for any experiment was rather large. In spite of these qualifications it seems clear that viral transformation of embryonic chick cells had not produced any significant alterations in the surface ionogenic groups. These authors also noted, as have others, that the detachment of cultured cells from glass by mild trypsinization reduced the electrophoretic mobilities of both cell types by about 10% relative to EDTA-detached cells. WEISS and SINKS [1970] in their study of normal and malignant lymphoid cells did find, however, that the net negative surface charge of cultured normal human lymphoid cells was significantly reduced by RN'ase treatment while cultured Burkitt lymphoma cells and circulating leukemic cells from a number of patients were not significantly reduced. The differences in RN'ase susceptibility between these normal and malignant cells could not be readily correlated with cell proliferation, and the authors tentatively suggested that the absence of RN'ase-susceptible material on the surfaces of malignant human lymphoid cells may have some diagnostic and therapeutic application, but cannot be considered as characteristic of malignant cells in general [WEISS and MAYHEW, 1966].

Many of the techniques employed for preparing cell suspensions for measurement of electrophoretic mobility involve separation of cells from each other or from a substrate by the use of enzymes, chelating agents or mechanical stress, all of which may lead to modifications of the ionogenic groups detectable at the cell surface. In addition, many of the solutions in which the cells have been suspended for the actual measurements of electrophoretic mobilities may alter the surfaces as the result of release of intracellular materials, and the measurements obtained in such solutions are difficult to relate to the conditions in which the cells exist *in vivo*. Other factors which may influence the surface charge of cells include their stage in the mitotic cycle. MAYHEW and O'GRADY [1965] examined parasynchronous cultures of cells and found that the electrophoretic mobility was highest during the peak of the mitotic phase. This increase of electrophoretic mobility during mitosis may be partly responsible for the increase in electrophoretic mobility of cells isolated from regenerating rat liver compared with their normal resting counterparts [BEN-OR *et al.*, 1960], and for the higher mobilities of embryonic compared with adult fibroblasts [SIMON-REUSS *et al.*, 1964]. It should be noted, however, that MACMURDO and ZALIK [1970] have found that gastrula cells of *Xenopus laevis* have higher mobilities than blastula cells although their division rate is lower.

V. Antigens

FRANKS [1968] has defined 4 groups of antigens which may be detected on the surfaces of cells.

1) Species antigens which characterize all members of a species and are absent from all other species.

2) Heterophil antigens which characterize all members of a species but may also be present in some other species, e.g., Forssman and Paul-Bunnel antigens.

3) Alloantigens (isoantigens) which are present in some but not all members of the same species, e.g., blood and transplantation antigens.

4) Cell specific antigens which are present only on certain cell types, e.g., tumor specific antigens, and tissue or organ specific antigens.

Neoplastic transformation of normal cells may be, but is not necessarily, accompanied by a variety of alterations in the antigens present at the cell surface. New antigens may appear, existing normal antigens may be lost or become exposed, antigens detectable in embryonic cells but not in adult cells may reappear at the surfaces of malignant cells, and even the new characteristic tumor antigens may be partly masked in certain situations.

A. Tumor Specific Antigens

These are antigens which cannot be detected in the normal cells. At least two types of tumor specific antigens have been found: the complement-fixing antigens of tumors induced by DNA and RNA viruses which have been located inside the cell by immunofluorescent staining [GILDEN *et al.*, 1965; POPE and ROWE, 1964], and antigens located primarily at the cell periphery which are responsible for transplantation resistance. HAUGHTON [1965] has shown that up to 97% of the transplantation antigens of normal and malignant lymphoid cells are present on the cell surface. The tumor specific transplantation antigens of the majority of tumors induced by chemical carcinogens such as methylcholanthrene [PREHN and MAIN, 1957], benzpyrene [OLD *et al.,* 1962] and dibenzanthracene [PREHN, 1960] or by physical factors such as implanted cellophane film [KLEIN *et al.,* 1963], show individual specificity, i.e., the tumor antigens of different tumors produced by the same agent in syngeneic animals, or even the same animal, do not cross react to any great extent [KLEIN *et al.,* 1960; PREHN, 1965]. It should be emphasized, however, that not all chemically induced tumors possess detectable tumor antigens [OLD *et al.,* 1962]. On the other hand, all tumors induced by a given virus appear, so far, to share the same tumor specific transplantation antigens [SJÖGREN, 1964a], regardless of the type of organ of origin of the tumor, even in different species of animals or with cells transformed in tissue culture [HABEL, 1962]. The malignant transformation of normal cells by DNA viruses is not, in general, accompanied by the continuous production of the virus by the viable tumor cells, although part of the virus genome persists and is presumably responsible for the tumor specific transplantation antigens which are not present in the complete virion.

In RNA virus induced tumors, unlike most DNA virus induced tumors, complete infectious virus continues to be produced. There is some evidence that, in addition to the type-specific viral envelope antigen present at the surface of the tumor cells, there may be new surface antigens similar in kind to those produced by DNA virus transformation. Thus, 'virus-free' sarcomas induced by avian leukosis virus in mammals still contain the specific antigen [JONSSON and SJÖGREN, 1966], and antiserum cytotoxic for Friend Leukemia cells has been found to retain cytotoxicity after exhaustive absorption with virus [STEEVES, 1968]. The 'spontaneous' tumors, so called because the transforming agent or process is still unknown, appear to lack tumor specific transplantation antigens [RÉVÉSZ, 1960; BALDWIN, 1966], or if present, they are very weak antigens [HIRSCH, 1958]. The failure to demonstrate such antigens may be due also to the small numbers of antigenic receptors on the cell surface and the induction of immunological tolerance in the host.

1. Evidence for Surface Localization

Evidence for the location of these new tumor specific antigens at the cell surface has been obtained by various techniques most of which involve the interaction of specific antibodies with the antigens. Cell lysis in the presence of antibody and complement has been demonstrated by vital dye staining [Slettenmark and Klein, 1962] and by release of ^{51}Cr [Haughton, 1965].

Fluorescent antibody staining of the tumor cell surface has provided a direct demonstration of the surface localization of the tumor antigens in cells transformed by SV40 [Tevethia *et al.*, 1965] and polyoma virus [Malmgren *et al.*, 1968], as has the use of mixed cell agglutination techniques [Barth *et al.*, 1967]. Other methods for demonstrating tumor specific antigens have been developed which provide less direct evidence for the surface localization of these tumor antigens. These methods involve the detection of interactions between sensitized lymphoid and target tumor cells. Specifically sensitized lymphocytes are capable of reducing the cloning efficiency of cultured tumor cells [Hellström and Sjögren, 1967], although normal control lymphocytes can cause apparently non-specific reduction of cloning. The release of substance(s) from the lymphocytes following interaction with specific antigens can inhibit the migration of macrophages [Bloom and Bennet, 1966] and this has been used by Oettgen *et al.* [1969] to detect tumor antigens. Habel [1969] has emphasized, however, that although these *in vitro* techniques undoubtedly detect tumor specific antigens, they are not neccessarily identical with those responsible for the rejection of viable tumor cells inoculated into specifically immunized hosts.

The actual size and location of the tumor specific transplantation antigens on the tumor cell surfaces are only just beginning to be investigated. Immunofluorescent studies of H-2 isoantigens of mouse cells [Cerottini and Brunner, 1967] and of organ specific antigens on the surfaces of human thyroid cells [Fagraeus and Jonsson, 1970] have indicated that these antigens occur as discrete patches whereas species antigens are more uniformly distributed over the surface. The resolution of this technique is limited, but the greater resolution obtained with the electron microscope combined with the use of rabbit hybrid antibody possessing dual specificity for mouse γ-globulin and ferritin has successfully demonstrated the patch-localization of isoantigens on the surfaces of mouse cells [Aoki *et al.*, 1969], and of the E antigen, which is probably a leukemia specific antigen occurring only in certain mouse leukemias [Aoki *et al.*, 1970].

2. *Masked Antigens*

Investigations by Häyry and Defendi [1970] have shown that transformation of normal cells by oncogenic viruses can result in the exposure of normal surface antigens which cannot be detected on the surfaces of normal cells unless the normal cells are first treated with certain proteolytic enzymes. These antigens, which they named S antigens, were detected by serological methods involving the use of a sensitive mixed hemagglutination technique and indirect immunofluorescence. They are probably not identical with tumor specific transplantation antigens. Häyry and Defendi [1970] demonstrated that the S antigens acquired by cells transformed with SV40 virus could be detected on the surfaces of primary cultures of normal embryonic cells, spontaneously transformed cells or cells transformed by polyoma virus after they had been subjected to mild treatment with trypsin or chymotrypsin but not following treatment with papain or neuraminidase. Incubation of test sera with N-acetylglucosamine and ovomucoid did not reduce the titer of the sera for the enzyme-exposed antigens, indicating that the exposed receptor sites did not involve N-acetylglucosamine which Burger [1969] has shown to be revealed on the surfaces of trypsin-treated normal cells or untreated transformed cells. Similarly, incubation of the sera with N-acetyl and N-glycolyl hematoside did not reduce the titer, indicating that the S antigen determinants were not identical with those identified on the surfaces of transformed or trypsin-treated normal cells by Hakamori and Murakami [1968] and Hakamori *et al.* [1968]. The S antigens were also shown not to cross react with Forssman antigens. Häyry and Defendi suggested that during transformation each tumor virus uncovers characteristic sites on the cell surface. These sites are present on the surfaces of normal cells but are unaccessible to antibodies unless trypsin-labile components are first removed.

B. Loss of Antigens

Decrease in the immunizing capacity of chemically induced tumors can occur rapidly following transplantation [Globerson and Feldman, 1964], although complete loss was not observed and the process may have involved selection of less antigenic cells. Although the tumor specific transplantation antigens of virus induced tumor appear to be more stable than those of chemically induced tumors [Sjögren, 1964b], the partial or complete loss of tumor antigens *in vitro* from polyoma virus transformed cells has been reported by RABINOWITZ and SACHS [1970]. This loss of antigenicity is accompanied by a decreased tumorigenicity *in vivo*. Other changes which may be related to antigenic loss include the masking of specific sites which bind a protein, concanavalin A [INBAR and SACHS, 1969], the reacquisition of contact inhibition of cell proliferation, and certain morphological alterations [RABINOWITZ and SACHS, 1969]. SACHS [1965, 1967] put forward the hypothesis that the presence of tumor specific transplantation antigens on the tumor cell surface was responsible for the loss of contact inhibition of proliferation, and the work of RABINOWITZ and SACHS [1970] appears to provide some experimental support for this hypothesis. Loss of normal antigens as a consequence of the carcinogenic process has been observed in a number of systems, but not all of the antigens have been shown to be located at the cell surfaces. Hepatoma cells from animals exposed to azo dyes have lost certain antigens present in the normal cells [WEILER, 1952; BALDWIN, 1963; ABELEV, 1965], and tissue-specific antigens have apparently been lost or greatly decreased in detectability in gastrointestinal tumors [NAIRN *et al.*, 1962] and in thyroid tumors [GOUDIE and MCCALLUM, 1963]. Loss of tumor specific antigens may also involve masking to varying extents by other groups at the cell surface. Thus, CURRIE and BAGSHAWE [1969] found that treatment of cells of primary methylcholanthrene-induced sarcomas with neuraminidase made them much more immunogenic.

C. Embryonic Antigens

GOLD and FREEDMAN [1965a, b] established by a number of serological techniques that nearly all adenocarcinomata arising from entodermally derived epithelium of the human digestive system contain identical tumor antigens, specific in the sense that they cannot be detected in adjacent normal control tissue or in other diseased and malignant adult human tissues. Examination of human embryonic and fetal organs revealed that components identical with these digestive system tumor specific antigens were present in the differentiating gut, liver and pancreas until the end of the second trimester of gestation, and that all other developing organs lacked these components. These tumor antigens were therefore named carcino-embryonic antigens (CEA). Later work by GOLD *et al.* [1968] revealed that suspensions of cultured tumor cells were agglutinated by anti-CEA sera and stained at their surfaces with fluorescein-labelled antisera, indicating that these antigens were chiefly located at the tumor cell surface. GOLD [1967] examined the sera of over 200 subjects, both normal volunteers and patients suffering from a variety of diseases and found antibodies specific for CEA in two groups, one group consisting of pregnant women and the other of 70% of patients bearing non-metastatic digestive system cancers. Patients in which these tumors had metastasized did not possess circulating anti-CEA antibodies, presumably because the larger masses of disseminated tumor had totally absorbed these specific antibodies, but induction of immune tolerant states was not excluded. HOLLINSHEAD *et al.* [1970] extended this work and prepared soluble membrane antigen fractions from these types of tumors and with these fractions obtained positive delayed hypersensitivity reactions in 17 out of 19 patients suffering from carcinoma of the colon and rectum. These authors emphasized that in order to elicit these delayed hypersensitivity reactions it was essential to avoid harsh methods for preparing the isolated soluble membrane antigens. From the behavior of the active antigenic fraction on sephadex G-200 columns the molecular weight appeared to be about 175,000.

O'NEILL [1968] adapted the immune adherence technique to the assay of antigens present on the surfaces of cultured hamster kidney cells before and after transformation with polyoma and Rous sarcoma virus. He found that Forssman antigen, which was absent from the normal cell lines, appeared on the surfaces of the transformed cell lines studied and that Rous virus transformed cells lost this antigen when they reverted to normal morphology. The Forssman antigen was also detectable on the surfaces of primary cultures of embryonic hamster kidney cells within 1 day of culturing, implying that transformation with oncogenic virus resulted in the appearance of an embryonic antigen absent from cells of neonatal and adult hamsters. Other factors apart from neoplastic transformation appear to be involved in the appearance or disappearance of the Forssman antigen on the surfaces of cultured cells. For example, FOGEL and SACHS [1964] using immunofluorescent techniques found that primary monolayer cultures of normal hamster embryo cells did not initially possess this antigen, and suggested that its subsequent appearance was due to induction of antigen synthesis resulting from the actual process of plating *in vitro.* Although, as O'NEILL [1968] later confirmed, they found Forssman antigen on polyoma virus transformed but not on normal BHK cells they also observed that this antigen was absent from a tumor-producing BHK line not transformed by virus. In view of the ambiguities associated with the conditions which may elicit the appearance of Forssman antigen on the surfaces of cultured hamster cells it would appear unjustified to assume that its appearance following transformation involves an embryonic antigen.

The production of embryonic serum α globulins by certain hepatomas of man, rat and mouse has been demonstrated by ABELEV [1963, 1968] and TATARINOV [1964], but this protein is secreted by the tumor cells and there is no evidence that it is associated with the cell surface except transiently while it is being released from the cells.

VI. Reduced Cell-Cell Adhesiveness of Malignant Cells

COMAN [1944] was the first to suggest from measurements of the forces required to separate pairs of cells that malignant cells have reduced mutual adhesiveness compared with comparable normal cells, and that the reduced adhesiveness might play an important role in tumor invasion and metastasis. Subsequently, MCCUTCHEON *et al.* [1948] using a more reproducible technique involving the agitation of fragments of tissue under standardized conditions found that more cells were released from carcinomas than from the corresponding normal tissues, and TJERNBERG and ZAJICEK [1965] obtained cells more easily from carcinomas than from benign epithelial tumors following aspiration with a syringe.

A. Structural Basis

Electron microscopy indicates that there may be a structural basis for the reduced mutual adhesiveness of tumor cells. EASTY and MERCER [1960] examined replicas of the surfaces of monolayer cultures of normal hamster kidney epithelial cells and of kidney tumors known to arise exclusively from the cortical tubular epithelial cells and observed that while the normal cells were closely apposed with a gap of 200 Å between them, the tumor cells formed far fewer contacts of this kind. Similar results were obtained by MARTINEZ-PALOMO *et al.* [1969] from a comparison of cultures of normal and transformed cells. These authors found that transformed cells were devoid of regions of fusion between adjacent cells, in contrast with the frequent presence of such junctions between normal cells. MARTINEZ-PALOMO [1970] extended these studies of the appearance of cell contacts to a number of tumors and comparable normal tissues *in vivo* using lanthanum hydroxide and uranyl and lead stains to enhance the contrast of the cellular membranes and to delineate the extracellular spaces for the identification of tight junctions and gap junctions. The tight junction (zonula occludens) represents a surface contact where the membranes of adjacent cells fuse and obliterate the intercellular space, forming a continuous seal around the cells. This junction is impermeable to macromolecules such as hemoglobin [MILLER, 1960] and to lanthanum compounds which are used as a stain to detect these regions [REVEL, 1968]. The gap junction, on the other hand, is formed by close approximation of cell membranes leaving a gap of approximately 20 Å between the membranes which is permeable to macromolecules and lanthanum. In most tissues the irregular distribution of the gap junctions and the absence of specific topographic localization with respect to luminal surfaces suggests that they represent focal regions of membrane apposition rather than continuous belts around the cells, as is the case with tight junctions. MARTINEZ-PALOMO [1970] found that tight junctions were completely lacking in the three tumors he examined, a rat hepatoma, a mouse skin carcinoma and a mammary carcinoma. In the mammary carcinoma apical junctions were observed which consisted of a series of desmosomes (maculae adherentes), the continuous seal running completely round each corresponding normal cell being absent. Occasional gap junctions were observed between the cells of these tumors but were much less frequent than between cells of the corresponding normal tissue. The lack of tight junctions in these solid tumors may be responsible for the reported decrease in intercellular adhesions of tumor cells since these junctions are known to be the most resistant to mechanical tension [FARQUHAR and PALADE, 1963]. As these junctions also effectively prevent the penetration of relatively small molecules in normal tissues, their absence in tumors creates a continuous extracellular space throughout the tumor, permitting the unrestricted passage of nutrients which may help to maintain the growth of the tumor. MCNUTT and WEINSTEIN [1969] also observed many fewer gap junctions between cells of cervical carcinomas than between cells of normal cervical epithelium.

B. Chemical Basis

1. Calcium

A chemical basis for the decreased mutual adhesiveness of malignant cells was suggested by COMAN [1944]. Calcium ions are known to be involved in the establishment and maintenance of adhesions between the cells of many tissues, and are generally considered to provide either a direct link between the apposed cell surfaces or to link each cell surface to an intercellular cement. WEISS [1967b] suggested that rupture of cell surfaces may occur when cells are separated, and observed that treatment of mouse sarcoma 37 cells with chelating agents which would remove calcium ions made these cells more easily deformable, which in its turn, may have facilitated cell separation. The observations of CURRUTHERS and SUNTZEFF [1944] of a decrease in the calcium content of normal epithelium following application of the carcinogen methylcholanthrene, followed by a further decrease when tumors developed, led COMAN to suggest that the lower calcium content of tumors was connected with or responsible for decreased mutual adhesion. More recent work by HICKIE and KALANT [1967] casts doubt on the generalization that malignant tissues are characterized by a lower calcium content. They found that the calcium and magnesium content of Morris hepatoma cells was actually higher than that of the corresponding normal rat liver cells. Analyses of whole tissues provide no information, however, concerning the concentration of calcium at the cell surface or in the extracellular spaces. It is possible that the surfaces of tumor cells are unable to bind as much calcium as those of normal cells, but the work of BANGHAM and PETHICA [1960] on the effect of calcium ions on the surface charges of Ehrlich ascites tumor cells and isolated liver cells revealed no significant differences in their calcium binding capacities. Similar results were obtained by PATINKIN *et al.* [1970b] from a comparison of the calcium binding capacities of normal chick embryo fibroblasts with their Rous sarcoma virus transformed counterparts, but as these authors emphasized, these results do not necessarily invalidate COMAN's [1944, 1953] suggestion of the role of calcium in reduced mutual adhesiveness of tumor cells, since calcium involved in mutual cell adhesion may be located in specific areas of the cell surface. It appears that the role of calcium in cell adhesion and separation remains obscure.

2. Surface Charge

The role of surface charge in cell adhesion has been reviewed by PETHICA [1961], WEISS [1967a] and CURTIS [1967]. It is generally conceded that the repulsive forces between cell surfaces originate in the ζ-potentials of the surfaces except at very close approaches, and AMBROSE [1965] has been the main proponent of the view that the higher surface charge densities of malignant cells are primarily responsible for the greater ease of separation of malignant cells. As indicated in Section IVC, the exceptions to the proposition that malignant cells have a higher surface charge than their normal counterparts are now too numerous to be completely ignored. Perhaps the strongest evidence relating surface charge density to the invasive and metastatic behavior of tumors is that of PURDOM *et al.* [1958], who measured the electrophoretic mobility of cells obtained from various sublines of the mouse MC1M sarcoma. These sublines vary progressively in their capacity to give rise to metastases following subcutaneous implantation and to give ascites on intraperitoneal inoculation. Thus, there may be a progression in the capacity of the cells of the various tumor sublines to separate from each other *in vivo*. PURDOM *et al.* [1958] found that these tumors demonstrated a progressive increase in electrophoretic mobilities which corresponded with increased ability to metastasize and invade. AMBROSE, in ABERCROMBIE and AMBROSE [1962], suggested that the progressive increase in surface charge accompanying increased malignancy was due to a reduced capacity to bind calcium, but FORRESTER [1969] found that addition of calcium ions to cell suspensions of the most invasive subline produced a pronounced decrease in electrophoretic mobility while the effect on the least invasive subline was very small, indicating that the increased surface charge of the most invasive cells was not due to a reduced calcium binding capacity. Support for the observations of PURDOM *et al.* [1958] came from SAKAI [1967] who found that several strains of rat ascites hepatoma cells differing in adhesiveness and metastatic frequency showed a progressive rise in electrophoretic mobility with increased ease of metastasis. The observations of WEISS and MAYHEW [1968] and PATINKIN *et al.* [1970a] of the very low electrophoretic mobility of the highly invasive macrophage and of the small amount of neuraminidase-susceptible neuraminic acid present on its surface must throw considerable doubt, however, on the suggestion that there is a simple first order correlation between surface charge density or the quantity of neuraminidase-susceptible surface sialic acid and the invasive capacity of cells.

3. Release of Hydrolytic Enzymes

Various proteolytic enzymes are widely used for the separation of cells from tissues or solid substrates, and it is obvious that if malignant cells are characterized by the capacity to release enzymes which may facilitate cell separation by degradation of cell surface or intercellular components these enzymes may play a significant role in the spread of cancer. Although much of the earlier histochemical work on hydrolytic enzymes associated with invasive tumors yielded conflicting results [reviewed by EASTY, 1966] the more recent work of SYLVÉN and his colleagues involving the capillary tube sampling of intercellular fluid from sterile granulomas and carefully characterized regions of tumors of different invasive capacities has provided strong support for the suggestion that invasive tumors are characterized by the ability to release hydrolytic enzymes into the extracellular environment [SYLVÉN and MALMGREN, 1957; SYLVÉN, 1968]. Inflammatory cells associated with the tumor undoubtedly contribute toward the enzyme content of tumor interstitial fluid, but the pattern of enzymes present in tumor interstitial fluid presents distinct differences from that presented by sterile or infected granulomas, notably in the presence of dipeptidase activity in the tumor fluid [SYLVÉN, 1968]. Moreover, SYLVÉN succeeded in demonstrating that the tumor fluid contained a trypsin-like protease, cathepsin B, with an optimum activity at about pH 5.6 in the presence of suitable activators. He considers that this enzyme plays a particularly important role in facilitating tumor cell separation and in the degradation of extracellular structures. Most of the remainder of the proteolytic activity appears to require lower pH for optimum activity and is probably associated with cathepsin D [SYLVÉN, 1965].

Using various pH indicator dyes in benzene-water emulsions, DEUTSCH [1928] demonstrated that the interfacial pH can differ from that of the bulk phase, and, subsequently, MCLAREN [1957], from a comparison of the pH activity curves of invertase attached to the surface of yeast cells and invertase present in solution, suggested that the enzyme was acting at a lower pH when bound to the surfaces of cells because of the negative surface charge associated with the cell surface. Similar factors could operate with mammalian cells, and enzymes such as cathepsin D, which may only acquire significant activity at pH values much lower than the surrounding interstitial fluid, may be active in close association with the surfaces of the cells. Acid environments may not be essential for significant activity of other enzymes which may degrade extracellular materials. For example, ROBERTSON and WILLIAMS [1969] have detected neutral collagenase activity associated with extracts of an invasive tumor and have also demonstrated cartilage-degrading activity in tumors invading the xiphisternum.

Direct evidence of the possible role of lysosomal enzymes in facilitating cell detachment from tumors has been provided by WEISS and HOLYOKE [1969] who examined the effects of hypervitaminosis A on the pulmonary metastatic pattern of spontaneous mammary tumors in mice. High doses of vitamin A are known to activate the lysosomal enzymes of many cells, and in addition to obtaining increased metastases in hypervitaminized animals WEISS and HOLYOKE found that cells were liberated in much greater quantities from the tumors of these animals than from control tumors when plugs of tumor tissue were agitated under standard conditions. Whether the lysosomes considered to be activated were present exclusively in the tumor cells or were associated with host cells such as macrophages is not known. It would be of considerable interest to know if the reverse effect, i.e., decrease in metastasis and cell separation, could be obtained by the use of lysosome stabilizing substances such as cortisone and chloroquine.

VII. Contact Regulation of Normal and Malignant Cells

Although WILLIS [1952] ascribed the invasiveness of malignant cells into surrounding normal tissues solely to pressure within the tumor resulting from cell proliferation, there is evidence that tumor cells from solid fragments of tumor may infiltrate fragments of adjacent normal tissues maintained under conditions which minimize cell proliferation [EASTY and EASTY, 1963]. The separation of tumor cells facilitated by release of enzymes is not a sufficient condition to produce the process of invasion in the absence of pressure. Cell locomotion must take place [ABERCROMBIE and AMBROSE, 1962].

A. Contact Inhibition of Movement *in vitro*

The *in vitro* studies of ABERCROMBIE and HEAYSMAN [1953, 1954], ABERCROMBIE *et al.* [1957], reviewed by ABERCROMBIE [1961, 1970] provided convincing evidence that absence of movement of the cells present in the majority of adult tissues may be determined by contact of cell surfaces with each other. In sparse monolayer cultures of normal fibroblasts the contact between the leading ruffled membrane of a migrating cell with the surface of another cell results in cessation of activity in the ruffled membrane and a contraction of the migrating cell which becomes arrested. Adhesion between the surfaces of the two cells in contact is frequently shown by the formation of extended cytoplasmic strands when the two cells subsequently move apart. Another factor which may be involved in cessation of movement of cells following contact is that the cell surfaces may not be sufficiently adhesive to permit migration of one cell over another. Thus, normal endothelial cells which in the intact organism are relatively non-adhesive, inhibit the migration of tumor cells [BARSKI and BELEHRADEK, 1965] and macrophages [ABERCROMBIE, 1970] which are not inhibited by normal fibroblasts. A consequence of contact inhibition of movement is that when a complete monolayer of cells is formed, movement of individual cells is severely restricted in any direction. The migration of cells from a number of sarcomas, particularly the highly anaplastic mouse sarcoma 180 [ABERCROMBIE *et al.,* 1957], is not inhibited by contact with normal fibroblasts, and this absence of contact inhibition of movement is probably intimately involved in the invasion of some sarcomas. Other sarcomas, however, are subject to contact inhibition of movement, either by normal cells or by each other [DEFENDI *et al.,* 1963; BARSKI and BELEHRADEK, 1965; PONTON and MACINTYRE, 1968]. Little work has been carried out on the migratory behavior of cultured carcinoma cells in contact with normal fibroblasts and epithelial cells, but in view of the influence of environmental factors such as the type of substrate [CARTER, 1965], nutrients and cell density [CURTIS, 1961] on the degree of contact inhibition between normal fibroblasts and the difficulties associated with the measurement of this phenomenon, this is not surprising.

Although ABERCROMBIE [1970] considers that contact inhibition of movement is essentially due to the contraction which develops on contact and paralyses cell movement, the nature of this interaction remains unknown. A careful study was made by FLAXMAN *et al.* [1969] of the nature of the surface contacts formed between pairs of cells which had been fixed at varying times after contact was seen to have been established between them. They found that tight junctions, involving the obliteration of the intercellular space, were formed within a few minutes of contact being established between cells which were exhibiting contact inhibition of movement. The authors emphasize the difficulties of defining the exact state of contact inhibition of the pairs of cells studied, and, hence, the difficulty of directly relating the formation of tight junctions to contact inhibition. These junctions are known to be very resistant to breaking and this may well be a factor involved in contact inhibition, but the possibility that transport of ions between the cells in the junctional region may alter cytoplasmic properties responsible for cell motility is also suggested by the authors. WOLPERT and GINGELL [1969] have proposed a detailed mechanism of immobilization of cells following contact. They suggest that the close approach of two cell surfaces may lead to an increase in surface potential which induces conformational alterations in the surface, which in turn activate adenyl cyclase leading to the production of cyclic AMP. This could alter the local calcium concentration [RASMUSSEN and TENENHOUSE, 1968] which may result in changes in the activity of microtubules situated just beneath the surface which are believed to be responsible for cell motility [GINGELL and PALMER, 1968].

B. Functional Junctions between Cells

It is known that in a variety of normal tissues cells in contact with each other have localized regions of increased permeability which permit the flow of inorganic ions [LOEWENSTEIN, 1966; BENNETT, 1966] and of larger molecules with molecular weights of up to 60,000 [LOEWENSTEIN, 1966; KANNO and LOEWENSTEIN, 1966]. These regions are also associated with electrical coupling between the cells [POTTER *et al.*, 1966; SHERIDAN, 1966; LOEWENSTEIN, 1967; BENNETT and TRINKAUS, 1970]. These low resistance junctions which interconnect adjacent cells have been detected in many tissues, both adult and embryonic, and also between cells in tissue culture. It has been suggested by numerous authors that these junctions permit cells to exchange control substances which influence functions such as movement, division, and the differentiation of cells within the tissue.

Malignant cells characteristically display to varying extents a lack of coordination and control in their growth, differentiation and movement both *in vivo* and *in vitro,* and observations have been made which suggest that this loss of control may be associated with the absence or reduced incidence of electrical coupling between the cells. Reduction in electrical coupling has been observed between cells of cancers of the liver [LOEWENSTEIN and KANNO, 1967], the thyroid [JAMAKOSMANOVIC and LOEWENSTEIN, 1968] and the stomach [KANNO and MATSUI, 1968]. This functional uncoupling associated with malignant transformation may also extend to the normal cells surrounding the tumor, particularly where the tumor is invading [LOEWENSTEIN and KANNO, 1967]. It does not appear to be associated with controlled cell proliferation such as that involved in normal liver regeneration or wound healing, and although cellular intercommunication is reduced between normal cells adjacent to a wound this is rapidly restored, often within minutes of wounding [LOEWENSTEIN and PENN, 1967]. These very important observations of LOEWENSTEIN and his colleagues have not been universally accepted, however. FURSHPAN and POTTER [1968] found no difference in the coupling between normal and tumor virus transformed cells in culture, while the degree of coupling of two cell lines derived from mouse sarcoma 180 depended on the time elapsed since renewal of the culture medium. These results could suggest that cancer cells may be coupled in tissue culture, provided the medium has not been recently renewed, but may be uncoupled in the animal. Further doubts were cast on the suggestion that cancer cells are functionally uncoupled, by the work of SHERIDAN [1970] who investigated the electrical coupling between cells of sarcoma 180, the Novikoff hepatoma and two Morris hepatomas. SHERIDAN concluded that electrical coupling in all these tumors was widespread, often extending between cells separated by many other cells, although he did not exclude the possibility that coupling may not have been as extensive as in the comparable normal tissue.

There is at the moment considerable disagreement about the structural basis of cell-cell coupling. Various ultrastructural features of the cell surface have been proposed as sites of coupling, but the issue is complicated by the possibility that there may be more than one type of cellular interconnection, e.g., junctions which are permeable to small ions may not be permeable to large molecules such as proteins. BULLIVANT and LOEWENSTEIN [1968] have suggested that coupling may be associated with septate desmosomes in several invertebrates. Electrical coupling between heart cells has, however, been demonstrated by MUIR [1967] following disruption of desmosome-like junctions at the intercalated discs. GOODENOUGH and REVEL [1970] appear to favor the gap junction as the site of electrical coupling but from a review of the results of other workers conclude that both desmosomes and zonulae occludentes may also be involved. The permeabilities between adjacent cells at the different types of structural junctions remain unknown, and the evidence that the junctions are indeed the sites of coupling remains largely circumstantial. The work of FLAXMAN *et al.* [1969] strongly suggests that tight junctions might be involved in contact inhibition of cell movement *in vitro,* but HARRIS [1970] examined cultures of 3T3 cells which demonstrate a high degree of contact inhibition of movement and division and found no evidence of tight junctions, gap junctions or desmosomes.

C. Contact Inhibition of Cell Proliferation

With the establishment of the phenomenon of contact inhibition of movement of normal fibroblasts in monolayer culture it was, perhaps, inevitable that a mechanism of contact inhibition of cell proliferation should be proposed since it was observed by many investigators that a culture of normal fibroblasts would not continue to grow indefinitely even when the medium was renewed frequently. Convincing evidence for the existence of a cell contact mechanism of control of cell proliferation *in vitro* came from the observations of TODARO and GREEN [1963] that the growth rate of the 3T3 line of mouse fibroblasts fell sharply when the density of a monolayer of these cells was such that a considerable degree of mutual contact was achieved, and that this inhibition was not due to an inhibitory factor present in the medium. Similar effects were observed by LEVINE *et al.* [1965] in cultures of human diploid fibroblasts. Many tumor cells and cells transformed *in vitro* will grow to higher saturation densities than normal cells and this suggested that they were less responsive to contact control of cell proliferation. Further convincing evidence of the role of loss of contact control of proliferation in tumor growth has come from the observations that cultures of mouse cells selected on the basis of increased susceptibility to contact control showed decreased tumorigenicity [POLLACK *et al.*, 1968], and that loss of tumorigenicity was accompanied by a reacquisition of susceptibility of contact control of proliferation *in vitro* [RABINOWITZ and SACHS, 1968].

STOKER [1964] and STOKER *et al.* [1966] observed that certain transformed cells which had largely escaped from mutual contact control retained the capacity to respond to the control exerted by contact with non-growing cells, and suggested that the transformed cells could receive and respond to the inhibitory stimuli but could not initiate or transmit. Subsequently, EAGLE *et al.* [1968] found that when a variety of normal or malignant cells were inoculated onto a monolayer of cells which had essentially ceased to grow, the ability of the inoculated cells to grow in contact with the stationary cells correlated very well with their inherent growth capacity, i.e., the higher the population density attained by cells in pure culture the less the susceptibility to inhibition of growth by other cell types. Unfortunately, many virus transformed cells did not demonstrate any clear relationship between inherent growth capacity in pure culture and susceptibility to inhibition by stationary cultures [EAGLE *et al.*, 1970]. Further complications were introduced by the discovery that factors other than density-dependent contact may operate in the control of proliferation of normal cells *in vitro.* Multilayers of normal fibroblasts were obtained by continuous perfusion of medium [KRUSE and MIEDEMA, 1965], and TODARO *et al.* [1965] found that addition of fresh serum to non-growing confluent cultures restimulated

DNA synthesis in a proportion of 3T3 cells. Transformed cells, however, seem to have a decreased requirement for serum for stimulation of growth [HOLLEY and KIERNAN, 1968]. Further complications arise from the reported release into the medium of growth inhibiting substances by cells [YEH and FISHER, 1969].

An additional growth promoting or releasing factor involving cell surface contact has been described by STOKER *et al.* [1968]. Normal fibroblasts which do not proliferate continuously when maintained in isolation in suspension in soft agar or methylcellulose were stimulated to grow by contact with small glass fibers present in the gel. Only those cells in actual contact with fibers proliferated and neighboring unattached cells did not proliferate, indicating that attachment was the stimulus and that a diffusable inhibitor was not operating. Attachment of the cell surface and subsequent spreading may alter surface or membrane function and favor cell division.

D. Metabolic Cooperation between Cells in Contact

SUBAK-SHARPE and his colleagues have shown that when cells are cultured *in vitro* there is an exchange of materials between cells in contact which can result in modifications of their metabolic processes. SUBAK-SHARPE [1965] isolated biochemical variants of the polyoma virus transformed hamster cell line which lacked detectable inosinic pyrophosphorylase (IPP^-) and adenylic pyrophosphorylase (APP^-) activity. The IPP^- cells were unable to incorporate hypoxanthine, while the APP^- cells were unable to incorporate adenine or adenosine. When IPP^- cells were cultured in the presence of IPP^+ cells (possessing inosinic pyrophosphorylase activity) in the ratio of 300 IPP^- cells to 1 IPP^+ cell in medium containing radioactive hypoxanthine, it was found that, in addition to the rare IPP^+ cell which was heavily labelled, there were other cells which were in direct or indirect contact with the isolated IPP^+ cells which were definitely labelled though to a lesser extent [SUBAK-SHARPE *et al.*, 1966, 1969]. IPP^- cells not in direct or indirect contact with IPP^+ cells did not display significant amounts of radioactivity. Indirect contact involves a contact between an IPP^- cell with another IPP^- cell which is itself in contact with an IPP^+ cell. Similar results were obtained with mixtures of APP^+ and APP^- cells cultured together in the presence of labelled adenine. When IPP^- cells were cocultured with equal numbers of APP^- cells under conditions of close cell contact in the presence of either labelled hypoxanthine or adenine, then virtually all the cells had incorporated label into their nucleic acid irrespective of which isotopically labelled purine had been provided [BÜRK *et al.*, 1968]. Thus, this type of metabolic cooperation involves a reciprocal exchange. Metabolic cooperation of this type involves contact between the cell surfaces, at least at a light microscope level. It is not known at present whether or not the biosynthetic ability apparently gained by this type of contact-associated metabolic cooperation is retained, and for how long, by the deficient cell after contact is broken between it and the competent cell. The nature of the materials exchanged between the cells is likewise unknown. Similarly, beyond the fact that the cells have to have been in microscopic contact for cooperation to occur, it is not known whether cooperation involves phagocytosis of the cytoplasm of one cell by the adjacent cell, the formation of specialized junctions permitting exchange of molecules or even the temporary fusion of the membranes of the cooperating cells with the establishment of direct cytoplasmic continuity. Although most of the work has been carried out using variants of transformed cell lines, metabolic cooperation has been observed between transformed and normal cells [STOKER, 1967], and also between human fibroblasts from normal individuals and fibroblasts which lack inosinic-guanylic pyrophosphorylase activity derived from patients suffering from Lesch-Nyhan disease [FIEDMANN *et al.,* 1968]. Thus, metabolic cooperation of this type appears to be common to both normal and malignant cells. Its possible relevance to cancer has been discussed by SUBAK-SHARPE [1969] who suggests that if cooperation occurs between normal and malignant cells *in vivo,* then tumors genetically sensitive to a therapeutic agent may acquire resistance from surrounding normal cells and the surrounding normal cells may acquire drug sensitivity from the tumor cells. Thus, reduction of contact-mediated metabolic cooperation might increase the selectivity of a chemotherapeutic agent.

VIII. Cell Contacts and the Formation of Metastases

During the embryonic development of multicellular animals there are extensive movements of cells within the embryo, and the adhesive properties of the cells are considered to play an important role in the final location and arrangement of these cells [see review by STEINBERG, 1964]. STEINBERG'S work strongly indicated that quantitative differences in the strength of adhesions between cells of different types, together with random cell movements were sufficient to account for these emergent patterns of cellular location. However, the observations of SIMON [1957a, b] and MEYER [1964] of the vascular migration of avian germ cells from the extra-embryonic mesoderm to the gonads appears to involve specificity of localization of these cells. GOWANS and KNIGHT [1964] studied the quantitative distribution of labelled large and small rat lymphocytes injected into the femoral vein of rats. Their results indicated that the small lymphocytes rapidly accumulated in the lymph nodes where they then entered the lymphatic circulation. These authors suggested that the selective lodging of these small lymphocytes in the post capillary venules of the lymph nodes might be due to a special affinity of these cells for the endothelium of these venules. WEISS and ANDRES [1952] had also suggested that adhesion among embryonic cells of different types involved a high degree of specificity on the basis of experiments which demonstrated that when chick embryo melanoblasts were injected intravenously into early non-pigmented embryos the cells homed to the sites they would have occupied in the donor embryo, giving rise to pigmented patches. BURDICK [1968] pointed out that these results could equally well be explained by differential survival of the injected cells rather than a highly specific lodgement involving specific cell adhesion mechanisms. He tested the hypothesis of WEISS and ANDRES by injecting tritium-labelled chick embryo cells from the mesonephros, limb and heart ventricle into early embryos and estimating the distributions of cells in the ventricle, limbs and mesonephroi by autoradiography. He found that the distribution of all the cells among the organs was identical for the three cell types. These results offered no support for the Weiss hypothesis, and similar results were obtained by HOLLYFIELD and ADLER [1970] who also studied the distribution of chick embryo cells of neural retina, optic lobe and liver in a variety of organs following intravenous injection into chick embryos.

The existence, or otherwise, of specific cell adhesions which may influence the site of lodgement of vascularly disseminated cells is highly relevant to the sites of lodgement of blood-borne tumor cells which may form metastases. COMAN [1953] has reviewed the mechanisms responsible for the distribution of blood-borne tumor cells and the sites of metastasis and concluded that they were mainly governed by the mechanics of circulation. There are indications from the experiments of KINSEY [1960] that sites of metastasis could not be accounted for solely on the basis of mechanical trapping. KINSEY placed grafts of various tissues subcutaneously in the thighs of mice which were subsequently injected intravenously with

S–91 melanoma cells. Metastases developed only in the animal's own lungs and in the grafts of pulmonary tissue. The quantitative distribution of tumor cells was not examined, however, and it cannot be excluded that the cells were distributed non-selectively throughout the organs of the host and that the lung tissue provided a very favorable site for tumor cell proliferation. EASTY *et al.* [1969a, b] attempted to correlate the quantitative distribution of a variety of heterologous tumor cells following injection of tritium-labelled cells into chick embryos with the extent of subsequent tumor growth in the organs. In general, they found that the most densely 'seeded' organs were the sites of most extensive tumor growth, but there were certain outstanding exceptions. The brain was one of the least efficient tumor cell trapping organs but was one of the best sites of tumor growth, while for the spleen the converse was found. Pronounced differences in the initial distribution of different tumors in certain organs were observed which could not be accounted for on the basis of cell size. Selective trapping involving adhesions of different strengths of the endothelium of different organs may have been involved. SUEMASU *et al.* [1970] carried out a similar study of the distribution of Cr-labelled cells and the pattern of organ metastasis using two rat tumors injected into the left ventricle. They demonstrated that the pattern of organ metastasis was different for the two tumors and concluded that the initial density of seeding in each organ corresponded with the incidence of subsequent metastasis in the organs, but examination of their data reveals certain prominent exceptions, e.g., the density of seeding of the Sato lung sarcoma in the lungs was 10 times higher than in the adrenals but the tumor incidence in the lungs was 3 times lower.

The importance of the cell surface in determining the sites of localization of intravascularly disseminated cells is frequently invoked, but direct evidence of its role is relatively weak. An indirect, but interesting approach to this problem is illustrated by the work of GESNER and his colleagues. They have studied the effects of preincubating rat lymphocytes with a variety of enzymes on their localization within various organs of the host following intravenous injection. GESNER and GINSBURG [1964] found that treatment with glycosidases decreased lymphocyte accumulation in the spleen and lymph nodes and increased it in the liver. The effect of these enzymes was specifically inhibited by the addition of L-fucose and N-acetyl-D-galactosamine to the incubation mixture, and the authors suggest-et that the enzymes removed carbohydrate moieties from the lymphocyte periphery, exposing new sites which led to their removal from the circulation by cells of the reticuloendothelial system. Alternatively, the integrity of the carbohydrate moieties may be essential for the lymphocytes to traverse the post-capillary venules of the lymphoid tissue. Similar results were obtained by GESNER and WOODRUFF [1967] following treatment of the lymphocytes with neuraminidase. Pretreatment with trypsin retarded the localization of lymphocytes in the lymph nodes, but, eventually, the nodes became populated with the labelled cells. The authors suggested that this may have been due to an *in vivo* regeneration of lymphocyte cell surface components altered or removed by pretreatment with trypsin [WOODRUFF and GESNER, 1968].

Very few studies of the effects of enzymes on the distribution and subsequent fate of blood-borne tumor cells, accompanied by appropriate analyses of the alterations of surface components, have been carried out. GASIC and GASIC [1962b] found that pretreatment of mouse tumor cells with neuraminidase under conditions which removed nearly all the Hale-positive staining material from their surfaces had no effect on the incidence or size of lung metastases following intravenous injection of enzyme-treated or untreated cells into mice. This lack of effect was thought to reflect the rapid *in vivo* regeneration of sialic acid components of the cell surface of the enzyme-treated cells. Intravenous injection of relatively large doses of enzyme removed the Hale-positive staining material from the surfaces of the vascular endothelium for several days, and injection of tumor cells into these enzyme-treated animals produced significantly fewer and smaller metastases than in untreated animals. GASIC and GASIC suggested that removal of sialic acid from the surfaces of the vascular endothelium in some way reduced the number or strength of adhesions between the tumor and endothelial cells and thereby reduced the incidence of metastatic growth. Later work by GASIC *et al.* [1968] indicated that the enzyme treatment of the host may also affect tumor cell attachment to endothelium in a less direct way. They found that when tumor cells were injected at varying times after injection of neuraminidase, the antimetastatic effect of the enzymes was most intense when the enzyme activity in the plasma had dropped considerably. In addition, they observed that neuraminidase produced a fall in blood platelet count which did parallel the antimetastatic effect of enzyme pretreatment. A similar antimetastatic effect was obtained by reducing the platelet count by the administration of anti-platelet serum, and they suggested that platelets might contribute to the establishment of metastases from blood-borne tumor cells by promoting attachment of these tumor cells to the vascular endothelium. CORMACK [1970] established in a number of carefully controlled experiments that treatment of Walker tumor cells with neuraminidase before bringing them into contact with rat mesothelium also produced a reduction in the subsequent growth of tumors and increased the survival time of the rats. He considered the effect was due to a reduction in the adhesion of tumor cells to the mesothelium as a result of removal of sialic acid from the tumor cell surface. Measurements of the numbers of tumor cells adhering to the membrane were not made, however. A direct demonstration of a reduction in the ability of cells to form mutual adhesions as a result of neuraminidase treatment has, however, been made by KEMP [1970] in a study of the aggregation of dissociated chick embryo muscle cells. The role of cell surface components in the adhesion, organ localization and extravascularization of both normal and malignant cells in the blood is only beginning, but should prove to be a very rewarding field of investigation.

References

ABELEV, G. I.: Unio Int. Contra Cancrum *19:* 80 (1963).

ABELEV, G. I.: Progr. exp. Tum. Res. (Basel), vol. 7. p. 104 (Karger, Basel/New York/Sydney 1965).

ABELEV, G.I.: Cancer Res. *28:* 1344 (1968).

ABERCROMBIE, M.: Exp. Cell Res., Suppl. *8:* 188 (1961).

ABERCROMBIE, M.: Europ. J.Cancer *6:* 7 (1970).

ABERCROMBIE, M. and AMBROSE, E.J.: Cancer Res. *22:* 525 (1962).

ABERCROMBIE, M. and HEAYSMAN, J.E.M.: Exp. Cell Res. *5:* 111 (1953).

ABERCROMBIE, M. and HEAYSMAN, J.E.M.: Exp. Cell Res. *6:* 293 (1954).

ABERCROMBIE, M.; HEAYSMAN, J. E. M. and KARTHAUSER, H. M.: Exp. Cell Res. *13:* 276 (1957).

AMBROSE, E.J.: In E.J. AMBROSE, Cell electrophoresis, p. 194 (Churchill, London 1965).

AMBROSE, E.J.; JAMES, A.M. and LOWICK, J.H.B.: Nature *177:* 576 (1956).

AOKI, T.; HÄMMERLING, U.; HARVEN, E. DE; BOYSE, E.A. and OLD, L.J.: J. exp. Med. *130:* 979 (1969).

AOKI, T.; STÜCK, B.; OLD, L.J.; HÄMMERLING, U. and HARVEN, E. DE: Cancer Res. *30:* 244 (1970).

ASHWORTH, L.A.E. and GREEN, C.: Biochim.biophys. Acta. *84:* 182 (1964).

BALDWIN, R.W.: Unio Int. Contra Cancrum *19:* 545 (1963).

BALDWIN, R.W.: Int. J. Cancer *1:* 257 (1966).

BANGHAM, A.D. and HORNE, R.W.: J. molec.Biol. *8:* 660 (1964).

BANGHAM, A.D. and PETHICA, B.A.: Proc. Roy. Phys. Soc., Edinb. *28:* 43 (1960).

BARKER, S.A.; STACEY, M. and TIPPER, D.J.: Nature *184:* 68 (1959).

BARLAND, P. and SCHROEDER, E.A.: J. Cell Biol. *45:* 662 (1970).

BARSKI, G. and BELEHRADEK, J.: Exp. Cell Res. *37:* 464 (1965).

BARTH, R.F.; ESPMARK, J.A. and FAGRAEUS, A.: J. Immunol. *98:* 888 (1967).

BENNETT, M.V.L.: Ann. N.Y. Acad. Sci. *137:* 509 (1966).

BENNETT, M. V. L. and TRINKAUS, J. P.: J. Cell Biol. *37:* 621 (1970).

BEN-OR, S.; EISENBERG, S. and DOLJANSKI, F.: Nature *188:* 1200 (1960).

Blaisie, J. K.; Dewey, M. M.; Blaurock, A. E. and Worthington, C. R.: J. molec. Biol. *14:* 143 (1965).
Bloom, B.R. and Bennett, B.: Science *153:* 80 (1966).
Branton, D.: Proc. nat. Acad. Sci. U.S. *55:* 1048 (1966).
Brenner, S. and Horne, R. W.: Biochim. biophys. Acta *34:* 103 (1959).
Bullivant, S. and Loewenstein, W.R.: J. Cell Biol. *37:* 621 (1968).
Burdick, M.L.: J. exp. Zool. *167:* 1 (1968).
Burger, M.M.: Proc. nat. Acad. Sci. U.S. *62:* 994 (1969).
Burger, M.M. and Goldberg, A.R.: Proc. nat. Acad. Sci. U.S. *57:* 359 (1967).
Carruthers, S. and Suntzeff, V.: Science *19:* 245 (1944).
Carter, S.B.: Nature *208:* 1163 (1965).
Cerottini, J.C. and Brunner, K.T.: Immunology *13:* 395 (1967).
Chapman, D. and Wallach, D.F.H.: In D. Chapman, Biological membranes, physical fact and function, p. 233 (Academic Press, New York / London 1968).
Chessin, L.N.; Bramson, S.; Kuhns, W.J. and Hirschhorn, K.: Blood *25:* 944 (1965).
Coman, D.R.: Cancer Res. *4:* 625 (1944).
Coman, D.R.: Cancer Res. *13:* 397 (1953).
Cook, G.M.W.: Biolog. Rev. *43:* 363 (1968).
Cook, G.M.W. and Jacobson, W.: Biochem. J. *107:* 549 (1968).
Cook, G.M.W.; Heard, D.H. and Seaman, G.V.F.: Nature *188:* 1011 (1960).
Cook, G.M.W.; Heard, D.H. and Seaman, G.V.F.: Nature *191:* 44 (1961).
Coombs, R.R.A.; Daniel, M.R.; Gurner, B.W. and Kelus, A.: Int. Arch. Allergy *19:* 210 (1961).
Cormack, D.: Cancer Res. *30:* 1459 (1970).
Cunningham, W.P. and Crane, F.L.: Exp. Cell Res. *44:* 31 (1964).
Currie, G.A. and Bagshawe, K.D.: Brit. J. Cancer *23:* 141 (1969).
Curtis, A.S.G.: J. nat. Cancer Inst. *26:* 253 (1961).
Curtis, A.S.G.: In The cell surface: its molecular role in morphogenesis (Logos Press / Academic Press, London / New York 1967).
Danielli, J.F. and Davson, H.J.: J. cell. comp. Physiol. *5:* 495 (1934–35).
da Silva, P.P. and Branton, D.: J. Cell Biol. *45:* 598 (1970).
Defendi, V.; Lehman, J. and Kraemer, P.: Virology *19:* 592 (1963).
Defendi, V.; Lehman, J. and Kaemer, P.: Virology *19:* 592 (1963).
Gier, J. de and Deenen, L.L. van: Biochim. biophys. Acta *84:* 294 (1964).
Deutsch, D.: Z. Physiol. Chem. *136:* 353 (1928).
Eagle, H.; Levine, E.M. and Koprowski, H.: Nature *220:* 266 (1968).
Eagle, H.; Foley, G.E.; Koprowski, H.; Lazarus, H.; Levine, E.M. and Adams, R.A.: J. exp. Med. *131:* 863 (1970).
Easty, G.C. and Mercer, E.H.: Cancer Res. *20:* 1608 (1960).
Easty, G.C. and Easty, D.M.: Nature *199:* 1104 (1963).
Easty, G.C.: In E.J. Ambrose and F.J.C. Roe The biology of cancer, p. 78 (Van Nostrand, London 1966).
Easty, G.C.; Easty, D.M. and Tchao, R.: Europ. J. Cancer *5:* 287 (1969a).
Easty, G.C.; Easty, D.M. and Tchao, R.: Europ. J. Cancer *5:* 297 (1969b).
Emmelot, P.; Bos, C.J.; Benedetti, E.L. and Rumke, P.: Biochim. biophys. Acta *90:* 126 (1964).
Fagraeus, A. and Jonsson, J.: Immunology *18:* 413 (1970).
Farquhar, M.G. and Palade, G.E.: J. Cell Biol. *17:* 375 (1963).

FINEAN, J.B. and RUMSBY, M.G.: Nature *197:* 1326 (1963).
FLAXMAN, B.A.; REVEL, J.P. and HAY, E.D.: Exp. Cell Res. *58:* 438 (1969).
FLEISCHER, S.; ROUSER, G.; CASU, A. and KRITCHEVSKY, G.: J. Lipid Res. *8:* 170 (1967).
FOGEL, M. and SACHS, L.: Exp. Cell Res. *34:* 448 (1964).
FORRESTER, J.A.: In E.J. AMBROSE Cell electrophoresis, p. 115 (Churchill, London 1965).
FORRESTER, J.A.: In G.E.W. WOLSTENHOLME and J. KNIGHT Homeostatic regulators, Ciba Found. Symp., p. 230 (Churchill, London 1969).
FORRESTER, J.A.; AMBROSE, E.J. and MACPHERSON, J.A.: Nature *196:* 1068 (1962).
FORRESTER, J.A.; AMBROSE, E.J. and STOKER, M.G.P.: Nature *201:* 945 (1964).
FRANKS, D.: Biol. Rev. *43:* 17 (1968).
FRIEDMANN, T.; SEEGMILLER, J.E. and SUBAK-SHARPE, J.H.: Nature *220:* 272 (1968).
FRYE, L.D. and EDIDIN, M.: J. Cell Sci. *7:* 319 (1970).
FUHRMANN, G.F.: In E.J. AMBROSE, Cell electrophoresis, p. 92 (Churchill, London 1965).
FURCHGOTT, R.F.: Cold Spr. Harb. Symp. quant. Biol. *8:* 224 (1940).
FURSHPAN, E.J. and POTTER, D.D.: In A.A. MOSCONA and A. MONROY Current topics in developmental biology, vol. 3. p. 95 (Academic Press, New York 1968).
GASIC, G. and GASIC, T.: Nature *196:* 170 (1962a).
GASIC, G. and GASIC, T.: Proc. nat. Acad. Sci. U.S. *48:* 1172 (1962b).
GASIC, G. and BERWICK, L.: J. Cell Biol. *19:* 223 (1963).
GASIC, G.; GASIC, T. and STEWART, C.: Proc. nat. Acad. Sci. U.S. *61:* 46 (1968).
GESNER, B.M. and GINSBURG, V.: Proc. nat. Acad. Sci. U.S. *52:* 570 (1964).
GESNER, B.M. and WOODRUFF, J.: J. clin. Invest. *64:* abstr. 1134 (1967).
GIER, J. DE and DEENEN, L.L. VAN: Biochim. biophys. Acta *84:* 294 (1964).
GILDEN, R.V.; CARP, R.I.; TAGUCHI, F. and DEFENDI, V.: Proc. nat. Acad. Sci. U.S. *53:* 648 (1965).
GINGELL, D. and PALMER, J.F.: Nature *217:* 98 (1968).
GLOBERSON, A. and FELDMAN, M.: J. nat. Cancer Inst. *32:* 1229 (1964).
GOLD, P.: Cancer *20:* 1663 (1967).
GOLD, P. and FREEDMAN, S.O.: J. exp. Med. *121:* 439 (1965a).
GOLD, P. and FREEDMAN, S.O.: J. exp. Med. *122:* 467 (1965b).
GOLD, P.; GOLD, M. and FREEDMAN, S.O.: Cancer Res. *28:* 1331 (1968).
GOODENOUGH, D.A. and REVEL, J.P.: J. Cell Biol. *45:* 272 (1970).
GOUDIE, R.B. and MCCALLUM, H.M.: Lancet *ii;* 1035 (1963).
GOWANS, J.L. and KNIGHT, E.J.: Proc. Roy. Soc. B. *159:* 257 (1964).
GRÖSCHEL-STEWART, U.; JONES, B.M. and KEMP, R.B.: Nature *227:* 280 (1970).
HABEL, K.: Virology *18:* 553 (1962).
HABEL, K.: In F.J. DIXON and H.G. KUNKEL Advances in immunology, vol. 10, p. 220 (Academic Press, New York 1969).
HAKOMORI, S. and MURAKAMI, W.T.: Proc. nat. Acad. Sci. U.S. *59:* 254 (1968).
HAKOMORI, S.; TEATHER, C. and ANDREWS, H.: Biochem. biophys. Res. Commun. *33:* 563 (1968).
HARRIS, A.: Amer. Zoologist *10*/3 Abstr. 324 (1970).
HAUGHTON, G.: Science *147:* 506 (1965).
HAYDON, D.H. and TAYLOR, J.: J. theoret. Biol. *4:* 281 (1963).
HÄYRY, P. and DEFENDI, V.: Virology *41:* 22 (1970).
HELLSTRÖM, I.E. and SJÖGREN, H.O.: J. exp. Med. *125:* 1105 (1967).
HICKIE, R.A. and KALANT, H.: Cancer Res. *27:* 1053 (1967).

HICKS, R.M. and KETTERER, B.: Proc. Roy. Microsc. Soc. *4:* 118 (1969).
HICKS, R.M. and KETTERER, B.: J. Cell Biol. *45:* 542 (1970).
HIRSCH, H.M.: Experientia *14:* 269 (1958).
HOLLEY, R.W. and KIERNAN, J.A.: Proc. nat. Acad. Sci. U.S. *60:* 300 (1968).
HOLLINSHEAD, A.; GLEW, D.; BUNNAG, B.; GOLD, P. and HERBERMAN, R.: Lancet *i:* 1191 (1970).
HOLLYFIELD, J.G. and ADLER, R.: Exp. Cell Res. *59:* 76 (1970).
INBAR, M. and SACHS, L.: Nature *223:* 710 (1969).
JAMAKOSMANOVIC, A. and LOEWENSTEIN, W.R.: J. Cell Biol. *38:* 556 (1968).
JERNE, N.K.: Cold Spr. Harb. Symp. quant. Biol. *32:* 591 (1967).
JONSSON, N. and SJÖGREN, H.O.: J. exp. Med. *123:* 487 (1966).
JORPES, E.: Biochem. J. *26:* 1488 (1932).
KANNO, Y. and LOEWENSTEIN, W.R.: Nature *212:* 629 (1966).
KANNO, Y and MATSUI, Y.: Nature *218:* 775 (1968).
KAWASAKI, H.: Tohoku J. exp. Med. *68:* 119 (1958).
KEMP, R.B.: J. Cell Sci. *6:* 751 (1970).
KINSEY, D.L.: Cancer *13:* 674 (1960).
KLEIN, G.; SJÖGREN, H.O.; KLEIN, E. and HELLSTRÖM, K.E.: Cancer Res. *20:* 1561 (1960).
KLEIN, G.; SJÖGREN, H.O. and KLEIN, E.: Cancer Res. *23:* 84 (1963).
KLENK, E. and UHLENBRUCK, G.: Hoppe-Seylers Z. Physiol. Chem. *319:* 151 (1960).
KRUSE, P.F. and MIEDEMA, E.: J. Cell Biol. *27:* 273 (1965).
LANGLEY, O.K. and AMBROSE, E.J.: Nature *204:* 53 (1964).
LANGLEY, O.K. and AMBROSE, E.J.: Biochem. J. *102:* 367 (1967).
LENARD, J.: Biochemistry *9:* 1129 (1970).
LENARD, J. and SINGER, S.J.: Proc. nat. Acad. Sci. U.S. *56:* 1828 (1966).
LEVINE, E.M.; BECKER, Y.; BOONE, C.W. and EAGLE, H.: Proc. nat. Acad. Sci. U.S. *53:* 350 (1965).
LOEWENSTEIN, W.R.: Ann. N.Y. Acad. Sci. *137:* 441 (1966).
LOEWENSTEIN, W.R.: Develop. Biol. *15:* 503 (1967).
LOEWENSTEIN, W.R. and KANNO, Y.: J. Cell Biol. *33:* 225 (1967).
LOEWENSTEIN, W.R. and PENN, R.D.: J. Cell Biol. *33:* 235 (1967).
LUCY, J.: In D. CHAPMAN Biological membranes, physical fact and function, p. 233 (Academic Press, London/New York 1968).
LUCY, J.A. and GLAUERT, A.M.: J. molec. Biol. *8:* 727 (1964).
MACMURDO, H.L. and ZALIK, S.E.: Experientia *26:* 406 (1970).
MADDY, A.H.: Biochim. biophys. Acta *88:* 448 (1964).
MADDY, A.H.: Int. Rev. Cytol. *20:* 1 (1966).
MADDY, A.H. and MALCOLM, B.R.: Science *150:* 1616 (1965).
MALMGREN, R.A.; TAKEMOTO, K.K. and CARREY, P.G.: J. nat. Cancer Inst. *40:* 263 (1968).
MARKHAM, R.; HITCHBORN, J.H.; HILLS, G.J. and FREY, S.: Virology *22:* 342 (1964).
MARTINEZ-PALOMO, A.: Lab. Invest. *22:* 605 (1970).
MARTINEZ-PALOMO, A.; BROILOVSKY, C. and BERNHARD, W.: Cancer Res. *29:* 925 (1969).
MAYHEW, E. and O'GRADY, E.A.: Nature *207:* 86 (1965).
MAYHEW, E. and WEISS, L.: Exp. Cell Res. *50:* 441 (1968).
MCCUTCHEON, M.; COMAN, D.R. and FONTAINE, B.M.: Cancer *1:* 460 (1948).
MCLAREN, A.D.: Science *125:* 697 (1957).
MCNUTT, N.S. and WEINSTEIN, R.S.: Science *165:* 597 (1969).

MERCER, E.H.: In R.W. RAVEN Progress in cancer, p. 66 (Butterworth, London 1963).
MEYER, D.B.: Develop. Biol. *10:* 154 (1964).
MILLER, F.: J. biophys. biochem. Cytol. *8:* 689 (1960).
MITCHISON, N.A.: Cold Spr. Harb. Symp. quant. Biol. *32:* 431 (1967).
MOOR, H.: Balzers Rep. *9:* 1 (1966).
MOOR, H. and MÜHLETHALER, K.: J. biophys. biochem. Cytol. *17:* 609 (1963).
MORA, P.T.; BRADY, R.O.; BRADLEY, R.M. and MCFARLAND, V.W.: Proc. nat. Acad. Sci. U.S. *63:* 1290 (1969).
MORGAN, W.T.J. and WATKINS, W.M.: Brit. med. Bull. *15:* 109 (1959).
MORTON, J.A.: Brit. J. Haematol. *8:* 134 (1962).
MOSCOVITCH, M. and CALVIN, M.: Exp. Cell Res. *3:* 33 (1952).
MUIR, A.R.: J. Anat., Lond. *101:* 239 (1967).
NAIRN, R.C.; FOTHERGILL, J.E.; MCENTEGART, M.G. and RICHMOND, H.G.: Brit. med. J. *i:* 1335 (1962).
NELSON, G.J.: J. Lipid Res. *8:* 374 (1967).
NOVIKOFF, A.B.: Cancer Res. *17:* 1010 (1957).
OETTGEN, H.F.; OLD, L.J.; MCLEAN, E.D.; BLOOM, B.R. and BENNETT, B.: In O. WESTPHAL, H. E. BOCK and E. GRANDMAN Current problems in immunology, Bayer Symp. 1, p. 153 (Springer, Berlin 1969).
OHTA, N.; PARDEE, A.B.; MCAUSLAN, B.R. and BURGER, M.M.: Biochim. biophys. Acta *158:* 98 (1968).
OLD, L.J.; BOYSE, E.A.; CLARK, D.A. and CARSWELL, E.: Ann. N.Y. Acad. Sci. *101:* 80 (1962).
O'NEILL, C.H.: J. Cell Sci. *3:* 405 (1968).
ONODERA, K. and SHERIDAN, R.: J. Cell Sci. *7:* 337 (1970).
PARASKEVAS, F.; SHO-TONE, L. and ISRAELS, L.G.: Nature *227:* 395 (1970).
PATINKIN, D.; SCHLESINGER, M. and DOLJANSKI, F.: Cancer Res. *30:* 489 (1970a).
PATINKIN, D.; ZARITSKY, A. and DOLJANSKI, F. Cancer Res. *30:* 498 (1970b).
PETERSON, J.A. and RUBIN, H.: Exp. Cell Res. *58:* 365 (1969).
PETHICA, B.A.: Exp. Cell Res., Suppl. *8:* 123 (1961).
POLLACK, R.E.; GREEN, H. and TODARO, G.J.: Proc. nat. Acad. Sci. U.S. *60:* 126 (1968).
PONTON, J. and MACINTYRE, E.H.: J. Cell Sci. *3:* 603 (1968).
POPE, J.H. and ROWE, W.P.: J. exp. Med. *120:* 121 (1964).
PORTER, K.R. and BRUNI, C.: Cancer Res. *19:* 997 (1959).
POTTER, D.D.; FURSHPAN, E.J. and LENNOX, E.S.: Proc. nat. Acad. Sci. U.S. *55:* 328 (1966).
POULIK, M. D. and LAUF, P. K.: Clin. exp. Immunol. *4:* 165 (1969).
PREHN, R. T.: Cancer Res. *20:* 1614 (1960).
PREHN, R. T.: Fed. Proc. *24:* 1018 (1965).
PREHN, R. T. and MAIN, J. M.: J. nat. Cancer Inst. *18:* 769 (1957).
PURDOM, L.; AMBROSE, E. J. and KLEIN, G.: Nature *181:* 1586 (1958).
RABINOWITZ, Z. and SACHS, L.: Nature *220:* 1203 (1968).
RABINOWITZ, Z. and SACHS, L.: Virology *38:* 336 (1969).
RABINOWITZ, Z. and SACHS, L.: Virology *40:* 193 (1970).
RAMBOURG, A. and LEBLOND, C.P.: J. Cell Biol. *32:* 27 (1967).
RAPPORT, M.M.: Ann. N.Y. Acad. Sci. *159:* 446 (1969).
RAPPORT, M.M.; GRAF, L.; SKIPSKI, V.P. and ALONZO, N.F.: Cancer *12:* 438 (1959).
RASMUSSEN, H. and TENENHOUSE, A.: Proc. nat. Acad. Sci. U.S. *59:* 1364 (1968).
REVEL, J. P.: Proc. 26th Meet. Electron Microscopy Soc. of America, p. 40 (1968).

REVEL, J. P. and KARNOVSKY, M. J.: J. cell Biol. *33:* C7 (1967).

RÉVÉSZ, L.: Cancer Res. *20:* 443 (1960).

ROBERTSON, D. M. and WILLIAMS, D. C.: Nature *227:* 259 (1969).

ROBERTSON, J.D.: Biophys. biochem. cytol. *3:* 1023 (1957).

ROSENBERG, S.A. and GUIDOTTI, G.: J. biol. Chem. *243:* 1985 (1968).

ROSENBERG, S.A. and GUIDOTTI, G.: J. biol. Chem. *244:* 5118 (1969).

ROUSER, G.; NELSON, G.J.; FLEISCHER, S. and SIMON, G.: In D. CHAPMAN Biological membranes, physical fact and function, p. 5 (Academic Press, London / New York 1968).

RUHENSTROTH-BAUER, G.: In E. J. AMBROSE Cell electrophoresis p. 66 (Churchill, London 1965).

SACHS, L.: Nature *207:* 1272 (1965).

SACHS, L.: Curr. Topics develop. Biol. *2:* 129 (1967).

SAKAGAMI, T.; MINARI, O. and ORII, T.: Biochim. biophys. Acta *98:* 111 (1965).

SAKAI, I.: Nagoya med. J. *13:* 51 (1967).

SEIFERT, H. and UHLENBRUCK, G.: Naturwissenschaften *52:* 190 (1965).

SHERIDAN, J.D.: J. Cell Biol. *31:* C.1 (1966).

SHERIDAN, J.D.: J. Cell Biol. *45:* 91 (1970).

SIMON, D.: C. R. Acad. Sci. *244:* 1541 (1957a).

SIMON, D.: C. R. Soc. Biol. *151:* 1010 (1957b).

SIMON-REUSS, I.; COOK, G.M.W.; SEAMAN, G.V.F. and HEARD, D.H.: Cancer Res. *24:* 2038 (1964).

SJÖGREN, H. O.: J. nat. Cancer Inst. *32:* 361 (1964a).

SJÖGREN, H.O.: J. nat. Cancer Inst. *32:* 661 (1964b).

SJÖSTRAND, F.S. and ELFVIN, L-G.: J. Ultrastruct. Res. *10:* 263 (1964).

SLETTENMARK, B. and KLEIN, E.: Cancer Res. *22:* 947 (1962).

SNEATH, J.S. and SNEATH, P.H.A.: Brit. med. Bull. *15:* 154 (1959).

SPRINGER, G.F.; NAGAI, Y. and TEGTMEYER, H.: Biochemistry *5:* 3254 (1966).

STECK, T.L. and WALLACH, D.F.H.: In H. BUSCH Methods in cancer research, vol. V, p. 93 (Academic Press, New York 1970).

STEEVES, R.A.: Cancer Res. *28:* 338 (1968).

STEIN, W. D. and DANIELLI, J.F.: Disc Farad. Soc. *21:* 238 (1956).

STEINBERG, M.S.: In M. LOCKE Cellular membranes in development, p. 321 (Academic Press, New York 1964).

STOKER, M. G. P.: Virology *24:* 165 (1964).

STOKER, M.G.P.: J. Cell Sci. *2:* 293 (1967).

STOKER, M.G.P.; SHEARER, M. and O'NEILL, C.: J. Cell Sci. *1:* 297 (1966).

STOKER, M.; O'NEILL, C.; BERRYMAN, S. and WAXMAN, V.: Int. J. Cancer *3:* 683 (1968).

SUBAK-SHARPE, J.H.: Exp. Cell Res. *38:* 106 (1965).

SUBAK-SHARPE, J.H.: In G.E.W. WOLSTENHOLME and J. KNIGHT Homeostatic regulators. Ciba Found. Symp., p. 276 (Churchill, London 1969).

SUBAK-SHARPE, J.H.; BÜRK, R.R. and PITTS, J.D.: Heredity *21:* 342 (1966).

SUBAK-SHARPE, J.H.; BÜRK, R.R. and PITTS, J.D.: J. Cell Sci. *4:* 353 (1969).

SUEMASU, K.; KATAGIRI, M.; SHIMOSATA, Y.; MIKUNI, M. and ISHIKAWA, S.: Gann. *61:* 7 (1970).

SYLVÉN, B.: Cancer Res. *25:* 458 (1965).

SYLVÉN, B.: Europ. J. Cancer *4:* 463 (1968).

SYLVÉN, B. and MALMGREN, H.: Acta radiol. Suppl. *154:* (1957).

TAKEUCHI, M. and TERAYAMA, H.: Exp. Cell Res. *40:* 32 (1965).

TATARINOV, J.: Vopr. Med. Khim. *10:* 90 (1964).
TEVETHIA, S.S.; KATZ, M. and RAPP, F.: Proc. Soc. exp. Biol. Med. *119:* 896 (1965).
TILLACK, T.W. and MARCHESI, V.T.: J. Cell Biol. *45:* 649 (1970).
TJERNBERG, B. and ZAJICEK, J.: Acta cytol. *9:* 197 (1965).
TODARO, G.J. and GREEN, H.: J. Cell Biol. *17:* 299 (1963).
TODARO, G.J.; LAZAR, G. and GREEN, H.: J. cell. comp. Physiol. *66:* 325 (1965).
VASSAR, P. S.: Lab. Invest. *12:* 1072 (1963).
VEERKAMP, J.H.; MULDER, I. and DEENEN, L.L.M. VAN: Z. Krebsforsch. *64:* 137 (1961).
WALBORG, E.F.; LANTZ, R.F. and WRAY, V.P.: Cancer Res. *29:* 2034 (1969).
WALLACH, D.F.H.: In W. ARBER Current topics in microbiology and immunology, p. 152 (Springer, Berlin 1969).
WALLACH, D.F.H. and EYLAR, E.H.: Biochim. biophys. Acta *52:* 594 (1961).
WALLACH, D.F.H. and GORDON, A.: Fed. Proc. *21:* 1263 (1968).
WALLACH, D.F.H. and ZAHLER, D.H.: Proc. nat. Acad. Sci. U.S. *56:* 1552 (1966).
WARREN, R.C. and HICKS, R.M.: Nature *227:* 280 (1970).
WARREN, L.; GLICK, M.C. and NASS, M.K.: J. Cell Physiol. *68:* 269 (1966).
WEILER, E.: Z. Naturforsch. *76:* 237 (1952).
WEISS, L.: In The cell periphery metastasis and other contact phenomena (North–Holland Publishing Co., Amsterdam 1967a).
WEISS, L.: J. Cell Biol. *33:* 341 (1967b).
WEISS, L.: In Biological properties of the mammalian surface membrane. Wistar Inst. Symp. Monogr., vol. 8, p. 73 (1968).
WEISS, L. and HOLYOKE, E.D.: J. nat. Cancer Inst. *43:* 1045 (1969).
WEISS, L. and MAYHEW, E.: J. Cell Physiol. *68:* 345 (1966).
WEISS, L. and MAYHEW, E.: J. Cell Physiol. *69:* 281 (1967).
WEISS, L. and MAYHEW, E.: Int. J. Cancer *4:* 626 (1969).
WEISS, L. and SINKS, L. F.: Cancer Res. *30:* 90 (1970).
WEISS, P. and ANDRES, G.: J. exp. Zool. *121:* 449 (1952).
WHITTAKER, V.P.: Ann. N.Y. Acad. Sci. *137:* 982 (1966).
WILLIS, R.A.: In The spread of tumours in the human body (Butterworth, London 1952).
WOLPERT, L. and GINGELL, D.: In G.E.W. WOLSTENHOLME and K. KNIGHT Homeostatic regulators. Ciba Found. Symp., p. 241 (Churchill, London 1969).
WOODIN, A.M. and WIENECKE, A.A.: Biochem. J. *99:* 493 (1966).
WOODRUFF, J. and GESNER, B. M.: Science *161:* 176 (1968).
WOODWARD, D. O. and MUNKRES, K. D.: Proc. nat. Acad. Sci. U.S. *55:* 872 (1966).
WU, H.; MEEZAN, E.; BLACK, P. H. and ROBBINS, P. W.: Biochemistry *8:* 2509 (1969).
YEH, J. and FISHER, H.W.: J. Cell Biol. *40:* 382 (1969).

Author's address: Dr. GERALD C. EASTY, Chester Beatty Research Institute, Institute of Cancer Research, Royal Cancer Hospital, Fulham Road, *London, S.W.3* (England)

Neoplasia and Cell Differentiation, pp. 234–278
(Karger, Basel 1974)

The Role of the Cell Surface in Embryonic Morphogenesis

M.F. COLLINS

Department of Anatomy, The University of Connecticut Health Center, Farmington, Conn.

Contents

I. Tissue Movements in Embryonic Morphogenesis

Vertebrate embryos accomplish profound transformations during development. Beginning as single cells, apparently simple, they progressively reveal a complexity of structure and function scarcely suggested by the modest zygote. Of the numerous processes and important events which take place as embryos mature, none is more striking than the morphogenetic changes which establish the animal's anatomy. The architecture of organ and tissue is the basis of the mature organism's identity and function. Those events which determine the precise positioning of embryonic cells and tissues are critical; the consequences of their perturbation are severe. Embryonic morphology clearly stems from the properties of the tissues composing the embryo. There is not only variety in the morphology of tissues, but also in the mechanisms underlying morphogenesis. Some tissues, or even individual cells, clearly change their associations with one another and move through the embryo as it develops. In other cases, morphogenesis is accomplished by deformation or expansion of tissue rather than actual translocation. Often the formation of an organ, and always the formation of an embryo, involves combinations of several processes. The literature contains many examples of tissue movements and alterations which contribute to morphogenesis. Brief consideration of a few will provide a context for subsequent material. One will find more complete coverage by Trinkaus [1969] and by DeHaan and Ursprung [1965].

A. Some Examples

Although the migration of individual cells for significant distances through the embryonic body is not typical, it does play several important roles in vertebrate morphogenesis. Primordial germ cells arise from extra-gonadal or even extra-embryonic locations and migrate to establish permanent residence in the prospective gonad. Additionally, numerous tissues derived from the neural crest are established only after neural crest cells emigrate to new locations. Several nervous, endocrine and even skeletal tissues owe their existence to the ability of their cellular precursors to detach themselves from the closing neural tube and find their way along certain paths to appropriate sites for differentiation [DuShane, 1935; Detwiler, 1936].

Development of the vertebrate heart illustrates several morphogenetic processes of general importance [DEHAAN, 1968]. This includes a well-documented case of small groups or clusters of cells migrating to establish an organ. These arise on both the left and right sides of the early embryo and migrate toward the midline where they are incorporated into the tubular heart [DEHAAN, 1965].

Gastrulation is a fundamental process in embryonic development. It establishes the three-layered structure of early embryos and thus roughs out the embryonic form which will follow. This process has been studied extensively in sea urchin embryos by GUSTAFSON and KINNANDER [1956, 1960]. Their films of numerous gastrulating sea urchin embryos indicate that the initial stages of gastrulation result from cell and tissue deformations at the vegetal pole of the blastula where invagination begins. They also demonstrated the important role in later gastrulation of secondary mesenchyme cells located at the tip of the invaginating archenteron. Those cells extend filopodia which contact and adhere to particular sites on the inner blastocoele surface. Once anchored, the filopodia contract, pulling in the archenteron. Many filopodia attach selectively at and around the prospective larval mouth region. Thus they direct the invaginating archenteron tip to the stomodeum [GUSTAFSON, 1963].

Vertebrate gastrulation does not require such conspicuous cellular structures. The flask cells at amphibian blastopores are greatly deformed, but it is not clear that such deformation is the cause rather than a consequence of the gastrulation process [but see BAKER, 1965]. Vertebrate gastrulation has been examined most completely with amphibian embryos. Here prospective endoderm and mesoderm move steadily into the embryonic interior, filling the blastocoele. As they enter, or soon thereafter, endoderm and mesoderm separate and follow different courses. A mantle of mesoderm spreads down across the inner surface of the ectoderm which remains as the embryo's outer layer. Simultaneously, a trough of endoderm spreads dorsad along the inner surface of the mesodermal mantle. The endoderm finally forms a closed gut by fusion in the dorsal midline. The mesoderm insinuates itself between ectoderm and endoderm and fuses in the ventral midline. These tissue movements are well illustrated by BALINSKY [1970]. Similar spreading of tissues over other tissues has been demonstrated in work *in vitro* to be covered in Section II, D.

The development of the pronephros exemplifies morphogenesis combining tissue growth and movement. The formation of this and subsequent kidneys depends upon the pronephric duct which originates from the nephrogenic mesenchyme in the anterior part of the vertebrate embryo and connects with the cloaca near the posterior end. This organ grows back just ventral to the somites to the cloaca. HOLTFRETER [1944] showed that the pronephric duct follows a particular pathway; various operations indicated that the pronephric duct favors particular tissue associations as it elongates.

Although cell and tissue movements are vital to normal development, they are by no means the only basis of morphogenesis. Neurulation and the formation of the olfactory, lens and ear placodes and vesicles involve cellular shape change more than cellular movement. Numerous other processes could be mentioned and sound investigations into their mechanisms could be reviewed. The object of this section, however, is simply to suggest the wide range of roles played by cell and tissue growth, deformation and movement in embryos.

B. Cellular Basis of Morphogenetic Movements

It is clear from many examples that the behavior of cells and tissues contributes substantially to embryonic morphogenesis. That is a most important point which has been established by experiments with embryonic components isolated from the intact embryo.

That analysis began with the classic paper of HOLTFRETER [1939]. He examined the interactions of individual tissues removed from amphibian embryos and cultured in balanced salt solutions. Toward one another these tissues exhibited what HOLTFRETER called positive and negative affinities. In the intact embryo, prospective neural ectoderm and prospective epidermal ectoderm from early neurulae are continuous with one another. Explanted into balanced salt solution they exhibit a positive affinity; that is, they adhere firmly to each other – the prospective neural tissue tends to be submerged in or covered by the prospective epidermis. After those tissues had been in culture some time the positive affinity gave way to negative affinity. The neural tissue was expelled from the interior of the epidermal tissue and the two masses of tissue then grew side by side. The relevance of this behavior to normal morphogenesis becomes clear as we consider that the change of affinity occurred at a time corresponding to a significant morphogenetic event. That event is the separation of neural and epidermal ectoderm from one another at the time of neural tube closure.

In vitro observations of this type have been widely interpreted as indicating that changes in tissues which constitute the embryo are in large measure responsible for morphogenetic events in the intact embryo.

Similar events have been observed at the cellular level. Experiments to be discussed in the next section demonstrate that groups of single embryonic cells, dissociated from one another in various ways, are capable of establishing complex patterns which reflect normal morphogenetic and histogenetic relationships. Such work has shifted the emphasis in morphogenesis to the cellular level, where substantial progress has been made.

II. Sorting Out: A Model for Morphogenesis

A. Correspondence to Normal Morphology

1. Sponge and Amphibians

Numerous 'sorting out' experiments have been performed with dissociated cells of different types in order to determine how those cells will interact with one another in tissue culture, and to learn what sort of organization dissociated cells can establish. Perhaps the earliest work of this kind was done by H.V. WILSON [1907]. He disrupted marine sponges, let them reaggregate to form small masses of tissue and examined their interactions and development. The small reaggregates eventually became complete new organisms.

WILSON's work was technically simple. By forcing small pieces of sponge through fine silk bolting cloth, he was able to dissociate them into very tiny fragments and single cells. When the dissociated cells were allowed to settle onto the bottom of a dish containing sea water their reaggregation began. Cells moved over the surface of the dish, made contact with one another and reaggregated. Once the reaggregates had formed they settled down and reestablished normal sponge morphology. These results invite two interpretations. First, cells within the reaggregates may lose their original differentiated characteristics and then redifferentiate in accord with their new locations in the tissue mass. By this process they would form tissues with proper relationships to one another. This view was considered favorably by WILSON himself. GALTSOFF [1923] suggested a second interpretation: that the cells do not redifferentiate, but rather sort out from one another and consequently find proper locations within the tissue mass. GALTSOFF held this view on the basis of direct observation of dissociated cells. He stated that each type retains its distinctive characteristics and remains identifiable on the basis of nuclear structure or cytoplasmic inclusions. He suggested that with the exception of archeocytes, the potency of these dissociated sponge cells was very limited. It is quite difficult to tell one type of dissociated sponge cell from another with certainty, so even today we cannot say for sure which of these alternatives is correct.

Others have done similar experiments with amphibians, birds and mammals and the results are beyond doubt. The first unequivocal demonstration of sorting out was accomplished with amphibian cells. HOLTFRETER [1944] extended his tissue affinity studies by examining the behavior of embryonic amphibian cells which he had dissociated by means of a brief exposure to high pH. He thus dissociated blastopore lip tissue, composed of mesoderm and head endoderm, stirred it and restored the ambient pH to normal. The cells aggregated, apparently at random, and formed firm spheres which were organized into well separated tissues. The first step in that organization which he observed directly was the prompt and complete disappearance of pigmented mesodermal cells from the surface of the reaggregates. Thus, sorting out was established for amphibian embryos.

2. Birds and Mammals

Dissociation and culture procedures for amniote tissues are more complex than similar procedures for amphibian or invertebrate tissues. Often dissociation is accomplished by enzymatic treatment followed by mechanical agitation. Proteases, including pronase, papain, collagenase and especially trypsin, have been the most successful agents for cell dissociation [see L. WEISS, 1958; STEINBERG, 1967; BANKS *et al.*, 1970]. Nonenzymatic procedures have used the chelating agents ethylenediaminetetraacetic acid (EDTA) and tetraphenylboron [RAPPAPORT and HOWZE, 1966]. At least one tissue can be dissociated simply by depriving it of bivalent cations followed by mechanical agitation [COLLINS, 1966b].

Once cells have been dissociated, they must be cultured to demonstrate and study sorting out. Amphibian material is readily maintained in balanced salt solutions [see JONES and ELSDALE, 1963; JACOBSON, 1967]. For birds and mammals, composition of the medium, temperature control and other culture conditions are more exacting. Tolerances for sorting out experiments are relatively wide, however, and several media have proven satisfactory [see STEINBERG, 1967].

Sorting out can be examined in several ways. The use of a gyratory shaker to hold flasks containing cell suspensions is often used [MOSCONA, 1961]. This technique has several advantages. It can be used to examine cellular reaggregation as well as sorting out; it produces many reaggregates; and it does not require cellular motility for aggregation because cells are carried into contact with one another by the movement of the medium. It excludes many extraneous factors which could affect the experiment. CURTIS [1969] has conducted aggregation experiments in a Couette viscometer. His procedure has advantages in common with the gyratory shaker procedure and also provides laminar flow in the suspending medium, which results in small aggregates of uniform size and permits the estimation of the total number of collisions between cells per unit time.

Additional culture methods have been used. For example, reaggregates can be formed and maintained in hanging drop cultures [STEINBERG, 1962] or on solid surfaces [MOSCONA and MOSCONA, 1952]. Additionally, dissociated cells may be centrifuged into pellets, and then cultured on chorioallantoic membranes just as readily as in liquid culture medium [WEISS and TAYLOR, 1960].

MOSCONA [1957] used trypsin-dissociated cells reaggregated on glass to show that sorting out occurs among avian and mammalian cells as well as amphibian cells. He took advantage of the different staining properties of mouse and chick cells. Having determined that in some cases cellular segregation within reaggregates is based not upon species differences but rather upon tissue differences, he combined chondrogenic cells from chick embryos with mesonephric cells from mouse embryos. Since it was clear that these cells could not redifferentiate into cells of another species, the only possible basis for the cellular segregation observed in the reaggregates was the sorting out of cells from one another on the basis of their tissue type.

The sorting out and histogenesis of cellular reaggregates have now been studied extensively. Where only tissue is involved, histogenesis often can be detected. For example, cells from the neural retina establish 'rosettes' which represent an approximation of the tissue architecture of the normal neural retina [MOSCONA, 1961]. When cells of several types are included, the organization of reaggregates is more complex. HOLTFRETER described the organogenesis of complete nephric systems in reaggregates composed of mesoderm and endoderm [HOLTFRETER, 1944]. WEISS and TAYLOR [1960] described extensive organogenesis by liver, skin, and kidney tissue from chick embryos. One feature which all reaggregates composed of two different tissues share is the early establishment of internal and external phases within the reaggregate. First, cells destined to segregate internally disappear from the reaggregate surface [HOLTFRETER, 1944]. This is followed by coalescence of the internally segregating cells and the establishment of islands of those cells which may be distributed throughout the interior of an aggregate. The pattern established between two particular tissues is generally repeatable. Many tissues have been examined in this way [STEINBERG, 1964, 1970]. Any of several combinations is of interest to one concerned with the mechanism of sorting out. Those combinations of tissues which normally come into contact during development display a correspondence to *in vivo* morphology. In such cases, the cells always establish their normal spatial relationship. A good example of this was provided by MOSCONA and MOSCONA [1952] who examined reaggregates composed of chondrogenic and myogenic cells derived from chick embryo limb buds. The aggregates formed by cells of these two types consisted of an inner phase of chondrogenic cells with an outer covering of myogenic cells. This corresponds to normal limb structure in which skeletal elements are surrounded by muscle tissue. Examples of this and of more complex histogenesis indicate that the forces which drive sorting out reactions *in vitro* most probably play an important role in the establishment of tissue relationships *in vivo*. Thus, there is a strong correspondence of reaggregate morphology to normal embryonic morphology. Efforts have been made to study this sorting out process in order to shed light upon the mechanisms which underlie normal embryonic development.

B. Directed Migration

The term 'directed migration' has long been in the literature of morphogenesis. In 1944, HOLTFRETER used the term with reference to sorting out by amphibian tissues. He used it descriptively to indicate that cells assumed positions within reaggregates which made sense in terms of normal structure. HOLTFRETER apparently did not mean to imply any particular mechanism or sequence of events. The term now implies the migration of cells toward a particular 'objective' which could be either a specific target or simply the interior or exterior of an aggregate or cell population.

A fully unified treatment of directed migration would be difficult because it has been reported in various systems and several mechanisms for it have been suggested. *In vivo* systems are of the greatest interest, but we know least about what happens in the intact embryo. The term 'directed migration' could be applied to neural crest cell migrations in embryos, but little is known about the underlying mechanism. On the other hand, the best understood case that can be called directed migration is contact inhibition *in vitro* of fibroblasts, but it is not clear just how to apply those observations to developing animals.

Directed migration suggests two questions: 'What happens?' and 'How does it work?'

1. In vivo

The growth or movement of cells toward particular sites *in vivo* is best illustrated by the nervous system. Nerve fibers grow from cell bodies to their end organs. Some axons reach their proper points of innervation with great precision [SPERRY, 1965]. Neural crest cells accurately select their new locations *in vivo,* but observation of their movements has revealed little about how they are accomplished. Work by WESTON [1963] and WESTON and BUTLER [1966] indicated that the orientation of the neural tube determines the direction of migration of neural crest cells away from it. However, prospective somite and intersomitic mesenchyme influence the rate of neural crest cells migrating through them. Our understanding of this is inadequate.

Primordial germ cells migrate to the gonads in various ways. In some embryos they move through solid tissue; while most interesting, developmentally, these are unfortunately not amenable to analysis and remain baffling. In bird embryos the transmission of the primordial germ cells via the vascular route suggests not actual directed migration but rather a trapping or homing mechanism. Thus it exemplifies the movement of cells to a particular target region, although the migration is fairly passive. Selective lodging of embryonic cells by 'homing' is discussed in Section II, D.

2. In vitro

Although homing or trapping may be considered a type of directed migration, that term is usually applied to cases like cell sorting in which the direction of movement, rather than the site of final localization, appears to be influenced. As with intact embryos, little more than direct observation and speculation about mechanisms has been possible. In reaggregates containing tissues of two types, internally segregating cells sometime assume positions in the very centers of the reaggregates; in such reaggregates it is as if cells of one kind actively migrated toward the center and/or cells of the other kind actively migrated toward the surface. Such reaggregates, composed of concentric spheres, have suggested that directed migration played a part in the cellular segregation.

What mechanism might account for such results? One explanation, widely considered and with some indirect support, is chemotaxis. Chemotaxis is a phenomenon requiring considerable effort and care to demonstrate convincingly; nevertheless, there is no question it does occur in nature. For example, chemotaxis is important for fertilization of some ferns and mosses. Chemotaxis has also been demonstrated unequivocally with cellular slime molds which exhibit cellular aggregation and morphogenesis. Substantial progress has been made in its analysis [BONNER, 1947; SUSSMAN, 1966; GERISCH, 1968].

Chemotaxis has not been proven for any vertebrate embryonic tissue, but there is one example in which it appears to be involved. TWITTY and NIU [1954] examined migration of propigment cells derived from salamander neural crests when these cells were explanted into narrow bore capillary tubes filled with coelomic fluid. Except near the end of the capillary tube which opened into the balanced salt solution outside, cellular behavior was as follows. Single cells which had been explanted were nearly stationary. Pairs of cells which were explanted close together moved apart, often up to considerable distances. When several cells were explanted near one another, they tended to move apart and to space themselves evenly along the length of the capillary tube. TWITTY and NIU interpreted these results as indicating that the cells were responding to gradients of diffusible substances which they released into the coelomic fluid. This was probably a case of negative chemotaxis.

How might chemotaxis result in sorting out? The idea is generally stated as follows. Cellular metabolism results in the production of acidic products which the cells release into their surroundings. Since reaggregates contain no circulatory system to remove waste products swiftly, metabolic products may accumulate in greater concentrations in the interiors of reaggregates than near their surfaces, where they might readily diffuse away into the medium. Thus, gradients of hydrogen ion concentration may be established in reaggregates with lower pH's in the interiors and higher pH's at the surfaces. If cells could detect and respond to such pH gradients, they might migrate up the gradients toward the centers or down the gradients toward the peripheries of reaggregates. Work by P. WEISS and SCOTT [1963] indicated that various cells, including chick embryonic heart and liver cells, often used in sorting out experiments, can respond to pH gradients. They established steep gradients *in vitro* and examined the behavior of dissociated cells placed in the gradients. The cells were affected locally; that is, each part of a cell behaved in accordance with the pH of its immediate vicinity. High pH caused cell surface contractions while low pH caused local

Either extreme effectively immobilized the affected cell surface area and the cell was moved by unilateral traction of the unaffected side. WEISS and SCOTT therefore concluded that pH gradients could polarize cell motility. A gradient of 24 pH units per millimeter was necessary for a positive effect. The authors suggested that the requisite steepness would not exist in open culture conditions but was quite plausible for the micro-environments of contiguous cells. They also observed that the external asymmetries, which influenced the direction of cell movement, appeared to involve an inhibition of the rear rather than a stimulation of the front of the cell.

In spite of this evidence showing the plausibility of pH gradient influences upon the direction of cellular movement, directed migration is inadequate to account for much of what is known about cell sorting within reaggregates. The actual sequence of events during sorting out is at variance with what would be expected from directed migration. STEINBERG [1962b] and TRINKAUS and LENTZ [1964] did not find internally segregating cells herded into the centers of reaggregates but rather their progressive accumulation in small clusters which were apparently distributed at random throughout the interiors of reaggregates. The clusters accumulated more cells and fused with one another. In some cases all clusters in a reaggregate would fuse to form a single, centrally-located mass of cells such as might be expected to result from directed migration, but this was not usually observed. CURTIS [1967] pointed out the possibility of a system in which directed migration accounts for the final patterning of cells within a reaggregate but in which multiple foci of aggregation appear throughout the reaggregate. It has not been possible to show how such foci would develop. One cannot exclude with complete certainty this modified directed migration hypothesis, but it still does not offer the best explanation for sorting out.

Another shortcoming of this hypothesis appears when one considers the hierarchy of tissues established by STEINBERG [1970] on the basis of their sorting out behavior. That hierarchy will be discussed in Section II, D; the important point here is that cells of most types can segregate either internally or externally, depending upon the type of cell with which they are combined. It is most difficult to reconcile these data with directed migration.

Mechanical contacts influence the direction of cellular migration in tissue culture. This can be contact between migrating cells and their substrata or contact among cells. Studies of the outgrowth of nerve fibers and of fibroblasts from tissue explants have led to the concepts of contact guidance and contact inhibition of cell movement. When fragments of neural tissue are explanted, under proper conditions, nerve fibers will grow out away from the explant. WEISS has examined this phenomenon in considerable detail [see P. WEISS, 1955]. His evidence indicated that growing axons will follow various kinds of mechanical guidelines, such as oriented fibrillar proteins in their environment. Such elements can establish the pathways and thus direct migration of growing nerve fibers. This contact guidance certainly must help to establish embryonic nervous systems, although it is by no means sufficient to provide for the many precise neural connections which must be made [see SPERRY, 1965].

Contact inhibition influences the direction and extent of migration of whole cells in tissue culture. ABERCROMBIE and HEAYSMAN [1953, 1954] observed the outgrowth of cultured chick fibroblasts. The fact that these cells tend to move away from the main explant had been the subject of speculation for many years. Chemotaxis had been suspected. The movement of individual cells turned out, however, to be random except when they made contact with their fellows. When two cells touched, their movement ceased temporarily; they usually moved apart. Since cells moving toward the explant would be more likely, on a statistical basis, to encounter movement-inhibiting contacts than cells moving away from the explant, the population of fibroblasts showed net movement away from the central explant. ABERCROMBIE and AMBROSE [1958] asked what happened to fibroblasts which contacted one another in culture and examined their behavior by means of interference microscopy. They reported that migrating fibroblasts were highly active at their front ends which possessed broad 'ruffled' membranes. When two migrating cells met, activity ceased as if the ruffled membranes had been paralyzed. Eventually, new ruffled membranes developed at other sectors of the cells and the cells then moved apart. P. WEISS and SCOTT [1963] suggested that local decreases in pH where cells touch may immobilize ruffled membranes and thereby inhibit motility. ABERCROMBIE [1964] and CURTIS [1967] have discussed contact inhibition and its possible mechanisms. It has been suggested that contact inhibition could be the basis for sorting out in cellular reaggregates. That is not, however, the best explanation. Contact inhibition has only been described in two-dimensional systems of cells migrating across a substratum in tissue culture; there is no evidence that ruffled membranes exist in three-dimensional tissues. It seems unlikely that a mechanism which underlies the dispersal of cells in tissue culture would cause cells to accumulate and cohere specifically in mixed reaggregates.

C. Timing

One mechanism for sorting out of dissociated embryonic cells which has been proposed by CURTIS [1960, 1961, 1962] may be referred to as the timing hypothesis. This was developed when considerable emphasis was being placed upon specific adhesions. Many believed that cells of one type simply would not adhere to cells of another type, or at least that cells were capable of some sort of self-recognition such that they would adhere much more firmly to their own kind than to cells of another type. CURTIS pointed out that one need not assume the existence of any specific adhesive molecules or specific patterning of adhesive units on cell surfaces—'areal specificity'—to account for the segregation of different cells within aggregates. As an alternative he proposed a temporal specificity, involving a sequential change of cellular properties subsequent to dissociation which would control the final locations assumed by cells within reaggregates. He noted that adhesive specificity would not necessarily result in sorting out to yield the consistent tissue patterns which are observed in reaggregates. In other words, if cell A adheres to its own kind more firmly than to cell B and cell B adheres to its own kind more firmly than cell A, that does not imply that either A or B would take up an internal position within a mixed reaggregate, or that there would be any consistency of result from one experiment to the next.

In 1960 cellular surfaces were thought to be covered by protein molecules as in the Davson-Danielli cell membrane model. This surface layer of protein would be important in any intercellular interactions. CURTIS considered data on the physical properties of protein monolayers to infer the probable organization and behavior of protein molecules at cell surfaces. He reported that when the area per molecule falls within a certain range, the protein monolayers exhibit properties of non-Newtonian liquids; that is, their viscosity is a function of applied shear. Shear forces are applied to cell surfaces as the cells move past one another or over solid substrata. Depending upon the condition of the protein layer at the cell surface, a shearing force can either increase or decrease the area per molecule and consequently the cellular adhesiveness. Thus, a reduction of shear force at a cell's surface could result in increased cellular adhesiveness, which could then restrict the cell from further movement throughout the reaggregate. This was thought most likely to happen at the external surface of the reaggregate where the exposure to a liquid medium, in contrast to motile cells, would provide an area of reduced shear. A modification of this theory

[CURTIS, 1967] allowed for the possibility of trapping of cells within the interior of reaggregates as well.

CURTIS stressed that dissociation alters cells and that they exhibit changing properties when recovering from dissociation procedures. He proposed that cells which recover their adhesiveness and other normal characteristics most rapidly would be trapped at reaggregate surfaces, that they in turn would provide relatively shear-free surfaces to trap others of their own kind, and that this would lead to an accumulation of cells of one type at the reaggregate surface. The result of such activity would be several layers of the rapidly recovering cells at the reaggregate surface, with slowly recovering cells collected in central aggregates or dispersed in smaller clumps throughout the reaggregate, depending upon the proportions of different cells. CURTIS reported that the surfaces of cells dissociated from *Xenopus laevis* embryos exhibited the non-Newtonian viscosity necessary for trapping at aggregate surfaces.

CURTIS [1961] reported experiments with *Xenopus* embryos supporting the timing hypothesis. He first determined the pattern of segregation achieved by dissociated cells from gastrulae. These corresponded to the normal morphology of the embryo; endoderm was internal to mesoderm, which was in turn surrounded by ectoderm, with the latter tissue generally forming a complete surface of the reaggregate. To test the timing hypothesis CURTIS reasoned that it should be possible to confuse the segregation process by allowing the different tissues different intervals for recovery after dissociation. If recovery of adhesive or other properties after dissociation were a causal factor in sorting out, then by giving one tissue a 'head start' the pattern of segregation might be altered.

Gastrula endoderm was dissociated and explanted for intervals of 4 or 6 hours. Endoderm was chosen since it segregates internally, perhaps because of a slow recovery rate. The endodermal cells were in contact with one another on the bottom of the culture dish and had an opportunity to begin recovery early. After 4 to 6 hours, a cohesive mass of endodermal tissue had been formed. Freshly dissociated ectodermal and mesodermal cells were placed on top of this endoderm which was then folded up to surround them completely. This established artificially a mass composed of three tissues with abnormal relationships. Those aggregates were cultured for 16 hours, fixed and examined microscopically. The following tissue patterns were observed: when the addition of ectodermal and mesodermal cells had been delayed for 4 hours, mesodermal cells intermingled with the endoderm while the innermost portion of the aggregate was occupied strictly by ectodermal tissue; when the addition had been delayed 6 hours, neither ectoderm nor mesoderm penetrated the surrounding endoderm; these two tissues were covered by the endoderm. A mass composed of ectoderm surrounded by mesoderm which was surrounded by endoderm resulted. This was just the reverse of the controls where endoderm had been given no head start. CURTIS proposed that interference with the relative timing of recovery of these tissues had caused the aberrant pattern and that sorting out observed by others was the result of different recovery rates for different tissues. The interpretation is consistent with CURTIS' results. Another interpretation – that ectodermal and mesodermal cells were simply trapped within the cohesive endoderm – would account for these results as well. A second experiment was also reported. In that experiment, dissociated ectodermal and mesodermal cells were placed on top of a mass of endodermal cells which had been cultured for 4 hours. CURTIS reported that the newly added cells penetrated the interior of the endoderm. That result cannot be explained by increased cohesion of endodermal cells. One cannot, however, consider these results as good support for the timing hypothesis. *Xenopus laevis* appears to be a poor animal for testing the hypothesis. *X. laevis'* development is extremely rapid, with complete gastrulation requiring little more than 5 hours. Thus, normal developmental changes in cellular properties could be confused with changes reflecting recovery from dissociation. CURTIS [1961] pointed out that all tissues were taken from midgastrulae (NIEUWKOOP's [1946] Stage 12). Endodermal cells cultured 4 hours would be the same age as the neural fold stage; normal development, rather than recovery, could be responsible for considerable changes during that interval. Similar results with another amphibian species would be more telling.

CURTIS approached this problem similarly with different organisms in 1962. He used four species of sponge which exhibited three different reaggregation rates as judged by the time required to form round reaggregates at 16–17°C. They were *Microciona sanguinea* (3 h), *Halichondria panicea* (8 h), *Suberites ficus* (15 h) and *Hymeniacidon perleve* (15 h). The species which formed reaggregates more quickly apparently did so because they recovered their adhesiveness more rapidly after dissociation. When *Microciona, Halichondria,* and *Suberites* cells were dissociated, taken in pairs and started reaggregating simultaneously, they formed separate and distinct aggregates. Such results had meant adhesive specificity to others. In contrast, *Suberites* and *Hymeniacidon perleve* cells, when treated similarly, exhibited complete or nearly complete intermingling. These last two species exhibited the same rate of reaggregation when tested alone. CURTIS reasoned from the timing hypothesis that these interactions could be modified by changing the timing relationships. Giving the cells 'slower' to reaggregate a head start should permit them to associate with 'faster' cells in reaggregates. This was observed. For example, when *Microciona* cells were added to *Suberites* cells which had been in culture for 10 hours, they formed common reaggregates with extensive intermingling. In addition, *Hymeniacidon* and *Suberites* cells, which do not sort out if started reaggregating simultaneously, can be made to sort out by delaying the addition of cells of either species to the mixed culture.

The timing hypothesis has considerable predictive value, at least for reaggregates of amphibian or sponge cells. It is not, however, an adequate explanation for cell sorting, especially when the differential adhesion hypothesis (Section II, D) can account for a wider variety of experimental results. Both the time course of events and the final product of sorting out are not as expected from the timing hypothesis [see STEINBERG, 1964; TRINKAUS, 1969]. The timing hypothesis as presented in 1961 required that cells which more rapidly recover their adhesiveness would establish a continuous layer over the entire surface of a reaggregate and provide a new stationary surface for trapping of cells of their own kind, thus leading to first a 'herding' of the second type of cell toward the interior of the reaggregates and yielding a final configuration with the internally segregating tissue contained in a single, central mass surrounded by an even layer of the externally segregating tissue. This is not observed. However, a clarification and elaboration of the timing hypothesis has been presented by CURTIS [1967, p. 248]. For example, the occurrence of several small regions of the internal tissue in some reaggregates may be explained by a relatively small number of internally segregating cells. Thus, islands of these cells might be isolated from one another as the externally segregating cells rapidly become trapped and stabilized, forming a continuous phase of non-motile cells. That would prevent complete migration of internally segregating cells to a single site in the interior. Reduced proportions of internally segregating cells cause those cells to localize in smaller and more numerous foci within reaggregates. This has been observed frequently. It has been systematically examined and interpreted by STEINBERG [1962].

Two difficulties with the timing hypothesis – of which its author is aware [CURTIS, 1967] – are the following. First, it does not account for the preferential spreading of fragments of tissue of one type over fragments of tissue of another type. These are not sorting out experiments but are closely related to them since they lead to the same results probably by the same mechanism. Tissue spreading also represents better the movements of tissues *in vivo.* Second, regardless of the value of the timing hypothesis as a means of explaining cell sorting under experimental conditions, it is not likely to have bearing on normal development, since recovery from dissociation procedures plays no role in intact embryos. The timing hypothesis does serve to emphasize possible changing adhesive properties which may occur during normal development and trigger important morphogenetic movements.

D. Selective Adhesion

1. Specific Adhesion

Cellular adhesive properties have often been thought to function decisively not only in reaggregates but also in normal morphogenesis. Cells were thought to select their neighbors on the basis of adhesive properties. Adhesiveness could be specific so that cells capable of 'self recognition' would adhere either exclusively or much more firmly to cells of their own kind than to different cells. This specific adhesion mechanism was considered to be most likely by LOEB [1922], TYLER [1940, 1955], and P. WEISS [1941, 1947]. They suggested that such specific adhesion might be mediated by sterically specific molecules located at the cell surface, whose role was to bind cells together. This is discussed below.

One example of specific adhesion which results in the isolation of cells from two different species has been investigated by HUMPHREYS [1963] and MOSCONA [1963]. They have found that, subject to some fairly minor technical reservations, cells from *Microciona prolifera* and cells from *Haliclona occulata* do not adhere to one another when cultured together. The cells appear to form aggregates composed exclusively of cells from one species or the other. Certainly this is what would be expected in the case of complete adhesive specificity. Additional investigation by CURTIS [1962] and by MACLENNAN and DODD [1967] suggested that this apparently complete specificity was the exception rather than the rule in sponge cell adhesiveness.

ROTH and WESTON [1967] and ROTH [1968] examined several embryonic vertebrate tissues for evidence of adhesive specificity. The probability that two cells will form and maintain an adhesion after a collision has occurred between them was taken to indicate the adhesive stability of the cells. ROTH and WESTON estimated the stability of adhesions between like and unlike cells as follows. Cells from one tissue were dissociated and allowed to form reaggregates of a given size. To the suspension of reaggregates were added tritium-labeled cells dissociated from the same or different tissues. The culture was maintained for an interval; then all reaggregates were fixed, sectioned, and examined by radioautography. The authors determined the relative abilities of reaggregates composed of cells of different kinds to remove various labeled cells from the medium by 'capturing' them. Some sort of specificity was involved in this process because aggregates composed of cells of any one type consistently captured more labeled cells of the same type than of a different type. Thus, while liver cell reaggregates captured liver cells better than retinal cells, aggregates of retinal cells captured retinal cells better than liver cells. While the results indicated an adhesion-related specificity, they did not show that adhesive specificity was involved in sorting out within mixed reaggregates or in intact embryos.

These results have been criticized, but not invalidated, by CURTIS [1970]. In 1969, CURTIS described a new method for conducting reaggregation experiments in a Couette viscometer, which permitted him to estimate the total number of collisions occurring between cells by counting the total number of particles at frequent intervals during a reaggregation experiment. The number of effective collisions was the number of collisions which resulted in adhesion between the colliding particles. Dividing the rate of effective collision by the rate of total collision gave what CURTIS defined as the stability ratio, which was undoubtedly related to cellular adhesiveness in some way.

In 1970, CURTIS applied this technique to the problem of adhesive specificity, by measuring the collision efficiency of cell suspensions containing cells of two types combined in different proportions. He used chick embryo liver and retinal cells prepared according to ROTH and WESTON [1967]. He found the collision efficiency of freshly-dissociated liver cells much greater than the collision efficiency of freshly dissociated retinal cells. Combinations of the two cells gave intermediate collision efficiencies. Thus, plotting collision efficiency as a function of percent retinal cells in suspension gave a curve with a negative slope. The curve's shape was important. In the absence of adhesive specificity, the curve would be linear. If adhesion were specific, the curve would be concave upward, because collisions between unlike cells would not be effective.

CURTIS [1970] plotted data obtained for liver and retinal cells in his figure 1. He fit a straight line to his data by means of a linear regression analysis and concluded that liver and retinal cells exhibited no adhesive specificity. His use of a linear regression analysis is questionable since there is no reason to assume a linear relationship. Visual inspection of that figure indicates that a concave upward curve passing exactly through five of the seven data points would give a much better fit than the straight line. Thus, an examination of CURTIS' figure 1 alone would tend to support rather than discredit ROTH and WESTON's claim of specificity. The data points in CURTIS' figure 2 can be fit well by neither a straight line nor a concave upward curve. Perhaps the proper conclusion to draw from CURTIS' figure 2 is that this particular method, imaginative as it is, lacks the precision necessary to answer the question put to it.

A dramatic piece of evidence which has been interpreted as supporting adhesive specificity is the phenomenon called 'homing'. In homing experiments, cells have been disseminated throughout embryos via the vascular route, and have appeared to come to rest specifically and permanently in appropriate locations. P. WEISS and ANDRES [1952, 1953] prepared suspensions of dissociated chick embryo cells which contained melanoblasts derived from neural crests of embryos from a pigmented strain. They injected these intravenously into 3- or 5-day chick embryos of a non-pigmented strain. Thus, cells were taken from pigmented donors and injected into non-pigmented hosts. Some host embryos which developed and hatched exhibited pigmentation of the donor type. The authors found differentiated pigment cells only in feathers and other regions normally populated by melanocytes. Their interpretation favored a preferential localization of injected cells in specific sites as if these cells, moving throughout the embryo, recognized target regions and formed stable adhesions only there. The authors suggested their results might reflect a general capacity for specific localization by embryonic cells, although neural crest cells have unusual migratory behavior. Technical difficulties with this work included the need to wait one or two weeks in order to determine the experimental results and the accompanying proliferation of cells during that interval. Other workers have observed that, in both amphibians and birds, melanoblasts differentiate and become pigmented only in the appropriate environment, suggesting that donor cells which might have localized in host regions which are not normally pigmented would not differentiate into melanocytes and hence would escape detection [see LEHMAN and YOUNGS, 1959].

The possibility of homing has been reinvestigated with more modern techniques by BURDICK [1968] and HOLLYFIELD and ADLER [1970]. BURDICK obtained donor tissue from chick embryos which had been incubated 5 1/4 days. He applied tritiated thymidine to donor embryos to label all their cell nuclei. Cells from mesonephros, limb bud, and heart ventricle were dissociated and injected into the circulatory systems of unlabeled host embryos, which were then sacrificed after intervals of 6 or 24 hours. Tissues taken from the host embryos were fixed and examined by means of autoradiography. The number of cells localized in a given organ was virtually identical for all three types of donor cells. For example, about two-thirds of all cells found in the three host organs were in the mesonephros, regardless of donor cell type; about one-fourth were in host limbs and 1–4% were in host ventricle. In other words, there was no evidence for preferential localization of injected cells, and the results did not support the existence of homing. Similar studies by HOLLYFIELD and ADLER [1970] confirmed these results and extended them by showing that polystyrene particles disseminated through the circulatory system localized in essentially the same way as did dissociated cells.

What kind of adhesive specificity and to what extent it is involved in sorting out is debatable. But adhesive specificity is not necessary to explain the results of sorting out experiments which have been performed. The differential adhesion hypothesis requiring only quantitative differences in adhesiveness among cells is adequate to account for the selectivity which is observed in cell sorting experiments.

2. Differential Adhesion

The most promising proposal to account for cell sorting and for some aspects of animal morphogenesis is generally called the differential adhesion hypothesis [STEINBERG, 1963, 1964, 1970]. Only a brief summary will be presented here; the original articles contain a more complete and rigorous treatment. The differential adhesion hypothesis requires few assumptions. It applies to any physical system consisting of discrete particles which move with respect to one another and which exhibit quantitative differences in adhesiveness. The hypothesis deals with the energies of adhesion but not the chemistry of adhesion. Therefore, it does not depend upon the chemistry of adhesion for its validity. STEINBERG stressed that the energies of adhesion reflect numbers of adhesive sites per unit surface contact area regardless of the molecular composition of those sites. Thus, one can constructively consider the morphological consequences of quantitative adhesive differences separate from the chemistry of adhesion. The hypothesis accounts for more of the sorting out behavior which has been observed in numerous experiments than do other proposals. The hypothesis does not, nor has it been claimed to, account for all of animal morphogenesis. It does not exclude other factors which may play a causal role in morphogenesis, such as local proliferation, cellular shape change, or even adhesive specificity [STEINBERG, 1970].

According to the differential adhesion hypothesis, cells of different types differ quantitatively in surface adhesiveness. When cells of two types are mixed together, the more adhesive ones should cohere in the interior of the reaggregate and force the others to the outside. Such behavior would establish an internal phase containing the more adhesive cells and an external phase containing the less adhesive cells. If the differences which determine sorting out among cells of different types are quantitative differences, there should exist a hierarchy of cells ranging from the most to the least adhesive. Also cells of one type should segregate internally with respect to less adhesive cells but externally with respect to more adhesive cells. The expected hierarchy does exist. STEINBERG [1970] reported that, when cells from six different chick embryo tissues are dissociated and recombined in binary combinations in culture, their sorting is predictable. For example, back epidermis cells segregate internally with respect to limb bud cells which in turn segregate internally with respect to pigmented retinal cells. The differential adhesion hypothesis suggests that epidermal cells are the most adhesive and should segregate internally when combined with pigmented retinal cells, and in fact they do. Six tissues permitted many binary combinations to be made. All were examined and the results established a consistent hierarchy. Some variability in the cellular behavior was discussed by STEINBERG [1970].

The differential adhesion hypothesis accounted for the sorting out hierarchy and several other experimental observations as well. STEINBERG [1970] listed the following: 1) Sorting out does in fact occur. 2) When irregular tissue fragments are cultured, they tend to round up and become spherical. 3) Sorting out follows a time course, characterized by the early disappearance of internally segregating cells from the surface layer of the reaggregate followed by the progressive accumulation of internally segregating cells into islands within the interior of the reaggregate. Once these islands are formed, they occasionally join one another to establish even larger islands [STEINBERG, 1962b; TRINKAUS and LENTZ, 1964]. 4) When two fragments of unlike tissue are fused with one another and then cultured, fragments of one type consistently spread over fragments of the other type. 5) Such spreading establishes an equilibrium configuration which corresponds closely to the configuration obtained when previously dissociated cells from those two tissues have sorted out.

An hypothesis placing critical emphasis upon quantitative differences in cellular adhesiveness requires an independent measurement, or at least estimate, of the values of adhesiveness for cells of different types. The importance of such a determination has been recognized by many investigators, and several ingenious attempts have been made to obtain such measurements. For example, it has been possible to measure the force required to separate two cells from one another with micromanipulators, or to measure the force required to remove cells from an artificial surface. L. WEISS [1962] has pointed out one drawback to such work – that there is no assurance that the force measured is actually the force required to separate the cell from its neighbor or substratum. It may, instead, measure the force necessary to rupture the cell periphery; these distraction experiments may result in the breakdown of cell parts instead of the clean separation of cells. STEINBERG [1964] has raised another objection. Cellular adhesiveness is more accurately expressed in units of work or energy rather than in units of force. Since force and energy are quite different, the force required to remove cells from a surface may not reflect the amount of energy used to remove the cells.

Other attempts to measure cellular adhesiveness have relied upon measurements of reaggregation kinetics under controlled culture conditions. In spite of improvements in this approach, which made it possible to study the kinetics of aggregation between cells of different types [ROTH and WESTON, 1967; ROTH, 1968; CURTIS, 1969], kinetic measurements do not bear upon the thermodynamic explanation of cell sorting [STEINBERG, 1970].

PHILLIPS and STEINBERG [1969] described a new method for equilibrium measurement of cellular adhesiveness capable of determining the energies of adhesion between cells. This was a modification of the sessile drop method used to calculate the specific interfacial free energies of liquid droplets from the shapes that the droplets assume in gravitational fields [see ADAMSON, 1960; DAVIES and RIDEAL, 1963; BUTLER and BLUME, 1966]. In general terms, the shape of a liquid drop or the shape of an aggregate of cells supported by a flat surface in a gravitational field will come to a configurational equilibrium reflecting the balance of two conflicting forces: 1) the tendency to minimize surface free energy and attain a spherical configuration, and 2) the gravitational force which tends to flatten the drop or aggregate upon the surface. Since the final state to be examined is one of equilibrium, it can be reached from either extreme, namely a spherical aggregate or a flat aggregate.

PHILLIPS and STEINBERG selected three types of chick embryo cells which tend to form homogeneous and spherical aggregates when cultured on gyratory shakers. These were 4-day heart ventricle cells, 4-day limb bud cells, and 5-day liver cells. These cells were dissociated by means of routine procedures. Spherical aggregates were assembled in gyratory shakers; flat aggregates were cut from sheets of cells which were made by centrifuging dissociated cells onto the bottoms of roller tubes, and then incubating them briefly to permit intercellular adhesions to form. The aggregates were maintained under controlled culture conditions and gravitational forces for specified intervals, at the end of which their profiles were photographed and compared.

The results were the following: aggregates which were initially flat rounded up, while aggregates which were initially round became flattened. The final shape of the reaggregate was independent of the initial shape. In other words, the aggregates all assumed a configuration characteristic of the tissue of origin and the gravitational force applied; hence, configurational equilibrium was attained. When the three tissues were compared on the basis of their profiles at the end of experiments with gravitational force held constant, limb bud was the most rounded, liver the least rounded, and heart intermediate. These results indicated that limb bud cells were more cohesive than heart cells, which were more cohesive than liver cells. That was exactly the result predicted on the basis of sorting out experiments with those three tissues; limb bud segregated internally with respect to heart or liver and heart segregated internally with respect to liver. PHILLIPS and STEINBERG [1969b] added chick embryo spleen to the hierarchy. They first determined its cohesiveness by the sessile drop method. From those results they predicted, and then confirmed by sorting out experiments, spleen's position in the hierarchy. Thus, an independent determination of the energies of adhesion for chick embryo cells from several organs was made and found to be in accord with cell sorting and tissue spreading results. That greatly enhanced the already considerable amount of evidence in support of the differential adhesion hypothesis.

E. Other Mechanisms

The cell and tissue rearrangements or morphogenetic movements which sorting out experiments have elucidated play a significant role in the establishment of animal morphology. They are especially important early in development when the most fundamental morphogenetic events, such as the establishment of the germ layers, occur. Experiments by HOLTFRETER [1939], TOWNES and HOLTFRETER [1955], and STEINBERG and KELLAND [1967] demonstrated the importance of differential adhesion in the establishment of ectoderm, mesoderm, and endoderm during amphibian gastrulation. However, other mechanisms play important causal roles in morphogenesis. While they are not the primary subject of this chapter, they should be mentioned to indicate the range of influences upon embryonic morphogenesis.

SAUNDERS and FALLON [1966] have reviewed the role of cell death in avian limb morphogenesis. The evidence which they presented indicated that, shortly after aggregation and growth of the mesenchymal precursors of the digits, those embryonic organs are further delineated by the onset of a precisely localized and programmed death of the cells between them. Thus, the digits become separate. SAUNDERS and FALLON discussed both the cytology and control of cell death and pointed out how its failure or absence resulted in webbing between the digits. They showed that widespread, well controlled cell death and tissue destruction were integral parts of normal development and that the timing and location of such events represented another way in which an organism's genome can regulate its phenotype. They cited other general reviews of the same topic, including ZWILLING [1964] and SAUNDERS [1966].

Changes in cellular shape have long been considered important in morphogenesis. A striking and well-known example involves the development of the lens of the eye from the optic vesicle. The cells of the medial wall of the optic vesicle elongate across the vesicle's lumen. Elongation continues until those cells contact the lateral wall of the vesicle, completely filling the lumen and establishing a solid cellular structure which will form the mature lens [see ROMANOFF, 1960]. Even a casual observation of the lens during these stages reveals that cellular elongation rather than proliferation accounts for the increase in tissue mass and morphogenetic change. Microtubules are probably involved in this process. If so, it would be possible to inhibit elongation with colchicine and demonstrate microtubules in the elongating cells by means of electron microscopy. Such a study is in progress.

Analysis of neural plate morphogenesis in the newt *Taricha torosa* by BURNSIDE and JACOBSON [1968] showed that the neural plate cells maintained contact with all of their neighbors during the extensive changes in neural plate shape which foreshadowed neurulation. Careful reconstructions from time-lapse movies of intact embryos, coupled with measurements of the thickness of the neural plate in different regions, indicated that the neural plate was distorted as a consequence of local thickenings in different sectors. These local thickenings reflected elongation of neural plate cells along their dorsal-ventral axes. Further examination showed that microtubules were abundant during these stages and that they were oriented along the axes of cellular elongation. It is most likely that microtubules are involved in these morphogenetic shape changes.

BAKER [1966] examined cells at the newly-formed dorsal lip of the blastopore in *Hyla* embryos. She detected a transition from cuboidal cells, to wedge-shaped cells, to flask cells during the early stages of invagination. She examined the ultrastructure of these cells and proposed a mechanism by which the changes in cell shape might be partially responsible for the invagination process. The cells were apparently held together firmly at their distal ends by tight junctions which were thought to coordinate the movements of cells into the archenteron. Microfilaments were seen in a dense layer at the apical ends of flask cells and probably contributed to their changes in shape.

BAKER and SCHROEDER [1967] approached the problem of neurulation similarly. They examined the ultrastructure of cells in *Hyla* and *Xenopus* neurulae. They found highly ordered arrays of 60 Å microfilaments in the constricted necks of neural plate cells. The filaments were arranged circumferentially, and by their contraction may have contributed to the asymmetric cell shape associated with the neurulation process.

SCHROEDER [1967] pursued the matter further with *Xenopus*. He presented an extensive model to explain neurulation movements, using specific changes in cellular configuration. Although the other factors have not been excluded, these works emphasized the importance of cell shape change in morphogenesis and implicated particular subcellular organelles.

III. Adhesive Mechanisms

Much insight into morphogenesis has been gained by considering events at the cell and tissue levels with no direct regard for the chemical or molecular basis for such events. Morphogenesis cannot be fully understood, however, until intercellular adhesion is also understood in molecular terms. Although the question of intercellular adhesive mechanisms has been studied diligently, it has not yet been answered satisfactorily. The complexity of cellular composition has hindered the identification of adhesive molecules. Just to identify some cell membrane components has been a substantial accomplishment [WARREN and GLICK, 1968; WARREN, 1969]. Purification and characterization of adhesive molecules involve a serious problem. The molecules cannot be assayed independently of the system from which they are derived. Since so many factors can influence the morphogenesis-related behavior of cells—for example, their geometry [LESEPPES, 1963]—it has been difficult to isolate adhesive molecules and be sure that their function was truly adhesive. Although work is progressing, no chemical identification of an adhesive molecule from the cell surface has resulted.

A. Intercellular Cement and Specific Molecules

Perhaps the earliest view of intercellular adhesion was that cells are held together by intercellular cementing substances. This would follow from the observations of tissues such as cartilage in which extracellular material is conspicuous and indeed must hold the cells together. For example, chondrocytes are embedded in the cartilage matrix. Although the matrix plays a supporting role for the entire organism, it also serves to hold the chondrocytes together. This example does not really pertain to embryonic development, however, because the cartilage cells are already in their proper positions before the matrix is elaborated. In fact, there is no case in vertebrate development where a gross intercellular cement of this type is involved in the adhesion and morphogenesis of embryonic cells. This does not exclude the possibility that molecules which exist outside the plasma membrane of an embryonic cell may have important adhesive properties which would contribute to the organism's morphogenesis. Whether such molecules are actually extracellular or better considered a part of the cell periphery seems to be one of semantics. It seems advisable to try to learn what molecules may be present at or near cell surfaces and what role they may play in intercellular adhesion.

When the examination of cell sorting began in earnest in the 1940's, before there was any explanation for selective adhesion on the basis of quantitative adhesive differentials, it seemed quite likely that some sort of adhesive specificity played a role both in sorting out and in normal morphogenesis. Cells were thought to be capable of discriminating between cells of the same and different types. Thus, they would cohere strongly with their own kind but fail to make firm attachments to cells of other types. LOEB [1922], TYLER [1940], and P. WEISS [1941] suggested that such specificity was the result of sterically specific molecules at cell surfaces. These molecules were thought to exert their specificity in the lock-and-key manner of antigen and antibody, a difficult proposal to test. SPIEGEL [1954, 1955] attempted a test. He prepared antisera against sponge cells and frog embryo cells and was able to demonstrate that these antisera would inhibit reaggregation of homologous dissociated cells. One possible interpretation of those results was that the antibodies bound specifically to antigenic cell surface macromolecules which were responsible for specific adhesion. However, another alternative, which SPIEGEL recognized, was that the antibodies simply bound to cell surfaces and covered up whatever sites might be involved in the adhesive process. Thus, SPIEGEL's results did not permit a definite conclusion. Most dissociated embryonic cells apparently adhere randomly when dissociated and cultured. This argues against adhesive specificity, or at least requires one to postulate a period of nonspecific adhesion preceding the onset of specificity. The results of sorting out experiments are more readily interpreted on the basis of differential adhesion according to which, cells of a more adhesive type will segregate internally with respect to cells of a less adhesive type.

One exception, and probably the best example of adhesive specificity, comes from work with two species of sponge *Microciona prolifera* and *Haliclona occulata* [see HUMPHREYS, 1967]. When these sponges are dissociated and mixed, they appear to form completely separate reaggregates: orange reaggregates of *Microciona* cells and white reaggregates of *Haliclona* cells. This is exactly the behavior that would result from adhesive specificity.

Attempts have been made to elucidate the chemical basis of this apparent specificity and to demonstrate an intercellular cement [see HUMPHREYS, 1967]. HUMPHREYS *et al.* [1960], HUMPHREYS [1963], and MOSCONA [1963] reported a 'chemical' dissociation procedure yielding sponge cells with properties different from those of mechanically dissociated cells. When fragments of sponge were treated with calcium- and magnesium-free sea water and then dispersed, their reaggregation proceeded normally at 18 to 20 °C, but was inhibited at 5 °C. The addition of supernatant from the chemical dissociation procedure to chemically dissociated cells promoted their aggregation at low temperature. It was thought that some factor normally involved in intercellular adhesion was released during the chemical dissociation procedure and taken up by dissociated cells permitting them to reaggregate. Several pieces of evidence have supported this view. The aggregation-promoting factors were removed from solution by homologous cells. The factors exhibited species specificity [HUMPHREYS, 1963]. Radioactive factor from *Microciona* was taken up by chemically dissociated *Microciona* cells [COLLINS, unpublished]. The species specificity of radioactive factor uptake has not yet been demonstrated. While the factor probably was absorbed at the cell surface, that has not been directly demonstrated. For both *Microciona* and *Haliclona* a comparison of chemically and mechanically dissociated cells revealed no significant differences in electrophoretic mobility tested over a wide range of pH [COLLINS, in preparation].

Efforts to isolate and characterize the aggregation-promoting factor have met with some success [HUMPHREYS, 1965; MARGOLIASH *et al.*, 1965]. It appeared to be a finely particulate substance with an *S* value of about 100 and a diameter of approximately 100 to 200 Å. It contained substantial amounts of carbohydrate, possibly the molecular basis for its specificity. It contained amino acids as well, but the work of HUMPHREYS [1965b] suggested that the factor did not contain a high molecular weight protein synthesized on ribosomes, since chemically dissociated cells of either species could survive and aggregate in a concentration of ethionine sufficient to reduce protein synthesis to less than 10% of controls. Repeated dissociation and aggregation of the same cells in the presence of ethionine argued against the possible storage and gradual use of large amounts of proteins.

Similar investigations with other species of sponges have given different results, and the case for adhesive specificity is somewhat limited. MACLENNAN and DODD [1967] examined several different kinds of sponges and found limited specificity. CURTIS [1962, 1967] examined some sponges which appeared to reaggregate specifically and found that by altering the timing of their recovery from dissociation he could modify their specificity accordingly. The *Microciona-Haliclona* specificity is not typical of all sponges, but it appears to be valid.

Sponges do not provide the only example of specific cell contact mediated by macromolecules. For example, CRANDALL and BROCK [1968] isolated complementary macromolecules from yeasts of opposite mating type. They reported that these were cell surface molecules that neutralize each other in the manner of antigens and antibodies. The authors began a chemical characterization of these molecules indicating that they are large mannose-containing glycoproteins.

Efforts to demonstrate specific adhesive molecules on vertebrate cell surfaces have not progressed as far as the work with invertebrates. The clearcut specificity leading to self isolation in some sponges does not exist among the vertebrate tissues. MOSCONA and his associates have tried to prepare aggregation-promoting factors from chick embryo cells and to demonstrate their tissue specificity. Neural retinal cells have been used more than others. MOSCONA [1962] reported a cell-free supernatant obtained by washing freshly dissociated neural retinal cells with appropriate media. The supernatant promoted the aggregation of freshly dissociated neural retinal cells, while it had much less effect upon the reaggregation of skin cells and liver cells and no effect at all upon kidney cells and limb bud mesenchyme cells. MOSCONA obtained some data indicating that the active supernatant contained protein, hexosomine, and hexuronic acid.

LILIEN and MOSCONA [1967] and LILIEN [1968] reported a similar material prepared from neural retinal cells grown for several days in monolayer culture in Eagle's basal medium with 20% serum. The medium used for cell aggregation was Eagle's medium without serum containing DNAase. Dissociated cells were cultured on gyratory shakers, and their aggregation was estimated by aggregate size after a given interval. The active supernatant added to neural retinal cell suspensions caused larger aggregates, but had no effect upon 10-day liver, 6-day heart, or 4-day limb bud cells. LILIEN reported that low temperature and cyclohexamide inhibited the accumulation of the active factor in the supernatant and that the activity was absorbed from the supernatant by homologous cells. He concluded that the active factor was probably a normal cell surface constituent serving to bind cells together in a specific way. However, serum-free media are generally unsuitable for avian cell culture and the need for DNAase in the culture medium indicated that considerable cell damage and loss of nuclear DNA to the medium occurred during these experiments. The active supernatants may have contained materials which bind the cells together, although they could have functioned in some other way.

If tissue-specific aggregation-promoting factors play a general role in animal cell adhesion, then it should be possible to produce specific factors from other tissues. KURODA [1968, 1969] showed that liver cells taken from young embryos and allowed to reaggregate for a standard time form larger aggregates than do cells from older embryos. He used 7-day cells which produce large aggregates and 18-day cells which produce very small aggregates. He prepared a supernatant fraction from 7-day liver cells which enhanced the aggregation of cells of the same stage at 28 °C but which had no effect upon 18-day cells at 28 °C. This supernatant did promote the aggregation of older cells at 38 °C if the experiments were continued for several days. The supernatants in these experiments appeared to contain considerable amounts of RNA and protein. The results seemed to indicate an aggregation-promoting factor, but there was no report of tissue specificity.

B. Ultrastructure

Areas of contact between adherent cells have been examined by electron microscopy in order to investigate the mechanism of intercellular adhesion and the possibility of an ultrastructural basis for adhesive selectivity. Cells of many kinds bear different surface coats external to their plasma membranes. FAWCETT [1966] illustrated filamentous material associated with surfaces of gut epithelial cells and concluded that the material was an integral part of the cell membranes. The nature of the association of the surface material with the cell membrane is not always clear, but such coatings do effectively constitute the outer surface of the cell and must be involved in intercellular adhesion.

Different epithelia exhibit several surface specializations thought to be adhesive sites. There are extensive areas of constant distance of separation between membranes of contiguous epithelial cells; it is believed that whatever maintains that distance also maintains the cellular adhesion [FAWCETT, 1966]. The space is occupied by polysaccharides, which have been demonstrated with staining procedures [PEASE, 1966; REVEL and KARNOVSKY, 1967; BEHNKE and ZELANDER, 1970]. In addition, specialized cell surface structures have been more firmly implicated in intercellular adhesion. Numerous epithelial tissues exhibit junctional complexes each one of which is composed of a zonula occludens, a zonula adherens, and a macula adherens or desmosome [FARQUHAR and PALADE, 1963, 1965]. These three components of the complex occupy characteristic positions near the apical ends of the epithelial cells. These specialized sites have been seen in non-epithelial tissue. For example, TRELSTAD *et al.* [1967] reported close and tight junctions between mesenchyme cells of gastrulating chick embryos. While the chemical composition of these specialized regions has not been determined, some evidence of their nature is available. DOUGLAS *et al.* [1970] demonstrated that desmosome plaques from two different epithelia could be removed by proteases, suggesting that protein or polypeptide constituted a major fraction of those organelles.

Specialized structures like desmosomes are generally accepted as adhesive sites. In 1925 CHAMBERS and RENYI noticed, in the course of microdissection of human epidermis, what appeared to be points of protoplasmic connections between cells which were being pulled apart. Of course, it was impossible for them to determine the exact nature of this unevenness of adhesion between epidermal cells, but it now seems likely that it is referable to the specialized structures associated with apposed plasma membranes seen in so many epithelial tissues. OVERTON [1962] noted that, in the chick blastoderm, the number of desmosomes increased with embryonic age as did the difficulty of dissociation. HILFER and HILFER [1966] reported that chick embryo thyroid cells which were being dissociated tended to remain attached by means of junctional complexes.

Specialized adhesive sites such as these undergo developmental changes in embryos. They first appear as close junctions and tight junctions between cells of the chick blastoderm. These connect mesodermal cells with one another and with the basal surfaces of epiblast and hypoblast cells. As development proceeds, they give way to more complex surface specializations of the desmosome type [TRELSTAD *et al.*, 1967]. Thus, embryonic cells change progressively in their adhesive properties, and their specialized surface structures themselves develop. OVERTON [1962] described a sequence of desmosome development in early chick blastoderm cells. The apperance of intercellular material was followed by the appearance of desmosomal plaques, which was in turn followed by the appearance of fibrillar material or tonofilaments. Although desmosome remnants remained after cellular dissociation, reassociating cells displayed a similar sequence of desmosome development.

LENTZ and TRINKAUS [1971] studied in detail ultrastructure of junctional complexes in surface cells of the developing *Fundulus* blastoderm. They examined these structures at several stages and reported a sequence of events comparable to that which occurs in chick blastoderms. They described junctional complexes similar to but not identical with those seen in more highly developed epithelia. The authors discussed evidence that these structures might function in cellular adhesion, in intercellular communication, and in contact inhibition.

OVERTON [1969], KAHN and OVERTON [1969a, b] reported that enzymatically dissociated chick embryo cells generally exhibited very little of the lanthanum-staining extracellular material seen in intact tissues, but that such material reappeared in abundance as cells reaggregated. The material associated with liver cells was largely amorphous; the material associated with limb bud cells was fibrillar; while that associated with cartilage cells was granular.

The authors proposed that the observable differences in surface materials may provide a basis for adhesive selectivity. They proposed this hypothesis with proper caution, however, since an adhesive role for lanthanum-staining materials has not been proven and it is not immediately clear how these materials would mediate adhesive selectivity. In the case of cartilage cells, there is no question but that the conspicuous extracellular material holds the cells together. The 8-day cartilage cells described by KAHN and OVERTON [1969a] had long been producing matrix and were already far past the stages during which morphogenetic movements established their locations in the embryo. Their extracellular products would not be typical of early embryonic cells.

Unlike cells frequently adhere to one another. Many examples were cited by ARMSTRONG [1970] who examined the incidence and morphology of adhesive cell junctions in heterotypic chick embryo cell aggregates. In these experiments, the junctions seen between like cells (heart–heart or pigmented retina–pigmented retina junctions) and unlike cells (heart–pigmented retina junctions) appeared to be identical. Thus, no basis for adhesive specificity in the morphology of these adhesive junctions has been demonstrated.

While ultrastructural examinations have provided little support for adhesive specificity, they might provide information bearing on the differential adhesion hypothesis. If intercellular adhesion is mediated largely by demonstrable adhesive specializations, one might demonstrate a correlation between the frequency of such junctions and the intercellular adhesion measured by the sessile drop method developed by PHILLIPS and STEINBERG [1969].

The conventional techniques of electron microscopy have shown in electron-lucid areas gaps of about 200 Å between membranes of adjacent cells. These gaps could be accounted for by the lyophobic colloid hypothesis of cell adhesion discussed in Section C. The matter has been complicated by the demonstration of stainable polysaccharide in the gaps.

C. Lyophobic Colloids

One proposal for intercellular adhesion which has attracted considerable attention was based upon the theory of stability of lyophobic colloids.

In this theory, the relative strengths of van der Waals forces of attraction and electrostatic forces of repulsion are decisive. CURTIS [1958] applied the theory to cells, calculated probable magnitudes for both forces, and determined that they should be in balance when the cells are separated by a distance of about 200 Å. That accounted for many of the electron-lucid areas seen between membranes of adjacent cells.

The hypothesis was in accord with other cellular properties. For example, cells bear negative charges at physiological pH; thus, there must be some electrostatic repulsion between the cells. Also, bivalent cations are essential for cell reaggregation. Many investigators have used the calcium-chelating compound EDTA to dissociate cells, and chick neural retinal cells can be dissociated simply by soaking the tissue in calcium- and magnesium-free solutions followed by mechanical agitation [COLLINS, 1966b]. While calcium ions might promote intercellular adhesion in other ways, they could do so by reducing the surface charge of cells and, hence, reducing the electrostatic repulsion between them. COLLINS [1966a, b, and unpublished] found that 1.8×10^{-3} M calcium ion could reduce electrophoretic mobility of various dissociated chick embryo cells by about 10%. CURTIS [1967] estimated that this was approximately the degree of charge reduction necessary to produce adhesion. CURTIS [1964] reported that monovalent cations promoted cellular adhesion, presumably by a reduction of electrostatic repulsion. Since those experiments concerned the adhesion of cells to glass rather than to other cells, they may reflect a mechanism entirely different from intercellular adhesion.

The lyophobic colloid hypothesis accounted for electron-lucid gaps often seen between cells in tissues. It did not predict the presence of stainable material between the membranes of adjacent cells [PEASE, 1966; REVEL and KARNOVSKY, 1967; BEHNKE and ZELANDER, 1970]. These results indicated that more than water and small molecules are present between the membranes. CURTIS [1967] argued in favor of the existence of a real gap between surfaces of apposed cells *in vivo*. He cited, among other things, the work of FARQUHAR and PALADE [1963] who demonstrated that hemoglobin can diffuse between the plasmalemmas of kidney tubules. He also cited the work of BRIGHTMAN [1965] who demonstrated that ferretin molecules, which have diameters of 100 Å, could penetrate quckly into the intercellular space and may actually be in molecular contact. CURTIS pointed out that such close contact was in accord with the lyophobic colloid hypothesis and would simply represent stabilization at the primary minimum.

CURTIS also reviewed recent evidence bearing upon that proposal and discussed some problems related to its acceptance. One of these concerned the role of enzymes in cellular dissociation. If enzymatic treatment of tissues increased net surface charge, then the effectiveness of enzymes in dissociation could be readily explained on the basis of the lyophobic colloid hypothesis. CURTIS noted that charge measurements of enzymatically treated and control cells have given variable results. Recent work with chick embryo cells [BARNARD *et al.*, 1969; MASLOW, 1970] indicated that enzymatic dissociation reduced neural retinal cell surface charge while increasing the charge on liver cells.

Electrophoretic measurement of cell surface charge provides an average charge density per unit surface area. It is quite possible that charged groups are distributed unevenly over cell surfaces; the actual charge density at points critical for intercellular adhesion may be quite different from that which can be measured by electrophoresis. There is, in fact, evidence for an uneven distribution of charged groups on the surfaces of *B. subtilis* cells [L. WEISS, 1963]. If one could examine the charge on selected sectors of dissociated cells, the results might serve better to test the applicability of the lyophobic colloid theory to cell adhesion.

In reviewing the lyophobic colloid hypothesis, TRINKAUS [1969] pointed out the excellent agreement between distances separating plasma membranes as predicted from CURTIS' calculations and as seen in electron micrographs. These applied to tight junctions as well as to the more widespread 200 Å separations. TRINKAUS noted, however, the existence of gap junctions in which plasma membranes were separated by 20–90 Å and which were not predicted from CURTIS' calculations.

Additional results bearing directly upon this hypothesis were provided by ARMSTRONG [1966], who examined the effects of bivalent cations in different concentrations on 1) the net surface charge densities of dissociated embryonic cells, and 2) the abilities of the cells to reaggregate. The ions enhanced reaggregation in a concentration-dependent way. Some ions were more effective than others. If a balance between van der Waals forces and electrostatic forces determines cellular adhesiveness, then concentrations of different cations which were equally effective in promoting reaggregation should reduce cell surface charge density to the same level. That was not so for the limb bud cells which ARMSTRONG tested. These results do not agree with the lyophobic colloid hypothesis, and they indicate that bivalent cations promote cellular adhesion in ways other than surface charge reduction.

D. Bivalent Cation Bridge

According to another view which has received theoretical support from PETHICA [1961], cells were bound together by means of bivalent cation bridges. These bridges would depend upon negatively charged sites on cell surfaces and the fact that anionic molecules (such as carboxylic acids) form slightly soluble salts with bivalent cations (such as calcium). Calcium ion, being bivalent, can bind simultaneously to two anionic sites. If those sites were located on the surfaces of contiguous cells, a bivalent cation bridge would be formed between the two cells. This bridge would tend to hold the cells together, and the number of such bridges per unit contact surface area would determine how firmly the cells adhered. This would account for the direct involvement of calcium in intercellular adhesion and for the dissociation of cells by calcium removal.

In spite of the sound chemical argument for this view, biological evidence for it is sparse. It does seem possible for aggregating cells to approach one another closely enough to establish an anionic bridge. LESSEPS [1963] demonstrated regions of reaggregating cells with low radii of curvature which appeared to be points of initial cellular contact. Electrostatic repulsion between cells could be reduced locally to permit close contact at these regions. Additionally, chick embryo cell surfaces bind calcium ions [ARMSTRONG, 1966; COLLINS, 1966a, b]. The pH mobility relationships and calcium-binding properties of cells selected from the adhesiveness hierarchy of STEINBERG [1970] have been determined [COLLINS, 1966a, and unpublished]. For freshly dissociated cells there was no significant correlation between the amount by which physiological concentrations of calcium ion diminished a cell's surface charge and that cell's position in the adhesion hierarchy. Additional experimentation will be required to exclude the bivalent cation bridge hypothesis, since enzymatic treatment appears to remove cell surface molecules which may be regenerated and may be involved in intercellular adhesion. Two possible approaches are: 1) to examine the surfaces of cells dissociated non-enzymatically, and 2) to examine the surface of enzymatically-dissociated cells which have had an opportunity to regenerate any lost surface material.

IV. Cellular Motility in Embryogenesis

A. Primordial Germ Cells

Vertebrate gonads are derived from two sources. First is the germinal epithelium, the cells of which differentiate *in situ* to form the somatic components of the gonads such as the Sertoli cells of testes or the follicle cells of ovaries. Second are the primordial germ cells which arise extragonadally or even extraembryonically; these migrate to the gonad and differentiate into gametes. In chick embryos the primordial germ cells are derived from extra-embryonic endoderm and, before migration, they are located mainly in the germinal crescent which surrounds the anterior end of the embryo [but see FARGEIX, 1969]. Several experimental procedures such as cautery or irradiation have been used to selectively destroy the crescent causing development of sterile gonads. Chick primordial germ cells migrate from the germinal crescent to the gonadal region via the circulatory system. MYER [1964] correlated the onset of circulation with the appearance of primordial germ cells within the embryo. DANTSCHAKOFF [1941] actually observed them in the blood stream. SIMON [1957] joined pairs of embryos in parabiosis; the germinal crescent of one had been destroyed. Primordial germ cells from the intact embryo colonized the gonads of the other embryo in all cases where the circulatory systems of the two blastoderms had fused.

Primordial germ cells of anurans are derived from endoderm also. They can be identified by their cytochemistry. Their paths of migration have been traced up the dorsal mesentery and then lateral to the primordial gonad. Embryos from which the endoderm has been removed are able to survive and differentiate quite well. In such cases, the gonads develop and differentiate but are sterile.

Mammalian primordial germ cells are also derived from endoderm. Their origin has been described by MINTZ [1960], who traced their route using their high alkaline phosphatase content.

Primordial germ cells, however, are not of endodermal origin. NIEUWKOOP [1946] exchanged endoderm between embryos of different species and showed that the primordial germ cells came from the species contributing mesoderm rather than the one contributing endoderm. In all of these cases, questions remain about the mechanism by which the primordial germ cells leave their sites of origin, pass through endothelium, migrate to presumptive gonad sites and, finally, select permanent residence there. Answers to these questions will require a better understanding of cell surfaces and the ways in which they change.

B. Neural Crest Cells

Perhaps no portion of the vertebrate embryo has so rich and diversified a fate as the neural crest. Derivatives of the neural crest include sensory ganglia, sympathetic ganglia, melanocytes, Schwann cells, adrenal medullae, and some visceral cartilage [DUSHANE, 1935; DETWILER, 1936]. These derivatives develop in widely separated parts of the embryo and their development is preceded by a dispersal of neural crest cells to different locations. This has been known for some time, and the migration of neural crest cells has been studied extensively [see TWITTY, 1966].

Until the use of autoradiography with tritiated thymidine to trace movements of individual cells, it was impossible to know by what route neural crest derivatives reached their new locations. It was not known that such routes even existed. Random dispersal and selective survival of neural crest elements were considered possible. Autoradiographic studies have shown, however, that neural crest cells migrate along quite specific pathways leading toward their final sites of localization [WESTON, 1963]. When chick embryo neural tubes containing neural crest cells were transplanted from labeled embryos into unlabeled hosts which were sacrificed at intervals thereafter, labeled cells were seen to have migrated from the dorsal side of the neural tube and to have followed two separate pathways. One pathway led directly into the ectoderm, where the cells spread ventrad to differentiate into pigment cells. The other pathway led through mesoderm, down past the sides of the neural tube, leaving some cells to form spinal ganglia, and on toward the dorsal aorta, near which the sympathetic ganglia would form.

The pathways followed by chick neural crest cells seemed to depend to a large extent upon the neural tube and crest themselves. For example, when neural tubes were implanted into host embryos with an abnormal orientation, the pathways followed by the migrating neural crest cells depended upon the orientation of the donor neural tube rather than the host tissue. That was shown by implanting the donor neural tube with abnormal orientation and observing that the direction of migration corresponded to the orientation of the neural tube.

This does not imply that the surrounding tissue has no effect upon the migration of neural crest cells. On the contrary, surrounding tissues do play a significant role. They seem to exert their effect, quantitatively rather than qualitatively; they influence rates of migration or prevent migration altogether. This effect is illustrated in the segmentation of the spinal ganglia. This metamerism corresponds to the segmentation of somites; DETWILER [1937] suggested that metamerism of the spinal ganglia was a direct consequence of mesodermal metamerism. The work of WESTON [1963] indicated a mechanism by which this neural segmentation may arise. Neural crest cells, which migrated directly through somitic tissue, exhibited an enhanced migration; they moved farther from the neural tube and moved more rapidly than neural crest cells which migrated between somites. Consequently, those latter cells were left behind in a pattern corresponding to the distribution of spinal ganglia. Segmentation of neural crest derivatives was not observed in the absence of mesodermal segmentation. Additionally neural crest derivatives could not migrate through all embryonic tissues, non could they follow their regular pathways of migration at all stages of embryogenesis. For example, labeled donor neural tubes were taken from young embryos before neural crest migration had begun and were implanted in more mature embryos in which considerable migration had occurred. When hosts were later examined autoradiographically, some neural crest derivatives – those farthest from the implanted neural tube – remained unlabeled, suggesting that the pathways of migration had been closed off as the host embryo matured.

C. Haptotaxis

Experiments by CARTER [1965, 1967] illustrated an extremely close relationship between cellular motility and cellular adhesion in an *in vitro* system. They provided examples of cells responding to adhesive differentials by moving in such a way as to maximize their adhesions.

The work was done with mouse fibroblasts (Earle's L. strain). These cells did not adhere firmly to cellulose acetate; consequently, they rounded up when cultured on that substratum. When the cellulose acetate had been covered with a layer of metallic palladium deposited from a vapor phase, however, the cells adhered to it tenaciously and could be seen to spread flat upon the surface. Increased palladium deposition caused greater adhesiveness as shown by greater flattening of the cells. CARTER was able to deposit the metal in a gradient, changing continuously from a heavy deposit to none at all. He examined the behavior of cells placed upon this gradient in contrast to cells on either side of it. Cells located on the gradient exhibited an intermediate degree of flattening and also different behavior, which was followed by tracing the path taken by individual cells. First, those on the gradient moved significantly greater distances in the same time interval and, secondly, their movements were not random but directed. The gradient induced a net movement of cells toward the heavier palladium deposit. In other words, the cells moved up the adhesion gradient, always selecting the more adhesive surface. This showed that cells are stimulated to move by the opportunity to increase adhesive contact and that the direction of their movement can be determined in this way.

This experiment dramatically illustrated the close and probably causal relationship between cellular adhesion and cellular motility. It supported the soundness of efforts to account for morphogenetic movements of embryonic cells on the basis of their adhesive properties, for it is most likely that factors which influence cellular motility *in vitro* would have a similar effect *in vivo.*

D. Changing Surface Properties

Progressive change is the hallmark of embryonic development. Consequently, progressive changes in cell surface properties are of descriptive interest both with respect to their timing and their physical basis. Such changes gain additional significance because cellular adhesion is so important in morphogenetic movements. If adhesive differences play an important role in animal morphogenesis, embryonic cells must differentiate with respect to surface adhesion and the timing of that differentiation must be regulated. This represents an area where the problems of morphogenesis and differentiation meet, an area through which the effects of the embryonic genome must be transmitted in order to influence embryonic morphology. In view of all this, there is a discouragingly small amount of information bearing directly upon that topic. This has resulted partly from the relatively recent appreciation of the importance of intercellular adhesion to morphogenesis and partly from our continuing uncertainty as to what physical and/or chemical properties are important to embryonic cell adhesion. Although much valuable information concerning composition, change and turnover of biological membranes is available [see MARCUS and SCHWARTZ, 1968; WARREN and GLICK, 1968; WARREN, 1969], it is not yet clear how this information will bear upon embryonic development.

While definite surface changes occur in cultured cells, changes which happen during the normal development or life cycle of organisms are of greater developmental interest. GARROD and GINGELL [1970] demonstrated a progressive change in surface properties of preaggregation cells of the slime mold *Dictyostelium discoideum.* They monitored surface charge density by means of electrophoretic mobility and found that mobility decreased progressively as cells neared the aggregation stage. BORN and GARROD [1968] had shown that cells suspended in distilled water could be made to adhere by adding sodium chloride or calcium chloride and that older cells required less salt to promote their aggregation. GARROD and GINGELL pointed out that their results appeared superficially to support the lyophobic colloid hypothesis of cell adhesion. But they also noted that the matter is complicated by the fact that EDTA and low temperature impaired cellular adhesion without having a significant effect upon the force of electrostatic repulsion between cells [GINGELL and GARROD, 1969]. The results of GARROD and GINGELL do not provide an easy explanation of aggregation of slime mold cells; even if they did, the explanation would not necessarily apply to vertebrate embryos. Those authors clearly demonstrated a change in cell surface physical properties during development, however, and they suggested a correlation between these changes and the adhesive properties of the cells involved.

Efforts have been made to study morphogenesis-related properties and their changes throughout development. Most tissues studied have been taken from relatively mature embryos. Most of the basic morphogenesis had been completed with the establishment of individual tissues and organs. The most important events are those occurring early in development when the isolation of cells from individual tissues is difficult and the numbers of cells are necessarily small.

The ease of dissociation of embryonic tissues has been examined since it might be related to cellular adhesiveness. For example, GROVER [1961] found old chick lung tissue more difficult to dissociate than younger tissue. Other tissues also become harder to dissociate. L. WEISS [1962], however, suggested that separation of cells from glass or from other cells represented a breakdown in the actual composition of the cell periphery rather than a detachment of adhesive sites. Cellular products such as collagen or other materials found among cells in tissues which are laid down after the major morphogenetic movements have occurred may simply fix the cells in locations previously obtained. Such materials would need to be digested in order to dissociate the tissue but would play no role in morphogenesis. Early chick blastoderms can be dissociated non-enzymatically, whereas more mature chick tissues require enzymes for dissociation. Perhaps the adhesive materials of more mature tissues, which require enzymatic treatment for breakdown, are different from the earlier adhesive mechanisms directly involved in the establishment of the basic embryonic structure.

The aggregation properties of embryonic cells also change progressively during development. Tissues taken from chick embryos at various ages, dissociated, and reaggregated formed aggregates of characteristic sizes. Large aggregates were formed by cells from young embryos; aggregate size declined with embryonic age, rapidly at first and then more gradually. Plots of aggregate size as a function of age in days resembled exponential decay curves for heart, liver, and neural retinal cells [MOSCONA, 1962; KURODA, 1962; GERSHMAN, 1970]. Younger cells also reaggregated more rapidly. While progressive changes do take place, there may be changes either in cell surface properties or in other factors influencing reaggregation.

From the standpoint of morphogenesis, changes which influence sorting out behavior of cells are more important than those which influence reaggregation rate. GERSHMAN [1970] demonstrated that the sorting out characteristics of neural retina, heart, and liver cells remained constant over most of the incubation period. When cells from 7-day livers labeled with tritiated thymidine were dissociated and mixed with dissociated cells from embryos incubated from 5 to 15 days, no sorting out was observed. The labeled 7-day cells remained intermingled with unlabeled cells from each of the other stages of development. When fragments of liver from different aged embryos were fused in culture, neither one spread over the other. Results with heart and neural retinal cells were the same except that they covered an even greater portion of the incubation period. Thus, whatever properties of the cell surface may change after 4 or 5 days of incubation, the adhesive properties involved in cell sorting remain unaltered. GERSHMAN also examined the electrophoretic mobilities of dissociated neural retinal cells incubated from 5 to 19 days. The pH–mobility relationship for neural retinal cells underwent complex changes with age. In the middle pH range, the mobility increased and then decreased as development proceeded. Mobility was not correlated with the constant sorting out characteristics which the cells maintain. The data indicated that surface charge density *per se* was not involved directly in intercellular adhesion. L. WEISS [1963, 1968] showed that average charge density over the cell surface does not give information concerning the distribution of that charge and that estimates of the average charge properties of the cell surface will not support predictions about contact phenomena in detail. GERSHMAN's results indicated that average charge density *per se* was not related to the cellular adhesive properties but did not exclude charged groups from an important role in adhesion. A subpopulation of negatively charged surface sites, such as ion-binding sites or sialic acid molecules, could participate in adhesion, but their relative abundance might be masked by the presence of other charged groups unrelated to adhesion.

If the molecular basis of intercellular adhesion were known, it would be much easier to monitor changes in significant cell surface properties during development. It may be possible to identify adhesive molecules with cells undergoing changes associated with morphogenetic events or which were, at least, demonstrably changing in their adhesive properties. Several approaches must be pursued before the roles of changing cell surfaces in animal morphogenesis can be understood. The differential adhesion hypothesis can be further refined and more widely applied, and sorting out systems can be analyzed more completely. Additional emphasis should be placed upon subcellular organelles known to be important for morphogenetic cellular shape changes. Surface components involved in intercellular adhesion must also be identified, and it is important to clarify the status of various cellular products or exudates which may participate in intercellular adhesion.

References

ABERCROMBIE, M.: in WILLMER Cells and tissues in culture, pp. 177–202 (Academic Press, New York 1964).

ABERCROMBIE, M. and AMBROSE, E.J.: Exp. Cell Res. *15:* 332 (1958).

ABERCROMBIE, M. and HEAYSMAN, J.E.M.: Exp. Cell Res. *5:* 111 (1953).

ABERCROMBIE, M. and HEAYSMAN, J.E.M.: Exp. Cell Res. *6:* 293 (1954).

ADAMSON, A. W.: Physical chemistry of surfaces (Interscience, New York 1960).

ANDRES, G.: J. exp. Zool. *122:* 507 (1953).

ARMSTRONG, P.B.: J. exp. Zool. *163:* 99 (1966).

ARMSTRONG, P.B.: J. Cell Biol. *47:* 197 (1970).

BAKER, P.C.: J. Cell Biol. *24:* 95 (1965).

BAKER, P.C. and SCHROEDER, T.E.: Develop. Biol. *15:* 432 (1967).

BANKS, B.E.C.; BANTHORPE, D.V.; LAMONT, D.M.; PEARCE, F.L.; REDDING, K.A., and VERNON, C. A.: J. Embryol. exp. Morph. *23:* 519 (1970).

BARNARD, P. J.; WEISS, L., and RATCLIFF, T.: Exp. Cell Res. *54:* 293 (1969).

BEHNKE, O. and ZELANDER, T.: J. Ultrastruct. Res. *31:* 424 (1970).

BONNER, J.T.: J. exp. Zool. *106:* 1 (1947).

BORN, G.V.R. and GARROD, D.: Nature *220:* 616 (1968).

BURDICK, M.L.: J. exp. Zool. *167:* 1 (1968).

BUTLER, J.N. and BLOOM, B.H.: Surface Sci. *4:* 1 (1966).

CARTER, S.B.: Nature *208:* 1183 (1965).

CARTER, S.B.: Nature *213:* 256 (1967).

CHAMBERS, R. and RÉNYI, G.S.: Amer. J. Anat. *35:* 385 (1925).

COLLINS, M.: J. exp. Zool. *163:* 23 (1966a).

COLLINS, M.: J. exp. Zool. *163:* 39 (1966b).
CRANDALL, M.A. and BROCK, T.D.: Science *161:* 473 (1968).
CURTIS, A.S.G.: Amer. Naturalist *94:* 37 (1960).
CURTIS, A.S.G.: Exp. Cell Res., Suppl. *8:* 107 (1961).
CURTIS, A.S.G.: Nature *196:* 245 (1962).
CURTIS, A.S.G.: J. Cell Biol. *20:* 199 (1964).
CURTIS, A.S.G.: The cell surface: its molecular role in morphogenesis (Logos Press/Academic Press, London 1967).
CURTIS, A. S. G.: J. Embryol. exp. Morph. *22:* 305 (1969).
CURTIS, A.S.G.: Nature *226:* 260 (1970).
DANTSCHAKOFF, V.: Der Aufbau des Geschlechts beim höheren Wirbeltier (Fischer, Jena 1941).
DAVIES, J.T. and RIDEAL, E.K.: Interfacial phenomena (Academic Press, New York 1963).
DEHAAN, R.L.: in R.L. DEHAAN and H. URSPRUNG Organogenesis, pp. 377–419 (Holt, Rinehart and Winston, New York 1965).
DEHAAN, R.L.: Develop. Biol., Suppl. *2:* 208 (1968).
DEHAAN, R.L. and URSPRUNG, H.: Organogenesis (Holt, Rinehart and Winston, New York 1965).
DETWILER, S. R.: Neuroembryology (Macmillan, New York 1936).
DOUGLAS, W. H. J.; RIPLEY, R. C., and ELLIS, R. A.: J. Cell Biol. *44:* 211 (1970).
DUBOIS, R.: J. Embryol. exp. Morph. *21:* 255 (1969).
DUSHANE, G.P.: J. exp. Zool. *72:* 1 (1935).
FARGEIX, H.: J. Embryol. exp. Morph. *22:* 477 (1969).
FARQUHAR, M.G. and PALADE, G.E.: J. Cell Biol. *17:* 375 (1963).
FARQUHAR, M.G. and PALADE, G.E.: J. Cell Biol. *26:* 263 (1965).
FAWCETT, D.W.: The cell (Saunders, Philadelphia 1966).
GALTSOFF, P.S.: Biol. Bull. *45:* 153 (1923).
GALTSOFF, P. S.: J. exp. Zool. *42:* 183 (1925a).
GALTSOFF, P. S.: J. exp. Zool. *42:* 223 (1925b).
GARROD, D. R. and GINGELL, D.: J. Cell Sci. *6:* 277 (1970).
GERISCH, G.: in A. MOSCONA and A. MONROY, Current topics in developmental biology, vol. 3.
GERSHMAN, H.: J. exp. Zool. *174:* 391 (1970).
GINGELL, D. and GARROD, D.R.: Nature *221:* 192 (1969).
GROVER, J.W.: Exp. Cell Res. *24:* 171 (1961).
GUSTAFSON, T.: Exp. Cell Res. *32:* 570 (1963).
GUSTAFSON, T. and KINNANDER, H.: Exp. Cell Res. *11:* 36 (1956).
GUSTAFSON, T. and KINNANDER, H.: Exp. Cell Res. *21:* 361 (1960).
HOLTFRETER, J.: Arch. exp. Zellforsch. *23:* 169 (1939) translated by K. KECK in B.H. WILLIER and J. OPPENHEIMER Foundations of experimental embryology pp. 186–225 (Prentice-Hall, New York 1964).
HOLTFRETER, J.: Rev. canad. Biol. *3:* 220 (1944).
HOLLYFIELD, J.G. and ADLER, R.: Exp. Cell Res. *59:* 76 (1970).
HUMPHREYS, T.: Develop. Biol. *8:* 27 (1963).
HUMPHREYS, T.: Exp. Cell Res. *40:* 539 (1965a).
HUMPHREYS, T.: J, exp. Zool. *160:* 235 (1965b).
HUMPHREYS, T.: in B.D. DAVIS and L. WARREN The specificity of cell surfaces, pp. 195–210 (Prentice-Hall, Englewood Cliffs 1967).
HUMPHREYS, T.; HUMPHREYS, S., and MOSCONA, A. A.: Biol. Bull. *119:* 294–295 (1960).

JACOBSON, A.G.: in WILT and WESSELS Methods in development biology, pp. 531–542 (Crowell, New York 1967).
JONES, K.W. and ELSDALE, T.R.: J. Embryol. exp. Morph. *11:* 135 (1963).
KHAN, T.A. and OVERTON, J.: Anat. Rec. *163:* 309 (1969a).
KHAN, T. and OVERTON, J.: J. exp. Zool. *171:* 161 (1969b).
KURODA, Y.: Exp. Cell Res. *49:* 626 (1968).
KURODA, Y.: in COWDREY and SENO Nucleic acid metabolism, cell differentiation and cancer growth, pp. 277–285 (Pergamon Press, Oxford 1969).
LEHMAN, H.E. and YOUNGS, L.M.: in M. GORDON Pigment cell biology, pp. 1–56 (Academic Press, New York 1959).
LENTZ, T.L. and TRINKAUS, J.P.: J. Cell Biol. *48:* 455 (1971).
LESSEPS, R.J.: J. exp. Zool. *153:* 171 (1963).
LILIEN, J. E. and MOSCONA, A. A.: Science *157:* 70 (1967).
LILIEN, J.E.: Develop. Biol. *17:* 657 (1968).
LOEB, J.: Science *55:* 22 (1922).
MACLENNAN, A.P. and DODD, R.Y.: J. Embryol. exp. Morph. *17:* 473 (1967).
MARCUS, P.I. and SCHWARTZ, V.G.: in L.A. MANSON Biological properties of the mammalian surface membrane, pp. 143–151 (Wistar Inst. Press, Philadelphia 1968).
MARGOLIASH, E.; SCHENCK, J.R.; HARGIE, M.P.; BAROKAS, S.; RICHTER, W.R.; BARLOW, G.H., and MOSCONA, A.A.: Biochem. biophys. Res. Comm. *20:* 383 (1965).
MASLOW, D. E.: Exp. Cell Res. *61:* 266 (1970).
MINTZ, B.: J. cell comp. Physiol. *56:* Suppl. *1:* 31 (1960).
MOSCONA, A.: Proc. nat. Acad. Sci. U.S. *43:* 184 (1957).
MOSCONA, A.: Exp. Cell Res. *22:* 455 (1961).
MOSCONA, A. A.: J. cell. comp. Physiol. *60* Suppl. *1:* 65 (1962).
MOSCONA, A. A.: Proc. nat. Acad. Sci. U.S. *49:* 742 (1963).
MOSCONA, A. and MOSCONA, M.H.: J. Anat., Lond. *86:* 287 (1952).
MYER, D.B.: Develop. Biol. *10:* 154 (1964).
NIEUWKOOP, P.D.: Arch. néerl. Zool. *8:* 1 (1946).
OVERTON, J.: Develop. Biol. *4:* 532 (1962).
OVERTON, J.: J. Cell Biol. *40:* 136 (1969).
PEASE, D.C.: J. Ultrastruct. Res. *15:* 555 (1966).
PHILLIPS, H.M. and STEINBERG, M.S.: Proc. nat. Acad. Sci. U.S. *64:* 121 (1969).
REVEL, J.P. and KARNOVSKY, M.J.: J. Cell Biol. *33:* C7 (1967).
ROMANOFF, A.L.: The avian embryo (Macmillan, New York 1960).
SAUNDERS, J.W.: Science *154:* 604 (1966).
SAUNDERS, J.W. and FALLON, J.F.: in M. LOCKE Major problems in developmental biology, pp. 289–316 (Academic Press, New York 1966).
SCHROEDER, T.E.: J. Embryol. exp. Morph. *23:* 427 (1970).
SIMON, D.: Soc. Biol. *141:* 1576 (1957).
SPERRY, R. W.: in R.L. DEHAAN and H. URSPRUNG Organogenesis, pp. 161–186 (Holt, Rinehart and Winston, New York 1965).
SPIEGEL, M.: Biol. Bull. *107:* 130 (1954).
SPIEGEL, M.: Ann. N.Y. Acad. Sci. *60:* 1056 (1955).
STEINBERG, M.S.: Proc. nat. Acad. Sci. U.S. *48:* 1769 (1962a).
STEINBERG, M.S.: Science *137:* 762 (1962b).
STEINBERG, M.S.: Science *141:* 401 (1963).

STEINBERG, M.S.: in M. LOCKE Cellular membranes in development, pp. 321–366 (Academic Press, New York 1964).
STEINBERG, M. S. and KELLAND, J. L.: Symp. Control Mechanisms in Morphogenesis, 134th Ann. Meet. Amer. Assoc. for Adv. Sci., New York 1967.
STEINBERG, M.S.: J. exp. Zool. *173:* 395 (1970).
SUSSMAN, M.: in A. MOSCONA and A. MONROY Current topics in developmental biology, vol. 1, p. 61 (Academic Press, New York 1966).
TRELSTAD, R. L.; HAY, E. D., and REVEL, J. P.: Develop. Biol. *16:* 78 (1967).
TRINKAUS, J.P.: Cells into organs (Prentice-Hall, Englewood Cliffs 1969).
TRINKAUS, J.P. and LENTZ, J.P.: Develop. Biol. *9:* 115 (1964).
TWITTY, V.C. and NIU, M.C.: J. exp. Zool. *125:* 541 (1954).
TYLER, A.: Proc. nat. Acad. Sci. U.S. *26:* 249 (1940).
TYLER, A.: in B.H. WILLIER, P. WEISS and V. HAMBURGER Analysis of development, pp. 556–573 (Saunders, Philadelphia 1955).
WARREN, L.: Curr. Topics Dev. Biol. *4:* 197 (1969).
WARREN, L. and GLICK, M.C.: in L.A. MANSON Biological properties of the mammalian surface membrane, pp. 3–15 (Wistar Inst. Press, Philadelphia 1968).
WEISS, L.: Exp. Cell res. *14:* 80 (1958).
WEISS, L.: J. theoret. Biol. *2:* 236 (1962).
WEISS, L.: J. gen. Microbiol. *32:* 331 (1963).
WEISS, L.: The cell periphery, metastasis and other contact phenomena (Wiley, New York 1967).
WEISS, P.: Growth, suppl. *5:* 163 (1941).
WEISS, P.: Yale J. Biol. Med. *19:* 235 (1947).
WEISS, P.: in B.H. WILLIER, P. WEISS and V. HAMBURGER Analysis of development, pp. 346–401 (Saunders, Philadelphia 1955).
WEISS, P. and ANDRES, G.: J. exp. Zool. *121:* 449 (1952).
WEISS, P. and SCOTT, B.I.H.: Proc. nat. Acad. Sci. U.S. *50:* 330 (1963).
WEISS, P. and TAYLOR, A.C.: Proc. nat. Acad. Sci. U.S. *46:* 1177 (1960).
WESTON, J.A.: Develop. Biol. *6:* 279 (1963).
WESTON, J.A. and BUTLER, S.L.: Develop. Biol. *14:* 246 (1966).
WILSON, H.V.: J. exp. Zool. *5:* 245 (1907).
ZWILLING, E.: in A. V. S. DE REUCK and M. P. CAMERON Cellular injury, pp. 352–368 (Churchill, London 1964).

Author's address: Dr. MICHAEL F. COLLINS, Department of Anatomy, The University of Connecticut School of Medicine, *Farmington, CT 06032* (USA)

Neoplasia and Cell Differentiation, pp. 279–318
(Karger, Basel 1974)

The Differentiation and Organization of Tumors *in vitro*

NELLY AUERSPERG and CYRIL V. FINNEGAN

Cancer Research Centre and Department of Zoology, University of British Columbia, Vancouver, B.C.

Contents

Introduction

The parallelism between the morphologic, metabolic and genetic abnormalities which characterize neoplastic and cultured cell populations appears to stem from the cells' adaptation, by rather similar means, to life in isolation. This adaptation leads to an increasing unresponsiveness to normal intercellular control mechanisms. As a result, cells in prolonged tissue culture become partly independent of their environment, and grow with similar characteristics under any adequate culture regime, while in the course of malignant progression *in vivo*, a conversion towards a common autonomous anaplastic tumor cell takes place.

Among the outstanding phenotypic characteristics of either neoplastic or cultured cells is their inability to maintain the normal differentiated state, a function which requires, in many adult tissues, a specific degree of supracellular organization to promote interactions among homotypic (like) as well as heterotypic (unlike) cells. This interdependence of differentiation and organization begins, in normal development, early in embryogenesis when the cells are first committed to a specific morphogenetic fate. Then epigenetic influences, based initially on the organization of the egg and subsequently on the immediate cellular micro-environment, progressively modify cytoplasmic functions, and thereby genetic activity and the cellular phenotype. As a result, cell interrelationships and the responsiveness of the cells to further external influences also change and thus, a constant interplay of epigenetic and genetic control mechanisms ensues through which increasingly restrictive commitments are imposed on the cells, commitments that are made manifest in the differentiation of tissue-specific structures and functions in the mature organism. It follows, as suggested by WOLPERT [1969], that the individual cell must come to know its exact position within any higher level of organization in order to undergo the correct molecular differentiation required to ensure that the proper histogenesis and organogenesis are accomplished and maintained. The means by which cells come to know their position in a developing system probably include the recognition of environmental and intercellular gradients [LOEWENSTEIN, 1968; CRICK, 1970], and the appearance of tissue-specific macromolecules on the cell surface [LILIEN, 1969]. It would seem logical that some of the important loci for the required cytoplasmic control mechanisms in such a system are at the region of the cell surface and cortical cytoplasm, an area which is immediately available to interact in the epigenetic events associated with, and controlling, development.

To have an effect on cellular activities, any environmental shifts must be within the interpretative capacity of the responding cells, for the organization and differentiation of a group of cells will be lost should the cells not be able to recognize their relative position in a system. Thus, some of the abnormalities which follow the dispersion of normal cells in tissue culture result from the inability of the cells to interpret this new drastically different position, because of the substitution of artificial, abnormal, organizational influences for the normal cues to which the cells had previously responded. On the other hand, the apparent inability of tumor cells *in vivo* to respond to the normal environment with which they are in contact seems to be based on an impaired capacity of these cells correctly to recognize and interpret their position, or to react to it, or both. Tumor cells, explanted into culture, should therefore respond as a blind person in a strange room; limited, because of an intrinsic defect, in their ability to react to an unusual situation or even to recognize it as such. One would expect then, that explanted tumor cells would respond to culture effects less consistently and less drastically than do normal cells.

Tumor cells can originate in cultures of normal cells as a result of transformation, which may occur spontaneously or be induced by viral, chemical or physical agents. In such a transformation to tumorigenicity, a hereditary defect which interferes with normal interactions with the environment is introduced into cells that have already undergone adaptive changes in response to the loss of normal intercellular control mechanisms *in vitro*. Thus, the sequence of events is at variance with that which occurs when tumors developed *in vivo* are explanted into culture, although it leads to a similar end result.

Unless otherwise stated it will be implied throughout this chapter that the tumors being discussed are malignant neoplasms. No satisfactory definition of malignancy in terms of cellular properties can be presented, since, as pointed out by FOULDS [1967], 'Malignancy is essentially an abstract concept based on medical practice: it is not a biological entity'. Malignancy of a cell population *in vivo* depends on certain combinations of a number of independently variable and dissociable characters, (such as uncontrolled proliferation, invasiveness, autonomy, or anaplasia), none of which by themselves provide a reliable criterion of malignant growth [FOULDS, 1969]. Thus, in reference to the behavior of a neoplasm, malignancy implies a tendency to be progressive, destructive and potentially lethal, in distinction to other less aggressive or more limited abnormal growths.

This chapter treats a comparison of the effects of the *in vitro* (tissue culture) environment on explanted normal tissues and on tumor tissues. Particular emphasis is placed on changes in differentiation and organization, and on the manner in which cellular changes during carcinogenesis *in vivo* may have modified subsequent *in vitro* growth and differentiation [see also HARRIS, 1964]. Some of the morphological and behavioral changes which are associated with *in vitro* transformation of normal cells, and which are also largely representative of the culture characteristics of explanted tumors, will be considered.

II. Culture of Normal Tissues

The maintenance of cytodifferentiation under adequate nutritional conditions *in vitro* is directly dependent on the degree of histotypic organization preserved in the explanted tissue. The preservation of this organization in turn is dependent on the presence of a cell mass of sufficient size, and is further enhanced in systems where, as in organ culture, cell migration is discouraged by the use of appropriate substrata. When the histotypic organization is maintained, complex morphological and functional differentiation may continue *in vitro* for considerable periods of time.

However, if culture conditions are such as to permit disorganization of the tissue to occur, as in cell-layer (monolayer or multilayer) culture, then cell migration begins rapidly and may be followed, within hours of explantation, by dedifferentation in the form of changes in cellular morphology, histotypic enzyme distribution and synthesis of DNA, RNA and protein, as well as in the loss of tissue-specific products and antigens [DAVIDSON, 1964; SAXÉN *et al.*, 1968]. It would appear that the signals from the immediate environment acting to maintain the differentiated phenotype are lost with migration, and the cells respond by a simplification of their morphology and physiology, i.e., by a shift to a less specialized state. That this initial limited dedifferentiation is frequently, if not always, only a modulation is demonstrated by its reversibility given the appropriate *in vitro* or *in vivo* modifications of nutritional, physical or mechanical environmental factors, or the reconstitution of a tissue mass in which the necessary cell interactions may occur [DANIEL and DE OME, 1965; CAHN, 1968; LAVIETES, 1970]. Variation in differentiation to cell types other than the originally explanted one has been reported; from neuroblasts to melanophores [COWELL and WESTON, 1970], from chondroblasts to fat cells [DAVIDSON, 1964] and from osteoblasts to chondrocytes [HALL, 1969]. While these variations from the initial phenotype may have been responses to some specific factor(s), it is also possible that a hierarchy of differentiation pathways, which are expressed according to an order of decreasingly exacting requirements, or which are selected in response to injury, are available to the cells. Indeed, squamous metaplasia of columnar epithelia *in vivo* may be based on a similar mechanism.

The levels of differentiation that are less dependent on undisturbed tissue organization for their expression, and which are maintained after cell migration in primary culture, vary according to tissue type. Terminal differentiation may continue in some cells, as in the formation of pigment [CAHN, 1968], cartilage [HOLTZER and ABBOTT, 1968], muscle fibers [BISCHOFF and HOLTZER, 1970] and stratified squamous epithelia [PROSE *et al.*, 1967; WILBANKS and SHINGLETON, 1970], but even so, some degree of intercellular contact, either retained or reconstituted, is required for optimal (sometimes even minimal) expression of the fully differentiated phenotype. Thus, initial loss of tissue-specific properties in culture appears to follow the disturbance of normal intercellular associations, rather than either genetic changes or the onset of mitotic activity, though the proliferative rate of those cells competent to divide does increase considerably with dispersion.

It is interesting to note that the sequence of events which follow explantation of a tissue fragment into culture, i.e., cell migration, dedifferentiation and proliferation, is strikingly similar to the cellular response to positioning at the edge of a wound *in vivo.* In fact, the cellular responses *in vitro* may represent a misinterpretation by the explanted cells of the positional information available to them, which leads to a response within the scope of their interpretative capacity and is represented by activities similar to those occurring in regeneration.

Regression of the level of differentiation continues with the passage of time in dispersed cell culture and the ability of the cells to redifferentiate appears to be increasingly curtailed, if not lost. This gradual reduction in tissue-specific properties is not simply a direct response to the artificial environment or to disaggregation *per se,* but rather, dedifferentiation appears to progress as the cumulative effect of repeated cell divisions in the absence of adequate intercellular contact. It follows then, that the rate, as well as the permanence, of dedifferentiation in culture may be controlled, to a degree, by alterations in either mitotic activity or in the extent of intercellular organization. Thus, isolated but non-dividing nerve cells or muscle fibers may remain differentiated under culture conditions in which similarly isolated, but dividing, cells dedifferentiate; while the confinement of proliferating cells to a limited space, where cell dispersal is reduced and intercellular contact enhanced, may result in the maintenance of a higher level of differentiation [ROSE, 1967]. Indeed, differentiation may be proportional to the rate of proliferation when, as in organ culture, maximal histotypic organization is sustained [FELL, 1957]. The dependence on cellular interactions for the maintenance of differentiation in proliferating cultured cell populations does seem to be to some extent tissue-specific and proportional to the normal contact requirements existing *in vivo*. Thus, for example leukocytes, fibroblasts and myoblasts are more apt to express, or remain competent to express, differentiated functions after periods of rapid proliferations in relative isolation from other cells, than are epithelial cells [MOORE and MCLIMANS, 1968; ELSDALE and FOLEY, 1969; RICHLER and YAFFE, 1970].

Standard culture regimens involve growth on a two-dimensional substratum with periodic cell-dissociation and dispersal. Normal cells will proliferate for varying periods when explanted to such conditions, but eventually the majority of cultures will degenerate and die out. While the average life span of cultures varies considerably, it is quite characteristic and predictable for many types of normal cells under a given set of culture conditions [HAYFLICK and MOORHEAD, 1961]. There is evidence accumulating that technical and, in particular, nutritional inadequacies may be largely responsible for the limited life span and for some of the dedifferentiation of normal cells *in vitro* [MACPHERSON, 1970; RICHLER and YAFFE, 1970; SPOONER and HILFER, 1971].

A small proportion of cells may survive this initial culture period and transform into an autonomous, 'permanent' cell line. There is now evidence that such transformations in response to prolonged growth *in vitro* might, in fact, take place spontaneously, i.e., without the influence of a demonstrable viral or chemical agent [RANADIVE *et al.*, 1968; SHARON and POLLARD, 1969; MACPHERSON, 1970; SANFORD *et al.*, 1972]. That an interference with normal relationships to adjacent tissues may be sufficient to cause an irreversible hereditary change to autonomy in normal cells and in their progeny, is indicated also by the transformations to malignancy which can be induced *in vivo* by growth either in diffusion chambers [SHELTON, 1963] or adjacent to plastic, asbestos or glass [CARTER and ROE, 1969; STANTON and WRENCH, 1972]. *In vitro*, the rate of success of transformations to culture autonomy is both tissue and species specific [MACPHERSON, 1970] and it requires a minimal time period of growth in culture before it takes place, the length of the period depending on the cell type and culture conditions [PARSHAD and SANFORD, 1972]. As this period may be shortened by growing the cells under suboptimal conditions [HAYFLICK, 1967] or, conversely, extended by providing special nutrients [MACPHERSON, 1970], it would appear that some cumulative sublethal damage is one of the factors leading to spontaneous transformations. While by definition, cell lines are considered as established or permanent when they have demonstrated the potential to grow indefinitely *in vitro* [FEDOROFF, 1967], in practice it may be difficult to determine at what point such a potential has been acquired, except with tissues having well known predictable growth characteristics as, for example, normal human fibroblasts [HAYFLICK and MOORHEAD, 1961]. The phenotypic changes of spontaneous transformation tend to appear gradually, in individual cells and in terms of the proportion of cells in the population, and tend to lead to the eventual acquisition by most lines of a relatively uniform set of characters in terms of growth rates, morphology, behavior, metabolism and cytogenetics. Tumorigenicity is acquired sometimes, but not invariably [SANFORD *et al.*, 1967] (see Section IV).

It now appears that much of the uniformity observed in the older established cell lines may, in fact, have been the result of the adaptation to, or the selection of cells for, standard tissue culture regimens, as recent modifications of *in vitro* techniques have permitted the establishment of cell lines with more individual characteristics [MOORE and MCLIMANS, 1968; RICHLER and YAFFE, 1970]. Indeed, low levels of differentiation appear to be maintained in spite of the persisting relative uniformity and apparent dedifferentiation in some older cell lines (iron fixation in hematopoietic cells, glycogen synthesis in liver cells, retention of tissue-specific antigens in others) [DAVIDSON, 1964]. Also, a competence to differentiate after long periods of proliferation has been demonstrated in some apparently undifferentiated cells, as, for example, in the selective stimulation of lipid accumulation in connective-tissue cells by hydrocortisone, the induction of collagen synthesis in fibroblasts following modifications of substrata and growth media, and the formation of contractile muscle fibers by myoblasts after crowding [DAVIDSON, 1964; RICHLER and YAFFE, 1970]. One of the more striking examples of competence retention in long-term culture of a normal tissue was that of an established line of bone marrow cells which, upon inoculation into irradiated animals, served as functionally competent hematopoietic stem cells [MCCULLOCH and PARKER, 1956]. No persistent competence for terminal differentiation, however, has been demonstrated in any existing cell lines of normal epithelial origin.

In the search for tissue-specific properties in cell lines it is important to rule out random variations occurring *in vitro* from having histotypic significance. For example, high rhodanase or arginase activity was found in a number of cell lines of diverse normal origins, while such activity *in vivo* would have been characteristic only of liver cells [WESTFALL, 1958].

In general, it would appear that the rate, the degree and the permanence of dedifferentiation occurring in cultured normal cells all vary with the tissue type of origin and, in particular, depend on the degree of intercellular organization that is required for the maintenance of differentiation by that tissue *in vivo*. Further, dedifferentiation is a function of the degree of tissue disorganization that takes place *in vitro*, of the time period during which the cells remain dedifferentiated, and of the mitotic activity of the population. Obviously, both the differentiation and the length of survival of a cell population are affected by modifications in the nutritional conditions of the culture, and undoubtedly the full capacity to differentiate of many cultured cells remains unknown because of the limiting environmental conditions *in vitro*.

The requirement for a minimal time of exposure to, and proliferation in, a particular environment, and the gradual, stepwise expression of the cellular characters associated with spontaneous transformation, are strikingly reminiscent of inductive phenomena in normal embryonic development.

III. Culture of Explanted Tumors

A. Organ Culture

Since the organization of tumor explants is optimally preserved in organ culture, this technique has been used extensively for the study of hormone effects on malignant tissues where the maintenance of a high degree of histologic integrity seems frequently required. Individual differences in hormone dependence or hormone-responsiveness which may persist among tumors within one histologic group can be analyzed in this system over brief periods of time [TURKINGTON and RIDDLE, 1970; HEUSON and LE GROS, 1972].

On occasions, the level of differentiation in organ culture exceeds that reached by the tumor *in vivo*, as illustrated by the keratinization of basal-cell carcinomas [FLAXMAN and VAN SCOTT, 1968] or the tubule-formation by anaplastic renal carcinomas [ELLISON *et al.*, 1969] upon explantation *in vitro*. Such excessive differentiation might take place in response to nutrient factors, or could result from the physical limitation of cell dispersal by the culture environment, which permits intercellular contact and organization to proceed further than *in vivo*. Alternatively, the response might be due to sublethal cell damage by sudden exposure to a relatively foreign environment and the result similar to the increase in differentiation sometimes observed *in vivo* in metastases, as well as in carcinoma cells upon transplantation [VAN SCOTT and REINERTSON, 1961], or during their initial unsuccessful attempts at invasion [WEED, 1968]. Finally, such increased differentiation following explantation could be the result of removing restraining host influences [FLAXMAN and VAN SCOTT, 1968] as is also sometimes observed in embryonic tissues in culture [STEINBERG, 1970].

B. Matrix Culture

Three-dimensional matrices provide a physical framework with closely confined spaces, where intercellular contact as well as the conditioning of the cellular micro-environment can reach levels that far exceed those possible in cell-layer culture. In a matrix, differences in proliferative capacity, differentiation and growth patterns between normal and malignant tissues as well as between individual tumors of a histologic group, may be maintained or partly restored following cell dispersal [LEIGHTON, 1954]. By using cellulose-sponge matrices as an artificial stroma, tumors previously grown as dedifferentiated cell-layers resume levels of organization and differentiation that approach their original histologic appearance *in vivo* [AUERSPERG and HAWRYLUK, 1962], and in association with a more adhesive, collagen-coated cellulose-matrix, carcinoma cells may form complex functional multicellular structures [LEIGHTON *et al.*, 1970]. When mixtures of malignant cells and embryonic tissues are allowed to sort out in this environment, the cancer cells invariably segregate on surfaces facing the culture medium, thus taking up the nutritionally optimal areas. It has been suggested that this might be due to their nondirectional migratory ability, and that a similar 'neoplastic blockade' of metabolically advantageous sites may take place *in vivo* [LEIGHTON, 1968].

In a series of tissues explanted into a fibrin foam matrix, malignant cells proliferated into adjacent spaces [KALUŠ *et al.*, 1968], and some of these outgrowths showed evidence of histogenesis (acinus-formation) and terminal differentiation (mucin synthesis), in the apparent absence of supporting connective tissue. It would seem, therefore, that these cells contained all the information required for histotypic organization and differentiation, given the establishment of sufficiently intensive cell interactions under the influence of the matrix framework. In contrast, fetal tissues such as liver or intestinal mucosa in the same culture system, formed organized outgrowths only in association with connective tissue elements, and benign tumors or normal adult tissues (with few exceptions) did not proliferate or spread into adjacent areas [KALUS and O'NEIL, 1968].

Thus, in three-dimensional culture, the histotypic and individual characteristics of explanted tumors are largely retained, and upon provision of an artificial stroma (matrix), cells that had dedifferentiated on two-dimensional substratum exhibit a considerable capacity to redifferentiate. The cells on occasion show self-organizing capacities which require heterotypic cell interactions in normal tissues, and they are able to proliferate under conditions inhibitory to normal tissues.

C. Short-Term Cell-Layer Culture

Malignant cells show considerable variation in their adaptability to migrate and proliferate from explants onto two-dimensional substrata. Therefore, the growth in primary cell-layer culture in a series of tumors from a single histologic source may range from no growth in some cases to extensive and sustained proliferation in others. In contrast, the growth in cell-layer cultures of most normal tissues is fairly consistent, but, particularly for epithelia, quite limited. Thus, if the growth potential is measured as a function of cases showing any outgrowth at all, the consistently growing normal tissues may appear at an advantage, though the average rate and duration of continued proliferation is frequently considerably greater in a group of tumors of the same tissue origin [SHERWIN *et al.*, 1967; PARK and KOPROWSKA, 1968]. The growth potential *in vitro* of malignant cells cannot be reliably predicted from their growth rate or growth pattern *in vivo* [WALKER *et al.*, 1965]. Rather, the survival of a tumor in culture seems to depend on the compatibility between the particular combination of cellular properties developed *in vivo* and the requirements for survival in culture, as well as on the capacity of the tumor cells to modify these properties in response to environmental influences *in vitro*.

In primary culture, tumor cells, like normal cells, exhibit phenotypic variations which reflect tissue-specific properties, but, in addition, the malignant cells retain some of the individual variations in adhesiveness, social behavior and differentiation which existed among the tumors *in vivo* [FOLEY and AFTONOMOS, 1965; WALKER *et al.*, 1965; SHERWIN, 1968].

Because of this variation on the relatively uniform and predictable theme shown by the normal tissue equivalent *in vitro,* the majority of tumor cultures deviate from normal in one or more of these parameters. Thus, normal squamous epithelium of the uterine cervix grows consistantly as a flat, cohesive cell layer [WILBANKS and SHINGLETON, 1970], but the intercellular organization in a series of carcinomas cultured from the same tissue may range from groups of single dispersed cells, to confluent monolayers, to highly cohesive multilayered colonies [AUERSPERG and WORTH, 1966]. This organization *in vitro* frequently correlates with the *in vivo* growth pattern of the individual tumor and similar correlations in growth patterns have been described for cultures of hepatomas [WATANABE and ESSNER, 1969], hypernephromas [ISRAELI and BARZILAI, 1970] and even for ascites tumors [SATO, H. *et al.,* 1968]. Varying degrees of tissue-specific intercellular organization may be maintained by groups of pulmonary carcinoma cells which have separated from the stroma of explanted tissue fragments. Cohesiveness within such cell groups varies considerably and does not seem to correlate with the adhesiveness of the tumor cells to stromal tissues. Similar isolated groups of homotypic cells, able to maintain a degree of histotypic organization, were never observed in cultures of normal lung tissue [SHERWIN *et al.,* 1967; SHERWIN, 1968].

Although malignant progression, as a rule, is associated with variable degrees of dedifferentiation, such tissue-specific properties as may have been retained *in vivo* by tumors appear to be expressed in primary cultures to a greater extent than are those present *in vivo* in normal tissues. Examples of terminal differentiation by malignant cells include the synthesis of histamine [FOLEY and DAVIS, 1965], melanin [KAHN and DONALDSON, 1970], thyrocalcitonin [GRIMLEY *et al.,* 1969] and mucin [REED and GEY, 1962]. While gradual changes in the morphology and organization of tumor cells in primary cultures may be indicative of cell degeneration, they may also indicate progressive *in vitro* differentiation, as, on a two-dimensional substratum, neuroblastoma cells appear to differentiate into mature nerve cells [GOLDSTEIN *et al.,* 1964] and epidermal or cervival carcinomas of the basal-cell type may undergo squamous differentiation [AUERSPERG and WORTH, 1966; FLAXMAN, 1972].

It would appear then, that, in short-term cell-layer cultures, tumor cells may dedifferentiate no more, and possibly less, than normal cells, and may show considerable growth potential. However, these characteristics are not consistent because malignant cell populations also retain individual differences in growth potential, intercellular organization and the capacity to differentiate. Carcinoma cells may undergo histogenesis in the absence of heterologous tissues; and, in comparison with normal tissues, tumor cells show a variable and possibly reduced effect of explantation on organization, and of disorganization on differentiation. This reduced effect of the culture conditions may reflect the fact that some degree of disorganization and dedifferentiation already exists in the tumors prior to explantation, as well as the fact that the responsiveness of malignant cells to environmental influences acting *in vitro* is impaired.

D. Long-Term Cell-Layer Cultures

Defining a malignant tissue in culture as a permanent cell line is made difficult by the diversity of the material and by the variable presence, already at the time of explantation, of properties like heteroploidy which serve as indicators of transformation to culture autonomy in cells from normal tissues. The metabolic and structural similarities of malignant cells *in vivo* to cell lines *in vitro* undoubtedly are of advantage to tumor cells in their adaptation to long-term culture [PARK and KOPROWSKA, 1968], and may account in part for the somewhat greater rate of success at which tumor cell lines are established as compared to their normal counterparts. The increased resilience and variability (both genetic and epigenetic) which cell populations acquire in malignant progression may also contribute to their ability to survive the major environmental changes encountered *in vitro*. However, most explanted malignant tumors, like their normal counterparts, do not proceed to grow indefinitely *in vitro* but survive only for a few cell generations. Therefore, while superficially similar, cell populations which evolve in carcinogenesis and in long-term culture are not identical and adaptation to one of these conditions does not *a priori* provide for survival in the other. This is made evident also by the observations that transformation to culture autonomy by normal cells does not necessarily make them tumorigenic [HAYFLICK, 1967; IOACHIM, 1969; MACPHERSON, 1970], that existing tumorigenicity may be lost by cells in prolonged culture [FOLEY *et al.*, 1965; ENG *et al.*, 1970] and that transformation to malignancy of cells *in vivo* does not directly result in their autonomy *in vitro* [DI PAOLO *et al.*, 1971].

Fluctuations in degrees of tumorigenicity and of differentiation *in vitro* by malignant tissues seem to take place independently [SHEIN, 1968; HU, 1969], but an interesting parallelism has been reported recently in the controls of these two cell characters [SILAGI and BRUCE, 1970]. Melanin formation and tumorigenicity in melanoma cell lines were both reversibly inhibited by BUdR (5-bromodeoxyuridine) at concentrations low enough to have little effect on the growth rates of the cultures. Thus, the regulation of these two cell characters was affected by BUdR differentially from the synthesis of products required for growth and division. Furthermore, BUdR seemed to affect differentiation and tumorigenicity similarly but independently, since the differentiation of normal embryonic cells [BISCHOFF and HOLTZER, 1970; COLEMAN *et al.*, 1970] and the tumorigenicity of amelanotic melanoma cells both could be inhibited by similar means. The inhibition of both characteristics was associated with changes in cell shape, reduced intercellular contact and increased contact inhibition of movement.

Long-term cell lines originating in tumors appear to differ from lines of normal origin in three important ways: 1) in their greater capacity to form multicellular organized structures in the absence of heterologous tissues, 2) in their greater capacity to maintain high levels of differentiation, and 3) in the relative ease with which the cells derived from malignant tissues adapt to long-term growth in suspension cultures. Each of these three characteristics will be considered separately.

1. Histotypic Organization in Long-Term Tumor Cultures

A limited number of tumor lines have retained the ability to form simple organized structures in the absence of heterologous tissues or supporting structures. In some cases this morphogenetic capacity is associated with the continuing synthesis of specific secretory products, such as hormones [PATILLO and GEY, 1968], or mucin [REED and GEY, 1962], while in other cell lines different functional or structural evidence of differentiation may be observed. Examples of the latter are seen in a cell line of renal origin forming multicellular blisters, reminiscent of renal tubular structure and activity, that seem to result from coordinated secretion with active fluid transport from the medium [LEIGHTON *et al.*, 1970]; in two other cell lines derived from a squamous carcinoma in which structures characteristic of the basal-cell and spinous-cell levels of differentiation in normal stratified squamous epithelium were observed [AUERSPERG, 1969a]; and in a cell line of hepatocarcinoma which retained the ability to form micropapillary and polypoid structures over several years although no liver-specific biochemical functions were identified [DAWE *et al.*, 1968b]. It is interesting to note that in the last case the epithelial cells were replaced on three separate occasions by rapidly growing spindle-shaped cells which did not form any organized structures, and which, since contamination appeared to be ruled out, seem to represent spontaneous transformations occurring in an already malignant cell line.

In all these cell lines the histogenetic capacity is increasingly expressed as the cultures become crowded so that differentiation seems to depend on the ability of the cells to establish autonomously a degree of density which permits the necessary interactions to take place. Apparently, other non-organizing tumor lines (as HeLa) may retain the competence to demonstrate tissue-specific organization if external support and closely confined spaces are provided, which may act to increase intercellular contact and to permit more intensive modifications of the cellular micro-environment [ROSE, 1967].

It should be noted that in the culture of most of these self-organizing lines, cells were always transferred in cohesive groups, since dissociation into single cells seemed detrimental to subsequent differentiation and sometimes also to subsequent proliferation. Thus, some of the control mechanisms associated with proliferation and differentiation (either their development or retention) were likely to be located in the cell periphery or cortical cytoplasm and became rapidly and irreversibly lost or nonfunctional in the absence of uninterrupted intercellular contact. It may be significant that among the lines mentioned, the only one which seems independent of continued intercellular contact for the long-term maintenance of its self-organizing capacity is of renal tubular origin [LEIGHTON *et al.,* 1970]. Here, completely dissociated cells retained the ability, apparently indefinitely, to reconstitute functional multicellular cysts and blisters. Similarily, in embryogenesis, kidney tubules may develop by the aggregation of previously independent mesenchymal cells which secondarily form an epithelium [SAXÉN *et al.,* 1968]. Therefore, the control mechanisms associated with this renal type of differentiation, in contrast to the other epithelia, may be either more stable or not irreversibly lost upon cell dissociation. The ability of malignant renal cells to proliferate and to organize in the absence of heterotypic influences could indicate that these tumor cells are independent of such influences, while, on the other hand, the dependence of the other carcinoma cells on continued intercellular contact could suggest that these malignant epithelial cells had themselves acquired some of the inductive properties of the originally associated mesenchymal (heterotypic) cells, and thus can exert 'heterotypic' influences on each other. This possibility is suggested also by the ability of carcinoma cells to form basement membrane without any contribution from stromal tissues [PIERCE, 1965], by their production of collagenase-sensitive material *in vitro* [AUERSPERG, unpublished], and by the capacity of some cultured epithelial cell lines to induce bone or cartilage *in vivo* [WLODARSKI, 1969]. Furthermore, the aggregation characteristics of carcinoma cells in gyratory shaker cultures seem to resemble those of mesenchymal embryonic tissues more than those of epithelial embryonic tissues [KITANO and KURODA, 1967; KURODA, 1968; REHM, 1968]. Finally in this regard, it may be significant to note that, by aggregation criteria, tumor cells are more cohesive than are their normal counterparts in the adult organism. In analogy, the capacity of embryonic cells to aggregate and to use other cells (rather than intercellular substances) as substrata, diminishes as they differentiate.

The most direct evidence that carcinoma cells may acquire inductive properties usually associated with mesenchymal cells comes from studies on the malignant transformation of embryonic mouse submandibular gland by polyoma virus, a transformation which, like embryonic induction, occurs *in vitro* in intact, or reconstituted, submandibular-gland rudiments but not in either isolated submandibular epithelium or mesenchyme [DAWE *et al.*, 1966]. Using mixtures of epithelial and mesenchymal components of submandibular-gland rudiments from two strains of mice, one of which possessed a chromosomal marker, it was possible to induce tumors with polyoma virus and to establish that all were epithelial in origin [DAWE *et al.*, 1971]. Yet these epithelial tumors could substitute for normal mesenchyme in supporting either normal epithelial morphogenesis in salivary glands or the malignant transformation by virus of the normal epithelium [DAWE *et al.*, 1968a]. The role of mesenchyme in supporting either epithelial differentiation [RUTTER *et al.*, 1968] or transformation [DAWE *et al.*, 1966] would seem to involve a capacity to induce proliferation, a capacity which the epithelial cells apparently may acquire in the course of malignant transformation.

The ability of epithelial cells mutually to support and stimulate their own organization in the continued absence of heterologous tissues is limited to malignant cells, and has led to the suggestion [LEIGHTON *et al.*, 1960] that aggregates of carcinoma cells *in vivo* may not just represent accumulations of independently multiplying cells, but replicating multicellular units within which cellular interactions may take place that largely determine the phenotypic characteristics of tumors [LEIGHTON, 1969].

Presumably, the cells from the majority of tumor cell lines, which grow singly, clone easily and form no multicellular structures, have lost any dependence on other cells for proliferative activity and have lost also any histogenetic responsiveness to intercellular contact.

2. Differentiation in Long-Term Tumor Cell Lines

Cell lines of tumor origin may, like those derived from normal tissues, retain the capacity to differentiate in response to the proper agent [DAVIDSON, 1964] or upon reconstitution of intercellular organization, either *in vitro* or *in vivo* [ROSENTHAL *et al.*, 1970]. A variety of different functions is also expressed under standard culture conditions, including such relatively non-specific activities as the formation of tonofibrils and desmosomes [SYKES *et al.*, 1970] or basement membrane [PIERCE *et al.*, 1964], as well as the more specific production of lipids [MORTON *et al.*, 1969], melanin [YASUMURA *et al.*, 1966b; HU, 1969], mucus [REED and GEY, 1962] or hormones [SATO *et al.*, 1970]. This increased capacity of some malignant cell types to proliferate in isolation over long periods and still remain competent to differentiate was dramatically demonstrated in the development of cell lines of teratocarcinoma which remained multipotential and, even after repeated cloning, retained the ability to form highly differentiated structures in aggregates or, to a degree, in two-dimensional culture [ROSENTHAL *et al.*, 1970]. As a comparison, it should be noted that the potential of very early embryonic cells to grow and differentiate in culture is minimal [BRINSTER and THOMSON, 1966; COLE *et al.*, 1966].

Although hormone secretion and hormone responsiveness by cells in culture were demonstrated long ago [GEY and THALHIMER, 1924; GEY *et al.*, 1938], it is only lately that stable cell lines with such functions have been established from tumors [SATO *et al.*, 1970]. This advance was largely due to the modifications of standard culture techniques designed to permit the explanted cells to adapt gradually to an *in vitro* environment, either by alternating periods of short-term culture and animal passages [BUONASSISI *et al.*, 1962], or by the establishment of transplantability *in vivo* followed by an *in vitro* regime designed to mimic initially the *in vivo* environment [PATILLO *et al.*, 1971]. In either system the cell lines adapted subsequently to standard culture conditions but remained highly differentiated, in some instances even upon cloning. It is possible that this kind of culture regime permits the adaptation to long-term culture of cell types that would otherwise not have survived even in a dedifferentiated state. Alternatively, the degree of dedifferentiation which took place may have been modified by the rate at which the environmental changes were introduced. Since the changes were gradual, they may have permitted the cells to respond by a stepwise development of increasing autonomy which enabled them to substitute cellular, or homotypic, controls for environmental, or heterotypic, ones in order to maintain the differentiated phenotype, and thus prevent irreversible dedifferentiation from occurring. It does appear as though tumor cells may have a greater ability to undergo such adaptive changes than do normal cells since similarly differentiated long-term cultures of normal endocrine tissues have not been established.

Cell lines established with these techniques include a steroid-secreting Leydig cell line, several pituitary-tumor cell lines which secrete ACTH or growth hormone [YASUMURA *et al.,* 1966b; SONNENSCHEIN *et al.,* 1970] and a hormone synthesizing human choriocarcinoma cell line which, since it produces both glycoprotein and steroid hormones, apparently has retained some multipotentiality [PATILLO *et al.,* 1971]. Other cell lines, obtained from adrenal tumors, appeared undifferentiated but still remained competent to respond to ACTH stimulation *in vitro* with steroid production [YASUMURA *et al.,* 1966a]. As this response is accompanied by a change in cell shape (i.e., rounding up), these cells may be used for a direct visual assay of ACTH activity [SCHIMMER *et al.,* 1968]. While this competence to respond to hormone induction was maintained for years *in vitro,* it was found to be reversibly lost if the cells were cultured in media with a low serum content. However, the minimal nutritional requirements of these cultures, whose full differentiation seems to depend upon environmental (nutritional) factors, appear to be similar to those of other undifferentiated cell lines, and the nutritional requirements for tissue-specific function could not be dissociated from those required for cell growth [CUPRAK and SATO, 1968].

To date, little information is available regarding prolonged growth in culture of hormone-*dependent* tumors other than the description [SATO *et al.,* 1970] of a method, employing selective destruction *in vitro* of hormone-independent proliferating cells with BUdR treatment, which led to the establishment of hormone-dependent cell lines.

3. Suspension Cultures of Tumor Cell Lines

Suspension cultures have been established from ascites tumors, malignant effusions and solid tumors [MOORE and KOIKE, 1964; MORGAN *et al.,* 1968], and, while representing the least degree of organization *in vitro,* the cells in some cases have continued to perform such specialized functions as mucin, hormone or melanin production [TOSHIMA *et al.,* 1968; BANCROFT and TASHJIAN, 1971; OBOSHI *et al.,* 1969], possibly in analogy with the differentiation of tumor cells in effusions *in vivo.*

A large number of cell lines in suspension culture are of hematopoietic origin [MOORE *et al.,* 1968], and there is an interesting similarity in the properties of cells grown from both healthy and leukemic donors [MOORE and MCLIMANS, 1968]. In cultures of either origin, the establishment of a permanent line is indicated by an increase in growth rate, increased acid production and the clustering of cells into large clumps, as though cell-surface changes had occurred which permitted increased cell contacts. It has been suggested that the lines were initiated by a small number of cells which survived the initial period of slow growth, and that these might be either dedifferentiated leukemia cells, or normal lymphocytes transformed by environmental factors[1]. Cells of normal and of malignant origin could not be distinguished by their life span or morphology since lines of normal lymphocyte origin have continued growth in culture for years without showing transformation, and since they, as well as cells from leukemic donors usually resemble lymphoblasts and both show a similar morphologic variation. Furthermore, cells from both sources may produce complete or incomplete immunoglobins. In addition, many of the cell lines derived from leukemias or lymphomas have normal karyotypes, making it difficult to ascertain their precise identity since, in contrast to most other tumors, human leukemia cells frequently remain diploid *in vivo* so that diploidy of cells *in vitro* is not necessarily indicative of their non-malignant origin.

Criteria which do distinguish human leukocytic cell cultures of normal and of malignant origin include the observations that only cell lines of malignant origin have been heterotransplanted successfully and that cells of malignant origin have a higher cloning efficiency in agar. This latter observation likely indicates a difference in requirements for direct or indirect cell interactions rather than a difference in anchorage dependence (see Section IV).

4. *Summary*

Both tissue-specific and malignancy-associated characters are retained to variable degrees by tumor cells in long-term culture.

While some of the properties acquired during carcinogenesis appear to be advantageous for the establishment of a permanent cell line, there is no consistent relationship between the degree of tumorigenicity of a cell population and its propensity to survive permanently *in vitro.* Differences in organization and differentiation between long-term cell lines of normal origin and of malignant origin are related to the independence from growth controls developed by the tumor cells *in vivo* and maintained to varying degrees in long-term culture. However, the autonomy acquired during carcinogenesis and the autonomy acquired in culture are not equivalent or cumulative. Rather, the impaired responsiveness which isolates tumor cells

1 It has been suggested recently that Epstein-Barr virus infection may be a prerequisite for the indefinite growth of lymphoblastoid cells in vitro [NILSSON *et al.,* 1971].

against normal host-control mechanisms *in vivo*, appears to render them less responsive to those tissue-culture influences that limit the life-span of normal cells and furthermore, may also isolate them against environmental influences *in vitro*, which lead to the disorganization and dedifferentiation of normal cells. Thus, contrary to possible expectations, tumor cell lines exhibit more phenotypic diversity and, on occasion, may retain more complex levels of organization and differentiation, than do lines of normal origin. The fact that this is not a consistent occurrence suggests that the autonomy acquired *in vivo* may vary at the time of explantation *in vitro*, and might also be altered or lost in culture.

IV. Malignant Transformation in vitro

A. General Introduction

In 1941, GEY reported that normal rat fibroblasts, after several months in culture, had acquired the ability to grow as malignant tumors when reimplanted into the animal strain of origin. This transformation, which occurred without apparent cause, was associated with an increase in growth rate, with changes in cell morphology, and with changes in cell behavior [GEY, 1941]. Subsequently, numerous other transformations to malignancy of normal cultured cells, all seemingly spontaneous, were reported [SANFORD, 1965], and the more recent discovery that similar transformations could be induced in a predictable manner by viruses, chemical carcinogens and, possibly, irradiation, has provided some of the most extensively studied models of experimental carcinogenesis [MACPHERSON, 1970]. The acquisition of tumorigenicity *in vitro* frequently seemed to accompany the transformation of normal cells to a permanent cell line, but the two processes are not equivalent. Rather, tumorigenicity behaves as one of several characteristics that may be acquired by established cell lines, characteristics such as an increase in growth rate and acid production, heteroploidy, and changes in antigenicity, growth requirements, cellular morphology and behavior.

Causal interrelationships between these cellular characteristics, which appear in various numbers and combinations following tansformation to culture autonomy, are largely unknown [SANFORD, 1967] and while some of them are definitely advantageous for survival in culture, tumorigenicity *in vitro* is a selectively neutral property that has no influence on the growth of the cells. The tumorigenicity *in vivo* of these same cells would, of course, depend on their antigenicity and the vigor of the host reaction as well as on the ability of the cells to utilize available nutrients and to initiate rapid proliferation with dispatch. Because of the mutual independence of transformation-associated cellular changes, the only adequate means of demonstrating malignancy in cultured cell lines is by their growth as destructive, invasive and potentially lethal tumors in immunologically compatible animals. In practice then, malignancy becomes a quantitative character for, while the *in vivo* endpoint of death is absolute, the capacity of the cultured cells to produce this result varies, as shown in differences among the latent periods and growth rates of the tumors produced and in the variations in minimum inoculum sizes required. Thus, failure of tumor formation by cultured cells is meaningful only on a comparative basis under identical experimental conditions.

Tumorigenicity tends to increase gradually during spontaneous transformation but more rapidly following viral or chemical transformation, and it may even increase in response to double transformation of already transformed, tumorigenic cells [DI PAOLO *et al.,* 1968]. Subsequently, the ability of cells in continuous culture to produce tumors may fluctuate or be lost completely [ENG *et al.,* 1970].

Some indication of the possible relationship between tumorigenicity and changes in the organization and differentiation of cultured cells can be made. The most common morphologic and behavioral characteristics of transformed cells include changes in cell shape, dedifferentiation, random growth, increased maximum cell density, and a greater independence from both substrata and nutritional factors [MACPHERSON, 1970]. Furthermore, it was found that when transformed tumorigenic cells show a reduction in malignancy it is frequently accompanied by the loss, to varying degrees, of these cell characters. Thus, in cells transformed by either RNA or DNA viruses, the loss of viral genome may result in a reduction of tumorigenicity accompanied by a change from random to more normally-oriented growth, and in a reduced ability to grow in suspension [MACPHERSON, 1966, 1970]. Similarly, when transformation is reversed by selective low-cell-density culture conditions in the continued presence of the viral genome [RABINOWITZ and SACHS, 1968], tumorigenicity of the cells is reduced and they show changes in shape, an increased adhesion to substrata and a reduced maximum cell density. Furthermore, when the tumorigenicity of polyoma-transformed cells is reversed, some of the cell-surface characteristics of normal cells are restored [INBAR *et al.*, 1969; RABINOWITZ and SACHS, 1970a]. When cells transformed by chemical carcinogens or irradiation undergo reversion, they regain, in addition to the above charcteristics of more normal cells, the property of a limited life span *in vitro* [RABINOWITZ and SACHS, 1970b].

Since these data indicate, though the correlation is by no means absolute [IOACHIM, 1969; RABINOWITZ and SACHS, 1969a], that some parallelism between degrees of tumorigenicity and transformation-associated changes in cell morphology and organization exists, the remainder of this chapter will explore in greater detail such phenomena as 1) contact inhibition of movement, 2) density-dependent inhibition of growth, 3) dependence on substrata and growth factors, 4) cell differentiation, 5) cell morphology, and 6) the structural basis for cellular changes in transformation.

B. Changes in Cellular Morphology and Behavior

1. Contact Inhibition of Movement

The change from an ordered monolayered growth pattern to one where the cells overlap and migrate at random, is among the more consistent modifications associated with transformation. This change in intercellular organization is usually interpreted as indicative of a reduction in contact inhibition of movement, i.e., a reduction in the proportion of collisions between migrating cells that result in a redirection of cellular movement [ABERCROMBIE, 1970]. On a population basis, such contact inhibition among either normal or malignant cells is not an all-or-none phenomenon [SANFORD, 1967] and quantitative studies of many non-neoplastic cell populations have shown considerable tissue, species, and age dependent variation in the degree of contact inhibition exhibited. In addition, the efficiency of this control mechanism depends on nutritional factors and other modifications of culture conditions [BAKER and HUMPHREYS, 1971], and is also directly proportional to the strength of cell-substratum adhesion [CARTER, 1965; BOYDE *et al.*, 1969]. The failure of contact inhibition of movement between transformed cells, which may be partially accounted for by a reduced adhesiveness to substrata, is probably mainly the result of cell-surface changes associated with transformation that interfere with the establishment of close intercellular contact. This contact is most likely required for the initiation of the cell membrane paralysis which preceeds the redirection of cellular migration, and appears to involve the formation of temporary tight junctions [FLAXMAN *et al.*, 1969], structures strikingly rare in transformed fibroblasts [MARTINEZ-PALOMO *et al.*, 1969] and in tumor cells in general.

The precise relationship of variations in contact inhibition to fluctuations in the tumorigenicity of cultured cells is not clear since, while increased randomness of cell movement does usually accompany the acquisition of tumorigenicity, sensitivity to contact inhibition may be retained [DEFENDI *et al.*, 1963] and malignant transformation may even occur without any observable change in cellular growth patterns [SANFORD, 1967; DI PAOLO *et al.*, 1968]. One of the initial changes in polyoma-transformed hamster embryonic fibroblasts was the change from oriented to randomly growing cells, but these cells were not tumorigenic until they acquired, subsequently, the ability to grow at high densities. Thus, in this instance, the loss of contact inhibition and the acquisition of tumorigenicity were not concurrent phenomena [VOGT and DULBECCO, 1963]. However, a similar change from parallel orientation to random growth did distinguish permanently transformed colonies from temporarily transformed populations of polyomavirus formed BHK 21 fibroblasts [STOKER, 1968].

In considering the possible effects of contact inhibition on growth patterns of tumor cells *in vivo,* it is necessary to take into account variations in their response to one another and to heterologous cells [ABERCROMBIE, 1967]. Contact inhibition among colliding malignant cells migrating *in vivo* would be expected to alter their direction of movement and thus promote their dispersal, but the lack of such inhibition might permit the formation of aggregates in analogy with the random piling up which takes place *in vitro.* On the other hand, the degrees of contact inhibition which may be expressed between tumor cells and other, normal, cells could influence the penetration of tumor cells into the surrounding heterologous tissue, the choice of paths of invasion along different stromal components, and some of the host-cell responses to the neoplasm.

2. Density-Dependent Inhibition of Growth

The number of cells that can be accommodated on a finite-sized surface is limited and normal cells cease to divide when a maximum number of cells is present per unit area of substratum, i.e., when their density reaches saturation. Under a given set of culture conditions, considerable tissue, species, and age dependent differences in maximum cell density are observed among normal cells. Usually, this density is increased upon malignant transformation, indicating that tumor cells are less sensitive than are normal cells to such growth control mechanisms. Furthermore, in contrast to normal cultures, tumor cultures may remain incompletely inhibited at high cell densities as cellular proliferation continues slowly and growth, as in suspension cultures, appears to be limited by nutritional factors.

The precise nature of the cell interactions required for density dependent growth inhibition is unknown. The process has some characteristics in common with contact inhibition of movement, namely, the requirement for close proximity of adjacent cells [STOKER *et al.*, 1966; SCHUTZ and MORA, 1968], cell-type specificity [BOREK and SACHS, 1966; EAGLE and LEVINE, 1967], and the tendency for both phenomena to be reduced in tumor cells. However, since the cell specificities are different, the type of intercellular contact has not been shown to be identical and independent variation is possible in the two [NJEUMA, 1971; LEVINSON *et al.*, 1971; STOKER and RUBIN, 1967], the basis for the two control mechanisms does not seem to be the same.

When cultures become crowded, the area of contact between adjacent cells increases, as the cells either become more columnar or they overlap and form multi-layers until further proliferation is inhibited [CASTOR, 1968]. The reduced sensitivity of tumor cells to growth inhibition, which may be expressed as a requirement for increased contact, might be due to 1) a change in intracellular growth-regulatory mechanisms which causes a diminished sensitivity to intercellular signals, or 2) a change at the cell surface resulting either in fewer receptors to such signals per unit surface-area, or a decrease in either the accessibility or the sensitivity of such receptors.

Extracellular diffusible substances also appear to be influential in this type of growth regulation, for the saturation density of a variety of cell types, including the characteristically monolayered 3T3 cells, increases considerably in response to increased supplies of culture medium, to high serum concentrations, and, in particular, to perfusion culture [HOLLEY and KIERNAN, 1968; KRUSE *et al.*, 1969; CASTOR, 1971; GRIFFITHS, 1972]. Several serum factors which limit the maximum density of normal cells, and which appear to have a reduced effect on transformed cells have been described [KOTLER, 1970], and one has been characterized as an acidic protein with insulin-like activity [TEMIN *et al.*, 1972]. Also, an inhibitor of RNA synthesis, possibly a polyamine, is released by confluent 3T3 cells into the culture medium, and both the ability to produce this inhibitor and the ability to react to it are apparently acquired only when the cells reach saturation density. Thus, possibly, intercellular contact must attain a critical degree to trigger the changes in cell membrane permeability necessary for growth inhibition by such diffusible factors to occur [YEH and FISHER, 1969].

It would appear that variations in saturation density more consistantly parallel changes in tumorigenicity than most other transformation-associated cell characteristics, so that they may be concurrently modified by the appropriate culture regime. Thus, in long-term mouse fibroblasts cultures, cell-transfer and maintenance at low cell densities produce non-tumorigenic cells highly sensitive to growth inhibition, while transfer and maintenance at high cell densities produces tumorigenic cells with a low sensitivity to growth inhibition. This suggests that the high cell density regime favors the selection of variant cells capable of multiplying even with extensive cell-to-cell contact, a property seemingly associated with tumorigenicity, while in low saturation cultures the variant cells with tumorigenic capacity are at a selective disadvantage [AARONSON and TODARO, 1968]. Furthermore, it appears that low-cell-density conditions not only act selectively, but may actually induce cellular changes which lead to contact-sensitive, less tumorigenic variants [RABINOWITZ and SACHS, 1969b]. In confluent transformed cultures treated with FUdR (5-fluoro-deoxyuridine), those cells capable of DNA synthesis can be killed, leaving cells more sensitive to growth inhibition and whose progeny demonstrates reduced tumorigenicity, increased sensitivity to growth inhibition by normal cells, increased cell flattening and increased adhesion to substrata, all suggesting changes in cell-surface properties [POLLACK *et al.*, 1968]. In a subsequent investigation it appeared that the decrease in tumorigenicity of such FUdR-selected variant cells was due to a reduced capacity of the cells to initiate tumors, but not to a change in the *in vivo* growth rate, the growth pattern, or the morphology once the tumors were established [POLLACK and TEEBOR, 1969]. Thus, it would appear that the reduced capacity to initiate tumors by the low-saturation-density-variants may result from their greater sensitivity to growth-control by juxtaposed normal cells, but once a tumor mass of sufficient size is formed this control is missing in the isolated central regions. Therefore, cells of different sensitivity to density-dependent growth-inhibition *in vitro* would differ in tumorigenicity because of variations in the efficiency with which they establish tumor masses of critical size [STOKER, 1967].

3. Dependence on Substrata and Growth Factors

Transformed cells, in contrast to normal cells, adapt with relative ease to growth in the absence of a solid substratum (suspension cultures), and this difference between normal and tumor cells persists when the cells are suspended in media rendered semi-solid by the addition of agar or methyl cellulose [MACPHERSON and MONTAGNIER, 1964]. Thus, when untransformed BHK 21 fibroblasts were suspended in semi-solid media they remained viable but divided only once and then ceased mitotic activity unless glass particles, larger than the cells and to which they could adhere, were provided. When silica fragments smaller than the cells were present growth was not stimulated, indicating that a cellular interaction with a rigid surface of a certain minimal size was required for proliferation to take place. It has been suggested that the underlying mechanism could involve cell membrane changes dependent on attachment to a substratum, or, following such attachment, could involve cell motility acting to release cell groups from any mutual growth inhibition [STOKER *et al.*, 1968].

The proliferative capacity of suspended BHK 21 cells could also be increased considerably upon the removal of acid polysaccharides from the agar [MONTAGNIER, 1968], and, conversely, colony formation was inhibited by the addition of heparin, dextran sulphate, ovomucoid, or chondroitin sulphate, in low concentrations. However, if native collagen was added, the colony forming capacity of these fibroblasts was enhanced 4–5-fold [SANDERS and SMITH, 1970]. While the basis of neither the inhibition by acid polysaccharides nor the enhancement by collagen is known, it would appear likely that the cell periphery is the site of activity, and the data also indicate that when provided with a solid substratum, the cells could overcome the inhibitory effects of the acid polysaccharides.

The anchorage dependence of the BHK 21 cells could be eliminated rapidly by transformation with polyoma virus, as the transformed cells grew into large colonies in the absence of substrata in semi-solid media [STOKER, 1968]. In addition, the transformed cells were unaffected in their proliferative capacity by either acid polysaccharide or collagen. Transformation in these cells, then, confers a high degree of independence from chemical growth-control factors and from any dependence on adhesion to a solid substratum. The latter change could indicate that the cells had been released from any mutual growth inhibition, or had become independent of those interactions with a substratum normally required for the initiation of division, or that the transformed cells had acquired the ability to use each other as substitutes for any necessary substratum. The changes which underly the unresponsiveness of transformed cells to the absence of solid substrata might well be related to the changes which allow these cells to spread and move indiscriminately over a variety of growth surfaces on which untransformed BHK 21 cells move and adhere differentially [AMBROSE, 1967]. The uniform response by the transformed cells to different growth surfaces suggests that they have lost the capacity of normal cells to recognize, and to react to, structural variations in substrata. It is interesting to note that this impairment is thought to be associated with the abnormal amount of undulating membrane activity in transformed cells

[AMBROSE, 1967], again implicating changes in the cell surface and cortical cytoplasm.

4. Differentiation of Transformed Cells

The dedifferentiation of normal cells in cell-layer culture begins immediately after explantation, in response to the loss of normal histotypic relationships, and proceeds during subsequent proliferation. Therefore, the effect of a malignant transformation at a later time is essentially one of accelerating a process already underway, in contrast to viral transformation *in vivo* or in organ culture [DAWE *et al.*, 1966] where proliferation and dedifferentiation are in fact initiated by the virus. While further dedifferentiation frequently does follow transformation in cell-layer cultures [ALBERT *et al.*, 1969], various levels of competence to differentiate under appropriate conditions may be retained [EVANS *et al.*, 1958; SHEIN, 1967; HEIDELBERGER and IYPE, 1968; DIAMANDOPOULOS and DALTON-TUCKER, 1969]. There is some evidence that the level of expressed differentiation following viral transformation of embryonic cells *in vitro,* may be directly proportional to the stage of embryonic development attained by the cells at the time of explantation [SHEIN, 1968].

Terminal differentiation may continue after viral transformation of cultured cells [DEFTOS *et al.,* 1968]. The persistence of differentiation in carcinogenesis is strikingly illustrated by cultured thyroid carcinoma cells that continued to secrete thyrocalcitonin after *in vitro* transformation by SV40 virus [GRIMLEY *et al.,* 1969], thus representing a population of rapidly proliferating epithelial cells that remained fully differentiated through carcinogenesis *in vivo,* subsequent growth in culture, and, finally, viral transformation. Such a high degree of stability in a differentiated function of a carcinoma cell is presumably the result of an equally high level of unresponsiveness to environment factors, attained early in carcinogenesis.

Specific functions may differentially increase, decrease, or not change following viral transformation. Thus, RSV[2]-transformed chick-embryo fibroblasts show an increase in hyaluronate production but not in collagen production, and transformed chick-embryo iris cells lose the capacity to synthesize melanin but gain the ability to produce hyaluronate [TEMIN, 1965]. Furthermore, the direction of change upon transformation depends in some unknown manner upon the particular combination of transforming agent and target cell, as illustrated by the reduced hyaluronate production

2 Rous sarcoma virus.

in DNA-virus transformed fibroblasts of mouse or man but the increased hyaluronate production in RSV-transformed chick fibroblasts [MACPHERSON, 1970].

Further indication of differential effects of transformation on various aspects of glycoprotein metabolism [BOSMANN *et al.,* 1968; BOSMANN, 1969] follows from the observation that, after viral transformation of fibroblasts, the activity of four glycosyl transferases (located on the smooth internal cell membranes and concerned with synthesis of cell-membrane glycoprotein) was increased, the activity of a glucosyltransferase (located on the plasma membrane and involved in collagen synthesis) was reduced, and the activities of several glycosidases (located in the lysosomes and involved in the breakdown of glycoproteins) were increased. Thus, transformation resulted in an increased synthesis of cell coat materials, a decrease in a differentiated function, and an increased capacity to break down glycoproteins, possibly including extracellular materials. Incidentally, the location of the glycosyl transferases is interesting in view of the prominence of Golgi complexes in so many tumor cells, including transformed fibroblasts [CORNELL, 1969a; MARTINEZ-PALOMO, 1969], since these organelles seem to be the normal site of cell-coat glycoprotein synthesis [RAMBOURG *et al.*, 1969; BENNETT and LEBLOND, 1970], and in view of the possibility that transformation-associated cell-surface changes might be responsible for some of the decrease in collagen synthesis.

5. Cellular Morphology

Changes in cellular morphology, similar to those of exfoliated tumor cells *in vivo,* are among the most consistent indicators of spontaneous malignant transformation *in vitro.* These changes, which take place gradually in transforming mouse embryo cell cultures [BARKER and SANFORD, 1970], occur in the sequence: increased basiphilia, increased number and size of nucleoli, increased nucleo-cytoplasmic ratio, retraction of cells from glass, and formation of cell clusters. These observations can be interpreted to indicate that initially in transformation considerable nuclear activity takes place, which effects and is affected by simultaneous cytoplasmic protein synthesis. All of this activity ultimately produces changes in the cell-surface and cortical cytoplasm, that are vividly demonstrated by the subsequent cell behavior. Cultures were invariably tumorigenic only if they had progressed to the point of cell retraction from the glass, i.e., where observable changes in adhesiveness or deformability, or both, had occurred. However, some transformed tumorigenic cultures showed only the increase in basiphilia and, sometimes, in nuclear size. Thus, while metabolic changes apparently involving RNA and protein synthesis are prerequisite and may be sufficient for the aquisition of malignancy, it would appear that the actual changes in the cell periphery must take place to ensure inevitable tumorigenicity in the cultured transformed cells.

The range of morphologic variation which can be recognized with the light microscope is quite limited and essentially the cells may be flattened or spherical, and symmetrical (epithelial) or asymmetrical (fibroblastic). In transformed cells, all variations between these forms have been observed [SATO, J. *et al.*, 1968; SHARON and POLLARD, 1969; MACPHERSON, 1966, 1970], and they all could be the result of changes in adhesiveness, deformability or the ability to maintain asymmetry (i.e., polarity). The changes taking place after spontaneous malignant transformation are particularly variable [SANFORD, 1967; SANFORD and HOEMANN, 1967; SATO, J. *et al.*, 1968; BAKER and SANFORD, 1970], while the form of virus-transformed cells is much more predictable, apparently being determined, depending on the system used, by either the target cells or the virus [MACPHERSON, 1970]. But even given the variation observed, viral transformation most frequently results in changes suggestive of diminished cellular adhesiveness. The physiologic state of the cells and possibly the state of *in vitro* differentiation (or dedifferentiation) at the time of transformation may have an influence on their subsequent shape, since human amnion cells, transformed at 10 days *in vitro* become fibroblast-like while those transformed at 34 days form cohesive epithelia [GAFFNEY *et al.*, 1970].

Transformed cells *in vitro* may alternate between epithelial and fibroblastic forms in response to different media [FOGH and GAFFNEY, 1970] or with time in culture [SANFORD *et al.*, 1961], and the growth pattern of carcinosarcoma cells cultured on chick-embryo chorioallantois may be sarcomatous or carcinomatous depending on the temperature of incubation [LEIGHTON *et al.*, 1967]. Another example of such morphological instability and variation is found in the observation that a direct correlation between epithelial or fibroblastic growth patterns of transformed cells *in vitro*, and their carcinomatous or sarcomatous growth pattern *in vivo* sometimes [BERWALD and SACHS, 1965], but not always [DIAMANDOPOULOS and DALTON-TUCKER, 1969; BARKER and SANFORD, 1970], exists.

While some tumors formed by cultured cells show definite tissue-specific characteristics [DIAMANDOPOULOS and DALTON-TUCKER, 1969], the distinction between sarcomas and carcinomas formed *in vivo* by transformed cells is usually based only on differences in cell shape and intercellular organization. When grown *in vivo* the majority of transformed cultures form masses of spindle-shaped cells, sometimes aligned in parallel array, resembling undifferentiated sarcomas. Some, which are also undifferentiated but consist of symmetrical round cells tending to form clumps, are classified as carcinomas. Unfortunately, in analogy with clinical histopathology, such observations frequently are taken to indicate the origin of the tumors from mesenchymal (sarcomas) or epithelial (carcinomas) cells and the predominance of sarcomas has been taken in turn to indicate the greater susceptibility of cultured fibroblasts to transformation. While this interpretation may be basically correct, its derivation from tumor histology seems unwarranted since the shape and the growth pattern of transformed cells *in vivo* are probably susceptible to modification by the same mechanisms which alter these characters *in vitro*. Epithelial cells may thus appear fusiform, i.e., fibroblast-like, upon transformation, while, conversely, fibroblasts may become round. That tumors histologically resembling sarcomas may originate in transformed epithelial cells is indicated by acid-phosphatase activity in fibrosarcoma-like tumors of prostatic origin [HEIDELBERGER and IYPE, 1968] and, perhaps more conclusively, by tumors with functional and histological characteristics of mesenchyme which have been cytologically identified as originating in salivary epithelium [DAWE *et al.*, 1971]. Finally, tumors of a mixed sarcomatous-carcinomatous morphology [DIAMANDOPOULOS and DALTON-TUCKER, 1969] might consist of a mixture of transformed epithelial cells and transformed fibroblasts, but, alternatively, adjacent normal cells of the same tissue could assume different shapes either upon transformation by two different transforming agents or as a result of physiologic differences at the time of transformation. The separation of such morphologically different but histogenetically similar cells into distinct groups within one tumor mass could be brought about through sorting-out processes based on differential adhesiveness [STEINBERG, 1970].

It would appear in general that, unless the origin of transformed cells is precisely known, or unless there is evidence of tissue-specific functions, sarcomatous or carcinomatous growth *in vivo* of transformed cells should be considered as no more than a morphological description.

6. *Structural Basis for Cellular Changes*

Most of the morphological and behavioral peculiarities of tumor cells described so far point to the cell surface and cortical cytoplasm as the probable region of important causal changes (see also chapter by EASTY).

Transformation *in vitro,* like carcinogenesis *in vivo,* is frequently accompanied by an increase in negative cell-surface charge density, apparently the result of an increased exposure of sialic acid residues, or increased production of acid mucosubstances, or both, at the cell periphery [MARTINEZ-PALOMO *et al.,* 1969; MACPHERSON, 1970; MONTAGNIER, 1970]. The resulting changes in the immediate ionic micro-environment may alter the availability of, and the membrane-permeability to, substances required for growth regulation, differentiation and other cellular functions; but equally important, the consequent reduction in cell deformability and the increased resistance to contact with other cells or substrata may have effects on cell shape, adhesiveness and contact-dependent controls of cell behavior [AMBROSE, 1967; WEISS, 1967].

Some changes in cell morphology and function associated with *in vitro* transformation seem to result from abnormalities induced in the protein moiety of the cell coat since similar changes can be produced temporarily in untransformed cells by treatment with proteolytic enzymes. Observations include the reduction in the duration of the G1 phase in BHK 21 cells by transformation or trypsin treatment [MONTAGNIER, 1970], the unmasking of glycolipids on the cell surface by either method [HAKOMORI *et al.*, 1968], and the agglutination by certain specific proteins of both tumor cells and untransformed cells, if the latter were first treated with protease [SHARON and LIS, 1972]. While the total number of agglutination sites on the cell surface seems similar in normal and transformed cells [INBAR and SACHS, 1969], the agglutinability of the cells varies in direct proportion to the saturation density and to the tumorigenicity of the cultures [POLLACK and BURGER, 1969]. Recent indications are that both transformation and protease treatment may produce the clustering of agglutination sites on the cell surface normally present in a dispersed form, and further, that any reversion of morphological and behavioral properties characteristic of transformed cells is associated with the reversion of cellular agglutinability [NICOLSON, 1972; INBAR *et al.*, 1972].

That abnormal protease activity may in fact be involved in transformation-induced alterations of the cell coat is suggested by changes observed in RSV-transformed chick-embryo fibroblast cultures [RUBIN, 1970]. Upon exposure to the virus, most of these cells are not permanently transformed but they are temporarily released from density-dependent inhibition of cell division and take on the appearance (round) and behavior (persistent rapid proliferation) of transformed cells. It appears that a non-dialyzable, heat-labile factor produced by the few permanently transformed cells is responsible for these temporary changes, and the effect of this factor can be mimicked by proteases. In this same vein, it may be significant that groups of hepatoma cells, which lack normal intercellular communication when invading *in vivo,* apparently induce a similar reduction in communication among the surrounding, histologically normal liver cells [LOEWENSTEIN and KANNO, 1966]. Furthermore, a similar interruption of ionic communication between normal cells can be produced directly by proteolytic enzymes acting at the cell surface [LOEWENSTEIN, 1968]. It is conceivable that neoplastic cells *in vivo* or *in vitro* could, by the release of proteases [SYLVEN, 1968], induce hyperplasia in the surrounding normal cells and thus predispose them to undergo malignant transformation.

Comparisons of normal and transformed cells at the ultrastructural level show, in addition to the well known nuclear and cytoplasmic changes associated with malignancy [CORNELL, 1969a] and rapid growth [HEINE *et al.,* 1969], considerable differences in the region of the cell-surface and the peripheral cytoplasm [MARTINEZ-PALOMO *et al.,* 1969]. The transformation of fibroblasts does not alter the appearance or the frequency of close intercellular junctions (120–200 Å intercellular spaces) or of incomplete desmosomes, but the frequency of tight junctions (fusion of plasma membranes) is greatly reduced. This change may account for the reduction in contact inhibition of movement [FLAXMAN *et al.,* 1969], in communication, and possibly in the exchange of growth regulatory substances [LOEWENSTEIN, 1968] observed in transformed cells. Transformation also seems to influence the appearance of surface microvilli. In culture, the number and the distribution of these structures depend on the physiological state of the cells HUMPHREYS and SINCLAIR, 1970], but in general their prominence is greatest when cells are round, reduced when the cells are deformed or flattened, and the microvilli are absent in regions of adhesion to substrata or of close intercellular junctions [FISHER and COOPER, 1967]. Since tumor cells frequently have prominent microvilli it is possible that a limited deformability impairs their ability to flatten and thus to adhere to substrata, and tends to reduce contact between the cells to one of interdigitating microvilli. However, it must be noted that, while such a limitation would probably interfere with the formation of tight junctions and with those functions dependent on them, incomplete desmosomes and dense junctions do form between adjacent microvilli and may result in strong adhesions between malignant cells [AUERSPERG, 1969a].

Within the cell, those organelles most closely associated with the maintenance and alterations of the cell shape are the cytofilaments and the microtubules. Submembrane accumulations of cytofilaments, both thin (40–50 Å) and larger (70–80 Å), are found adjacent to the culture medium in normal cells either in discrete bundles (stress fibers) or as a continuous layer [BUCKLEY and PORTER, 1967]. They seem to represent a contractile structure involved in membrane movements (ruffling, phagocytosis, cytokinesis),in cellular motility [FRANKS *et al.,* 1969; GOLDMAN and FOLLETT, 1969], in changes of cell shape [CLONEY, 1966; SPOONER and WESSELS, 1970], and in the maintenance of tension among cohesive cultured epithelial cells [AUERSPERG, 1969b] and fibroblasts [JAMES and TAYLOR, 1969]. Possibly, they also influence cellular deformability and the passage of some materials through the cortical cytoplasm. The larger filaments (80–120 Å) appear to be involved in the movements of particulate organelles and in the compartmentalization of the cytoplasm, either independently or in conjunction with the microtubules [GOLDMAN and FOLLETT, 1969; FREED and LEBOWITZ, 1970]. The latter would seem to be relatively rigid structures since they appear to play an important role in the maintenance of cell shapes [BRANSON, 1968; TILNEY and GIBBINS, 1969].

The oriented submembrane cytofilament structure present in normal fibroblasts disappeared in the irregularly shaped transformed cells [AMBROSE *et al.*, 1970], and similarly, the SV40 transformation of cultured cells produced a reduction in polarity (toward increased symmetry) which was also associated with the loss of stress fibers and of filament condensations in the cortical cytoplasm [GRIMLEY *et al.,* 1969]. On the other hand, when malignant transformation did not disturb the fusiform shape, no change in submembrane filaments was noted [CORNELL, 1969a]. If the adhesion of cultured cells to neighboring structures is altered or interrupted [AUERSPERG, 1969b], then rearrangements of cytofilaments, associated with changes in cell shapes and cell interrelationships, take place, indicating that these cellular characteristics depend on an interplay of adhesive forces at the cell surface with intracellular contractile and tensile forces.

It is not known how actions of cytofilaments and microtubules are controlled, but since these structures also regulate the movements and distribution of particulate organelles, it is possible that both the loss of normal cell shape and the metabolic disorder which occurs following malignant transformation could result, in part, from abnormalities in their functions. Indeed, it would seem to be particularly important to ascertain the relationship of cytofilaments and microtubules to surface factors since such an interaction might well represent an epigenetic control mechanism which is modified in a controlled manner in normal differentiation, and which may be disturbed in malignant cells[3].

V. Comments

In general, the loss of organization and differentiation by explanted tumor tissues parallels that of explanted normal tissues, with differentiation being gradually diminished as intercellular organization is reduced from organ culture, to aggregate and three-dimensional matrix culture, to cell-layer culture. In addition to any tissue-specific properties, cancer cell populations *in vitro* may retain some of the variations of morphology and behavior which existed *in vivo* among individual tumors, and which are recognized as deviations from the regular growth pattern of normal tissues. These variations also behave *in vitro* as expressions of differentiation in that they diminish with time in culture and with any disorganization of the system.

There are indications that the age and the developmental stage of normal cells in culture, at the time of transformation to tumorigenicity, may decidedly influence the subsequent phenotype (i.e., the level of differentiation) of their malignant progeny [SHEIN, 1968; GAFFNEY *et al.*, 1970]. This raises the interesting possibility that the age or the developmental stage of the target cells *in vivo,* at the initiation of carcinogenesis, may be similarly responsible for some of the unexplained phenotypic variation which exists among tumors of identical histologic origin. Indeed, such differences existing prior to explantation have been proposed recently as underlying the subsequent variations in competence observed in normal myoblast cell lines [RICHLER and YAFFE, 1970].

When compared to normal cells, malignant cells seem to demonstrate a reduced requirement for intercellular contact to support the differentiated state and the capacity to proliferate *in vitro*. In this respect, malignant tissues are reminiscent of several invertebrate embryonic systems, in that they behave as self-differentiating (self-regulating) sub-units that are capable of expression unhindered by many environmental influences; and particularly those influences imposed frequently, in more dependent systems, by juxtaposed cells. Such biological sub-units seem to be determined early, specifically and irreversibly; indeed, they appear to be so rigidly determined

3 It has recently been reported that microtubules may influence the agglutinability of cells by concanavalin A [BERLIN and UKENA, 1972].

as to be without the more pliable controls, or the freedom of response, present in less determined (i.e., more regulative) cell groupings. Thus, both the neoplastic and the embryonic systems are more independent of immediate external circumstances, possibly because of changes in the cell-surface/cortical region rendering these self-contained cells less responsive to external cues.

On the other hand, cancer cells apparently may acquire the competence to exert epigenetic influences similar to heterologous induction in normal development, i.e., a capacity to induce proliferation (followed by differentiation) in adjacent normal tissues, and possibly also among themselves [DAWE *et al.,* 1971]. Thus, within groups of interacting cancer cells, mutual stimulation of mitotic activity, even to levels beyond those permitted by total unresponsiveness to normal restrictive growth control mechanisms, could take place.

Once given an increased cell mass, in both normal and neoplastic tissues, the requisite interactions between the cells can occur. It is then, in a developing system, that the subtle epigenetic cues are probably most prevalent and most effective. And it is at precisely this level that normal and tumor cells differ, both in their responsiveness to, and in their dependence on, normal regulatory mechanisms of proliferation and differentiation. Therefore, attempts to control cancer proliferation at the level of intercellular interactions, as suggested by LEIGHTON [1969], would seem to be very appropriate, and the observations implicating the cancer cell-surface in tumor destruction [SHOHAM *et al.,* 1970] and in the restoration of a normal growth pattern in previously transformed fibroblasts [BURGER and NOONAN, 1970] are in support of this interpretation.

There exist interesting considerations of histogenesis based exclusively on biomolecular processes, as in the discussion by CARTER [1968] of the possible role of cell membrane instability, rather than any external cell-mediated cues, in density-dependent inhibition of growth, the stimulation of proliferation, and the migration of cells. However, it is difficult to interpret the morphogenetic events described for both neoplastic and embryonic systems solely in terms of these processes. For, while the characteristics of the individual cells will obviously have something to say about the identifying characteristics of the tissue in which they reside, it does not seem to be equally certain that they will say all that need be said; indeed, that the understanding of cytogenesis can be fully accomplished without also understanding histogenesis, and perhaps even organogenesis, seems doubtful. It would appear that the development of the properties of the lower levels of organization in a hierarchical system is controlled by higher levels so that the former differentiates as an incomplete reflection of the latter. Perhaps malignancy may be regarded as a condition in which the reponsiveness of the lower level of organization to the directional cues imposed by the higher levels is impaired, a condition brought about in part by changes in the peripheral portion of the individual cell's organization. The studies of cell populations *in vitro* considered in this chapter could be taken to indicate that this unresponsiveness to epigenetic factors can be established independently of those genetic changes that lead to the production of a malignant cell type.

Acknowledgement

We wish to thank Dr. J.W. JULL, Cancer Research Centre, The University of British Columbia, for his helpful and critical comments.

References

AARONSON, S. A. and TODARO, G. T.: Science *162:* 1024–1026 (1968).
ABERCROMBIE, M.: In Vitro *6:* 128–142 (1970).
ABERCROMBIE, M.: In P. DENOIX Mechanisms of invasion in cancer, pp. 140–144 (Springer, Berlin/Heidelberg 1967).
ALBERT, D. M.; RABSON, A. S.; GRIMES, P. A. and VON SALLMAN, L.: Science *164:* 1077–1078 (1969).
AMBROSE, E. J.: In P. DENOIX Mechanisms of invasion in cancer, pp. 130–139 (Springer, Berlin/Heidelberg, 1967).
AMBROSE, E. J.; BATZDORF, U.; OSBORN, J. S. and STUART, P. R.: Nature *227:* 397–398 (1970).
AUERSPERG, N.: J. nat. Cancer Inst. *43:* 151–173 (1969a).
AUERSPERG, N.: J. nat. Cancer Inst. *43:* 175–190 (1969b).
AUERSPERG, N. and HAWRYLUK, A. P.: J. nat. Cancer Inst. *28:* 605–627 (1962).
AUERSPERG, N. and WORTH, A.: Int. J. Cancer *1:* 219–238 (1966).
BAKER, J.B. and HUMPHREYS, T.: Proc. nat. Acad. Sci. U.S. *68:* 2161–2164 (1971).
BANCROFT, F.C. and TASHJIAN, A.N.J.: Exp. Cell Res. *64:* 124–128 (1971).
BARKER, B. E. and SANFORD, K. K.: J. nat. Cancer Inst. *44:* 39–63 (1970).
BENNETT, G. and LEBLOND, C. P.: J. Cell Biol. *46:* 409–416 (1970).
BERLIN, R.D. and UKENA, T.E.: Nature N.B. *238:* 120–122 (1972).
BERWALD, Y. and SACHS, L.: J. nat. Cancer Inst. *35:* 641–661 (1965).
BISCHOFF, R. and HOLTZER, H.: J. Cell Biol. *44:* 134–150 (1970).
BOREK, C. and SACHS, L.: Proc. nat. Acad. Sci., U.S. *56:* 1705–1711 (1966).
BOSMANN, H. B.: Exp. Cell Res. *54:* 217–221 (1969).
BOSMANN, H. B.; HAGOPIAN, A. and EYLAR, E. H.: J. Cell Physiol. *72:* 81–88 (1968).
BOYDE, A.; GRAINGER, F. and JAMES, D. W.: Z. Zellforsch. *94:* 46–55 (1969).

BRANSON, R. J.: Anat. Rec. *160:* 109–122 (1968).
BRINSTER, R. L. and THOMSON, J. L.: Exp. Cell. Res. *42:* 308–315 (1966).
BUCKLEY, T. K. and PORTER, K. R.: Protoplasma *64:* 349–380 (1967).
BUONASSISI, V.; SATO, G. and COHEN, A.I.: Proc. nat. Acad. Sci. U.S. *48:* 1184–1190 (1962).
BURGER, M. M. and NOONAN, K. D.: Nature *228:* 512–515 (1970).
CAHN, R. D.: In H. URSPRUNG The stability of the differentiated state, pp. 58–84 (Springer, New York 1968).
CARTER, S. B.: Nature *208:* 1183–1187 (1965).
CARTER, S. B.: Nature *220:* 970–974 (1968).
CARTER, R. C. and ROE, F. J. C.: Brit. J. Cancer *23:* 401–407 (1969).
CASTOR, L. N.: J. Cell Physiol. *72:* 161–172 (1968).
CASTOR, L.N.: Exp. Cell Res. *68:* 17–24 (1971).
CLONEY, R. A.: J. Ultrastruct. Res. *14:* 300–328 (1966).
COLE, J. R.; EDWARDS, R. G. and PAUL, J.: Develop. Biol. *13:* 385–407 (1966).
COLEMAN, A. W.; COLEMAN, J. R.; KANKEL, D. and WERNER, T.: Exp. Cell Res. *59:* 319–328 (1970).
CORNELL, R.: J. nat. Cancer Inst. *43:* 891–906 (1969a).
CORNELL, R.: Exp. Cell Res. *57:* 86–94 (1969b).
COWELL, L. A. and WESTON, J. A.: Develop. Biol. *22:* 670–697 (1970).
CRICK, F.: Nature *225:* 420–422 (1970).
CUPRAK, L. J. and SATO, G. H.: Exp. Cell. Res. *52:* 632–645 (1968).
DANIEL, C. W. and DE OME, K. B.: Science *149:* 634–636 (1965).
DAVIDSON, E. H.: In E. W. CASPARI and J. M. THODAY Advances in genetics, vol. 12, pp. 143–280 (Academic Press, New York 1964).
DAWE, C.J.; MORGAN, W.D. and SLATICK, M.S.: Int. J. Cancer *1:* 419–450 (1966).
DAWE, C. J.; MORGAN, W. D. and SLATICK, M. S.: In R. FLEISCHMAJER and R. E. BILLINGHAM Epithelial-mesenchymal interactions, pp. 293–312 (Williams and Wilkins, Baltimore 1968a).
DAWE, C. J.; WHANG-PENG, J.; MORGAN, W. D.; O'GARA, R. W. and KELLY, M. G.: J. nat. Cancer Inst. *40:* 1167–1193 (1968b).
DAWE, C. J.; WHANG-PENG, J.; MORGAN, W. D.; HEARON, E. C. and KNUDSEN, T.: Science *171:* 394–397 (1971).
DEFENDI, V.; LEHMAN, J. and KRAEMER, P.: (1963). Virology *19:* 592–598 (1963).
DEFTOS, L. J.; RABSON, A. S.; BUCKLE, R. M.; AURBACH, G. D. and POTTS, J. T.: Science *159:* 435–436 (1968).
DIAMANDOPOULOS, G. T. and DALTON-TUCKER, M. F.: Amer. J. Path. *56:* 59–77 (1969).
DI PAOLO, J. A.; RABSON, A. S. and MALMGREN, R. A.: J. nat. Cancer Inst. *40:* 757–770 (1968).
DI PAOLO, J.A.; NELSON, R.L. and DONOVAN, P.J.: J. nat. Cancer Inst. *46:* 171–181 (1971).
EAGLE, H. and LEVINE, E. M.: Nature *213:* 1102–1106 (1967).
ELLISON, M. L.; AMBROSE, E. J. and EASTY, G. C.: Exp. Cell Res. *55:* 198–204 (1969).
ENG, C. P.; KLEINE, L. P. and MORGAN, J. F.: J. nat. Cancer Inst. *45:* 235–242 (1970).
EVANS, V. J.; HAWKINS, N. M.; WESTFALL, B. B. and EARLE, W. R.: Cancer Res. *18:* 261–266 (1958).
FEDOROFF, S.: J. nat. Cancer Inst. *38:* 607–611 (1967).
FELL, H. B.: J. nat. Cancer Inst. *19:* 643–650 (1957).
FISHER, H. W. and COOPER, T. W.: J. Cell Biol. *34:* 569–576 (1967).
FLAXMAN, B.A.: Cancer Res. *32:* 462–469 (1972).

FLAXMAN, B. A. and VAN SCOTT, E. J.: J. nat. Cancer Inst. *40:* 411–422 (1968).
FLAXMAN, B. A.; REVEL, J. P. and HAY, E. D.: Exp. Cell Res. *58:* 438–443 (1969).
FOGH, J. and GAFFNEY, E. V.: J. nat. Cancer Inst. *44:* 215–223 (1970).
FOLEY, J. F. and AFTONOMOS, B. T.: J. nat. Cancer Inst. *34:* 217–229 (1965).
FOLEY, J. F. and DAVIS, R. B.: Nature *205:* 785–786 (1965).
FOLEY, G. E.; HANDLER, A. H.; LYNCH, P. M.; WOLMAN, S. R.; STULBERG, C. S. and EAGLE, H.: Cancer Res. *25:* 1254–1261 (1965).
FOLLETT, E. A. C. and O'NEILL, C. H.: Exp. Cell Res. *55:* 136–138 (1969).
FOULDS, L.: In discussion of: HAYFLICK, L., Nat. Cancer Inst. Monogr. *26:* 355–385 (1967).
FOULDS, L.: Neoplastic development, vol. 1 (Academic Press, New York 1969).
FRANKS, L. M.; RIDDLE, P. N. and SEAL, P.: Exp. Cell Res. *54:* 157–162 (1969).
FREED, J. J. and LEBOWITZ, M. M.: J. Cell Biol. *45:* 334–354 (1970).
GAFFNEY, E. V.; RAMOS, L., and FOGH, J.: Cancer Res. *30:* 871–879 (1970).
GEY, G. O.: Cancer Res. *1:* 737 (1941).
GEY, G. O. and THALHIMER, W.: J. Amer. med. Ass. *82:* 1609 (1924).
GEY, G. O.; SEEGAR, G. E., and HELLMAN, L. M.: Science *88:* 306–307 (1938).
GOLDMAN, R. D. and FOLLETT, E. A. C.: Exp. Cell Res. *57:* 263–276 (1969).
GOLDSTEIN, M. N.; BURDMAN, J. A. and JOURNEY, L. I.: J. nat. Cancer Inst. *32:* 165–199 (1964).
GRIFFITHS, J.B.: Exp. Cell Res. *75:* 47–52 (1972).
GRIMLEY, P. M.; DEFTOS, L. J.; WEEKS, J. R. and RABSON, A. S.: J. nat. Cancer Inst. *42:* 663–680 (1969).
HAKOMORI, S.; TEATHER, C. and ANDREWS, H.: Biochem. biophys. Res. Commun. *33:* 563–568 (1968).
HALL, B. K.: Life Sci. *8:* 553–558 (1969).
HARRIS, M.: Cell culture and somatic variation (Holt, Rinehart and Winston, Toronto 1964).
HAYFLICK, L.: Nat. Cancer Inst. Monogr. *26:* 355–385 (1967).
HAYFLICK, L. and MOORHEAD, P. S.: Exp. Cell Res. *25:* 585–621 (1961).
HEIDELBERGER, C. and IYPE, P. T.: In H. KATSUTA Cancer cells in culture, pp. 351–363 (University of Tokyo Press, Tokyo 1968).
HEINE, U.; LANGLOIS, A. J.; RIMAN, J. and BEARD, J. W.: Cancer Res. *29:* 442–458 (1969).
HEUSON, J.C. and LEGROS, N.: Cancer Res. *32:* 226–232 (1972).
HOLLEY, R. W. and KIERNAN, J. A.: Proc. nat. Acad. Sci., U.S. *60:* 300–304 (1968).
HOLTZER, H. and ABBOTT, J.: In H. URSPRUNG The stability of the differentiated state, pp. 1–16 (Springer, New York 1968).
HU, F.: G. ital. Derm. *44–110:* 496–503 (1969).
INBAR, M. and SACHS, L.: Nature *223:* 710–712 (1969).
INBAR, M.; BEN-BASSAT, H. and SACHS, L.: Nature N.B. *236:* 3–16 (1972).
IOACHIM, H.L.: J. Nat. Cancer Inst. *42:* 101–113 (1969).
ISRAELI, E. and BARZILAI, D.: Cancer *25:* 824–834 (1970).
JAMES, D. W. and TAYLOR, J. F.: Exp. Cell Res. *54:* 107–110 (1969).
KAHN, L. B. and DONALDSON, R. C.: Cancer *25:* 1162–1169 (1970).
KALUS, M. and O'NEAL, R. M.: Arch. Path. *86:* 52–59 (1968).
KALUS, M.; GHIDONI, J. J. and O'NEIL, R. M.: Cancer *22:* 507–516 (1968).
KITANO, Y. and KURODA, Y.: Exp. Cell Res. *48:* 350–360 (1967).
KOTLER, M.: Cancer Res. *30:* 2493–2496 (1970).
KRUSE, P. F.; WHITTLE, W.; MIEDEMA, E. D.: J. Cell Biol. *42:* 113–121 (1969).
KURODA, G.: GANN *59:* 281–288 (1968).

LAVIETES, B. B.: Develop. Biol. *21:* 584–610 (1970).
LEIGHTON, J.: J. nat. Cancer Inst. *15:* 275–293 (1954).
LEIGHTON, J.: In H. KATSUTA Cancer cells in culture, pp. 143–156 (University of Tokyo Press, Tokyo 1968).
LEIGHTON, J.: Cancer Res. *29:* 2457–2465 (1969).
LEIGHTON, J.; IAMMARINO, R. M. and MARK, R.: In P. DENOIX Mechanisms of invasion in cancer, pp. 212–217 (Springer, Berlin/Heidelberg 1967).
LEIGHTON, J.; ESTES, L.W.; MARSUKHANI, S. and BRADA, Z.: Cancer *26:* 1022–1028 (1970).
LEIGHTON, J.; KALLA, R. L.; TURNER, J. M. and FENNELL, R. H.: Cancer Res. *20:* 575–586 (1960).
LEVINSON, W.; HEILBRON, D. and JACKSON, J.: J. nat. Cancer Inst. *46:* 323–335 (1971).
LILIEN, J. E.: In A. A. MOSCONA and A. MONROY Current topics in developmental biology, vol. 4, pp. 169–195 (Academic Press, London 1969).
LOEWENSTEIN, W. R.: Perspect. Biol. Med. *11:* 260–272 (1968).
LOEWENSTEIN, W. R. and KANNO, Y.: J. Cell Biol. *33:* 225–235 (1967).
MACPHERSON, I.: In W. H. KIRSTEN Malignant transformation by viruses, pp. 1–20 (Springer, New York 1966).
MACPHERSON, I.: In G. KLEIN and S. WEINHOUSE Advances in cancer research, vol. 13, pp. 169–215 (Academic Press, New York 1970).
MACPHERSON, I. and MONTAGNIER, L.: Virology *23:* 291–294 (1964).
MARTINEZ-PALOMO, A.; BRAISLOVSKY, C. and BERNHARD, W.: Cancer Res. *29:* 925–937 (1969).
MCCULLOCH, E. A. and PARKER, R. C.: Proc. Canad. Cancer Res. Conf. *2:* 152–167 (1956).
MONTAGNIER, L.: C.R. Acad. Sci. *267:* 921–924 (1968).
MONTAGNIER, L.: Bull. Cancer *57:* 13–22 (1970).
MOORE, G. E. and KOIKE, A.: Cancer *17:* 11–20 (1964).
MOORE, G. E. and MCLIMANS, W. F.: J. theor. Biol. *20:* 217–226 (1968).
MOORE, G. E.; GERNER, R. E. and EITARO, I.: In H. KATSUTA Cancer cells in culture, pp. 60–72 (University of Tokyo Press, Tokyo 1968).
MORGAN, J. F.; BOWMAN, B. and LIPINSKI, E.: In H. KATSUTA Cancer cells in culture, pp. 48–59 (University of Tokyo Press, Tokyo 1968).
MORTON, D. L.; HALL, W. T. and MALMGREN, R. A.: Science *165:* 813–816 (1969).
NICOLSON, G.L.: Nature N.B. *239:* 193–197 (1972).
NILSSON, K.; KLEIN, G.; HENLE, W. and HENLE, G.: Int. J. Cancer *8:* 443–450 (1971).
NJEUMA, D.L.: Exp. Cell Res. *66:* 244–249 (1971).
OBOSHI, S.; SEIDO, T.; SHIBATA, H. and SUGANS, H.: GANN *60:* 419–424 (1969).
PARK, H. Y. and KOPROWSKA, I.: Cancer Res. *28:* 1478–1489 (1968).
PARSHAD, R. and SANFORD, K.K.: J. nat. Cancer Inst. *49:* 1155–1163 (1972).
PATILLO, R. A.; GEY, G. O.; DELFS, E.; HUANG, W. Y.; HAUSE, L.; GARANCIS, J.; KNOTH, M.; AMATRUDA, J.; BERTINO, J.; FRIESEN, H. G. and MATTINGLY, R. F.: Ann. N.Y. Acad. Sci. *172:* 288–298 (1971).
PIERCE, G.B.; BEALS, T.F.; SRI RAM, J. and MIDGLEY, A.R., jr.: Amer. J. Path. *45:* 929–961 (1964).
PIERCE, G. B.: Cancer Res. *25:* 656–669 (1965).
POLLACK, R. E. and BURGER, M. M.: Proc. nat. Acad. Sci., U.S. *62:* 1074–1076 (1969).
POLLACK, R. E. and TEEBOR, G. W.: Cancer Res. *29:* 1770–1772 (1969).
POLLACK, R. E.; GREEN, H. and TODARO, G. J.: Proc. nat. Acad. Sci., U.S. *60:* 126–133 (1968).
PROSE, P. H.; FRIEDMAN-KIEN, A. E. and NEISTEIN, S.: Lab. Invest. *17:* 693–716 (1967).

PUGH-HUMPHREYS, R.G.P. and SINCLAIR, W.: J. Cell Sci. *6:* 477–484 (1970).
RABINOWITZ, Z. and SACHS, L.: Nature *220:* 1203–1206 (1968).
RABINOWITZ, Z. and SACHS, L.: Virology *38:* 336–342 (1969a).
RABINOWITZ, Z. and SACHS, L.: Virology *38:* 343–346 (1969b).
RABINOWITZ, Z. and SACHS, L.: Virology *40:* 193–198 (1970a).
RABINOWITZ, Z. and SACHS, L.: Int. J. Cancer *6:* 388–398 (1970b).
RAMBOURG, A.; HERNANDEZ, W. and LEBLOND, C.P.: J. Cell Biol. *40:* 395–414 (1969).
RANADIVE, K. J.; WAGH, U. V.; VERNEKAR, S. D.; BHISEY, A. N. and BOSE, S.: In H. KATSUTA Cancer cells in culture, pp. 299–318 (University of Tokyo Press, Tokyo 1968).
REED, M. V. and GEY, G. O.: Lab. Invest. *11:* 638–652 (1962).
REHM, M.: Exp. Cell Res. *51:* 237–257 (1968).
RICHLER, C. and YAFFE, D.: Develop. Biol. *23:* 1–22 (1970).
ROSE, G. G.: J. Cell Biol. *32:* 89–112 (1967).
ROSENTHAL, M. D.;WISHNOW, R. M. and SATO, G. H.: J. nat. Cancer Inst. *44:* 1001–1014 (1970).
RUBIN, H.: Science *167:* 1271–1272 (1970).
RUTTER, W. J.,; CLARK, W. R.; KEMP, J. D.; BRADSHAW, W. S.; SANDERS, T. G. and BALL, W.D.: In R. FLEISCHMAJER and R. E. BILLINGHAM Epithelial-mesenchymal interactions, pp. 114–131 (Williams and Wilkins, Baltimore 1968).
SANDERS, F. K. and SMITH, J. D.: Nature *227:* 513–515 (1970).
SANFORD, K. K.: Int. Rev. Cytol. *18:* 249–311 (1965).
SANFORD, K.K.: Nat. Cancer Inst. Monogr. *26:* 387–418 (1967).
SANFORD, K. K.; WESTFALL, B. B.; CHU, E. H. Y.; KUFF, E. L.; COVALESKY, A. B.; DUPREE, L. T.; HOBBS, G. L. and EARLE, W. R.: J. nat. Cancer Inst. *26:* 1193–1219 (1961).
SANFORD, K.K.; HANDLEMAN, S.L.; HARTLEY, J.W.; JACKSON, J.W. and GANTT, R.R.: J. nat. Cancer Inst. *49:* 1177–1189 (1972).
SANFORD, K.K. and HOEMANN, R.E.: J. nat. Cancer Inst. *39:* 691–703 (1967).
SATO, H.; KUROKI, T. and GOTO, M.: In H. KATSUTA Cancer cells in culture, pp. 35–47 (University of Tokyo Press, Tokyo 1968).
SATO, J.; NAMBA, M.; OSNI, K. and NAGANO, D.: Jap. J. exp. Med. *38:* 105–118 (1968).
SATO, G.; AUGUSTI-TOCCO, G.; POSNER, M. and KELLY, P.: In E. B. ASTWOOD Recent progress in hormone research, vol. 26, pp. 539–546 (Acad. Press, New York 1970).
SAXÉN, L.; KOSKIMIES, O.; LATHI, A.; MIETTINEN, H.; RAPPOLA, J. and WARTIOVAARA, J.: In M. ABERCROMBIE, J. BRACHET and T. J. KING Advances in morphogenesis, vol. 7, pp. 251–293 (Acad. Press, New York 1968).
SCHIMMER, B. P.; UEDA, K. and SATO, G. H.: Biochem. biophys. Res. Commun. *32:* 806–810 (1968).
SCHUBERT, D.; HUMPHREYS, S.; DE VITRY, F. and JACOB, J.: develop. Biol. *25:* 514–546 (1971).
SCHUTZ, L. and MORA, P. T.: J. Cell Physiol. *71:* 1–6 (1968).
SHARON, N. and LIS, H.: Science *177:* 949–959 (1972).
SHARON, N. and POLLARD, M.: Cancer Res. *29:* 1523–1526 (1969).
SHEIN, H. M.: Arch. ges. Virusforsch. *22:* 122–142 (1967).
SHELTON, E.; EVANS, V. F. and PARKER, G. A.: J. nat. Cancer Inst. *30:* 377–391 (1963).
SHEIN, H. M.: Science *159:* 1476–1477 (1968).
SHERWIN, R. P.; RICHTERS, V. and RICHTERS, A.: Cancer *20:* 1–22 (1967).
SHOHAM, J.; INBAR, M. and SACHS, L.: Nature *228:* 512–515 (1970).
SILAGI, S. and BRUCE, S. A.: Proc. nat. Acad. Sci., U.S. *66:* 72–78 (1970).

SONNENSCHEIN, C.; RICHARDSON, U. I. and TASHJIAN, A. H.: Exp. Cell Res. *61:* 121–128 (1970).
SPOONER, B.S. and HILFER, S.R.: J. Cell Biol. *48:* 225–234 (1971).
SPOONER, B. S. and WESSELS, N. K.: Proc. nat. Acad. Sci., U.S. *66:* 360–364 (1970).
STANTON, M.F. and WRENCH, C.: J. nat. Cancer Inst. *48:* 797–821 (1972).
STEINBERG, M.: J. exp. Zool. *173:* 395–434 (1970).
STOKER, M. G. P.; SHEARER, M. and O'NEILL, C.: J. Cell Sci. *1:* 297–310 (1966).
STOKER, M.G.P.: In P. DENOIX Mechanisms of invasion in cancer, pp. 193–203 (Springer, Berlin/Heidelberg 1967).
STOKER, M. G. P.: Nature *218:* 234–238 (1968).
STOKER, M. G. P. and RUBIN, H.: Nature *215:* 171–172 (1967).
STOKER, M. G. P.; SHEARER, M. and O'NEILL, C.: J. Cell Sci. *1:* 297–310 (1966).
STOKER, M. G. P.; O'NEILL, €.; BERRYMAN, S. and WAXMAN, V.: Int. J. Cancer *3:* 683–693 (1968).
SYKES, J. A.; WHITESCARVER, P.; JERNSTROM, P.; NOLAN, J. F. and BYATT, P.: J. nat. Cancer Inst. *45:* 107–122 (1970).
SYLVÉN, B.: Europ. J. Cancer *4:* 463–474 (1968).
TEMIN, H. M.: J. nat. Cancer Inst. *35:* 679–693 (1965).
TEMIN, H.M.; PIERSON, Jr., R.W. and DULAK, N.C.: In H. ROTHBLAT and V.G. CRISTOFALO Growth, nutrition and metabolism of cells in culture, vol. 1 pp. 49–81 (Acad. Press, New York 1972).
TILNEY, L. G. and GIBBINS, J. R.: J. Cell Biol. *41:* 227–250 (1969).
TOSHIMA, S.; MOORE, G. E. and SANDBERG, A. A.: Cancer *21:* 202–216 (1968).
TURKINGTON, R. W. and RIDDLE, M.: Cancer Res. *30:* 127–132 (1970).
VAN SCOTT, E. J. and REINERTSON, R. P.: J. Invest. Derm. *36:* 109–131 (1961).
VOGT, M. and DULBECCO, R.: Proc. nat. Acad. Sci., U.S. *49:* 171–179 (1963).
WALKER, D. G.; LYONS, M. M. and WRIGHT, J. C.: Europ. J. Cancer *1:* 265–273 (1965).
WATANABE, H. and ESSNER, E.: Cancer Res. *29:* 631–644 (1969).
WEED, G.: Lying-in, J. Reprod. Med. *1:* 117–146 (1968).
WEISS, L.: The cell periphery, metastasis and other contact phenomena (North Holland Publ. Co., Amsterdam 1967).
WESTFALL, B. B.; PEPPERS, E. V.; EVANS, V. J.; SANFORD, K. K.; HAWKINS, N. M.; FIORAMONTI, M. C.; KERR, H. A.; HOOBS, G. L. and EARLE, W. R.: J. biophys. biochem. Cytol. *4:* 567–585 (1958).
WILBANKS, G.D. and SHINGLETON, H.M.: Acta. cytol. *14:* 182–186 (1970).
WLODARSKI, K.: Exp. Cell Res. *57:* 446–448 (1969).
WOLPERT, L.: J. theor. Biol. *25:* 1–47 (1969).
YASUMURA, Y.; BUONASSISI, V. and SATO, G. H.: Cancer Res. *26:* 529–535 (1966a).
YASUMURA, Y.; TASHJIAN, A. H. and SATO, G. H.: Science *154:* 1186–1189 (1966b).
YEH, J. and FISHER, H. W.: J. Cell Biol. *40:* 382–388 (1969).

Authors' address: Dr. NELLY AUERSPERG and Dr. C. V. FINNEGAN, Cancer Research Centre and Department of Zoology, University of British Columbia, *Vancouver 8, B.C.* (Canada)

Neoplasia and Cell Differentiation, pp. 319–349
(Karger, Basel 1974)

Differentiation and Organization in Teratomas

J.A. GAILLARD
Department of Pathology, Hospital Center, Evreux

Contents

Introduction

For a long time, the study of teratomas was enriched only by morphological data. Facts concerning the initial stages were lacking. Thus, these stages had to be reconstructed from later observations, necessarily different from the first manifestations of the process. Interpretation of these tumors was possible only by analogy, or by extrapolating, with greater or lesser justification, and often prematurely, from the advances in diverse biological areas, notably, histology and embryology. For this reason, explanatory theories on their histogenesis have multiplied, with a view toward coordinating what we know with what we ignore. It is always difficult, if not illusory, to coordinate in space and time the evolution of a system, each of whose elements has its own destiny, with but topographical and architectural observations as guide lines, the latter, arbitrarily 'petrified' by the histological technique. There is always the lack of precise facts concerning the changes that occur in the appearance of a phenomenon once it has begun. Our intellect can furnish only a schematic model, static and approximate.

For the past several years, however, the study of teratomas has benefited from experimental biological material, which gives hope for eventual isolation and identification of the stem cells, as well as for the verification of their potentialities, either in tissue cultures or grafts, or under other experimental conditions.

I have attempted to review the latest developments in the study of these tumors, and to present a synthesis of the often contradictory results of human histo-pathological observations and animal experimentation. It was, therefore, not a question of compiling a summary of all knowledge on this subject, but rather of evaluating what knowledge is still needed to obtain a clear picture of the histogenesis of teratomas, and especially to determine the extent to which their development is comparable to that of an embryo, as is often asserted.

There is a vast amount of literature on teratomas. I have been forced to make an arbitrary choice from the mass of publications, giving preference to those which represented general reviews, or those providing original ideas. I have especially cited those underlining points which are, in my opinion, important, and which are usually neglected because they are not in agreement with one of the generally-accepted concepts. In this manner, I trust that this work conforms with the spirit of experimental science—as conceived by CLAUDE BERNARD.

I. Definition and Nosology

In various parts of the organism, particular neoformations can appear, consisting of tissues resembling those consecutively formed during ontogenesis, i.e., from the first segmentations of the egg, up to and through the fetal and adult stages. These tissues, being foreign to the structure of the organ in which they develop, give rise to heterologous tumors. These growths should be distinguished from homologous tumors, which retrace the history of the development of a given organ.

It would be tedious to describe all the possible localizations of these complex, multi-tissued neoplasms, some of which are very rare [WILLIS, 1958]. It can be noted, however, that they are distributed according to a certain topographical order, around a median or para-median line running from the pelvis to the cranial arch. They exist in both deep and superficial forms. The deep ones are found chiefly in the gonads, but also in the epiphysis [MAIER, 1861]; the anterior mediastinum [see COURY, 1945], and the retro-peritoneum [see PORTUGAL, 1919]. Among the superficial ones, the better-known develop in the sacro-coccygeal region. Teratomas have been described in many species: horse, guinea-pig, fowl, mouse [see STEVENS, 1967]. The author has observed an ovarian teratoma in a hare.

These tumors are identified by an abundant and highly varied nomenclature, intended to be both descriptive and pathogenic. It is always difficult, however, to find the exact word capable of defining a phenomenon of hypothetical origin whose structure varies greatly with the particular case. Thus, these neoformations are burdened with terms whose meanings vary according to different authors, the more so, since they were already ambiguous from an etymological standpoint. Moreover, one was often obliged to add a supplementary adjective, in order to provide a complementary description, or to indicate the degree of malignancy – since we are dealing with a question of real tumors.

In the English-language literature, the term 'teratoma' or 'teratomata' is generally used. In French literature, the term 'dysembryoma' is often preferred. Both terms can be criticized: 'teratoma' implies the questionable notion of malformation, while 'dysembryoma' does not provide a clear distinction between a possible dysgenesis carried over from the embryogenetic period, and a possible post-embryonic dysgenetic development realized at the expense of totipotential cells. Because this work is being published in English, the first of the two expressions, 'teratoma' will be used. This, in order to conform with common usage; without, however, attributing to it either, an overly precise pathogenic meaning, or the capacity to indicate different degrees of structural complexity.

II. Structure of Teratomas

A. Analytical Structure

Since the time of JOHNSON [1856], established practice has been to classify the tissues found in teratomas according to the germ layers from which they are derived. This procedure is useful for analyzing the architecture of the most complex tumors. However, it is less interesting for the simplified forms. One gets a better picture of the observable facts, by studying their structures according to the degree of maturation of the tissue-components, particularly since certain teratomas actually contain the primitive germ-layers.

1. Immature Differentiation

By definition, an immature tissue is one that has not completed its morphological evolution. Thus, for a given tissue, the period of immaturity spans from the first stages of egg development to the complete adult stage. Immature proliferations generally have a high potential malignancy, and often erase the multi-tissued structure characteristic of teratomas. These immature proliferations correspond to the different segments of the egg during its embryonic stage, when 6 layers instead of 3, are in fact laid down [MARIN-PADILLA, 1968]. Three of these layers are embryonic in nature: The ectoblast, the endoblast, and the mesoblast; while 3 are extra-embryonic: the trophoblast, the extra-embryonic mesoblast (or primitive mesenchyma), and the yolk-sac. These layers have been more especially studied and identified in genital teratomas, with variations according to whether the testicle is involved (predominating ectoblastic and trophoblastic layers); or, the ovary (predominating extra-embryonic entoblast).

Immature proliferations in teratomas are often difficult to analyze for several reasons. First of all, in histo-pathology, different neoplastic tissues, especially those involving primitive germ layers, can show similar patterns (trabecular, papillary, vesicular, glanduliform, reticulated etc.). Secondly, during gastrulation, the 3 embryonic layers show a rather similar morphological aspect, and at the moment when the primitive streak appears, they are identified more by their reciprocal topographic situation inside

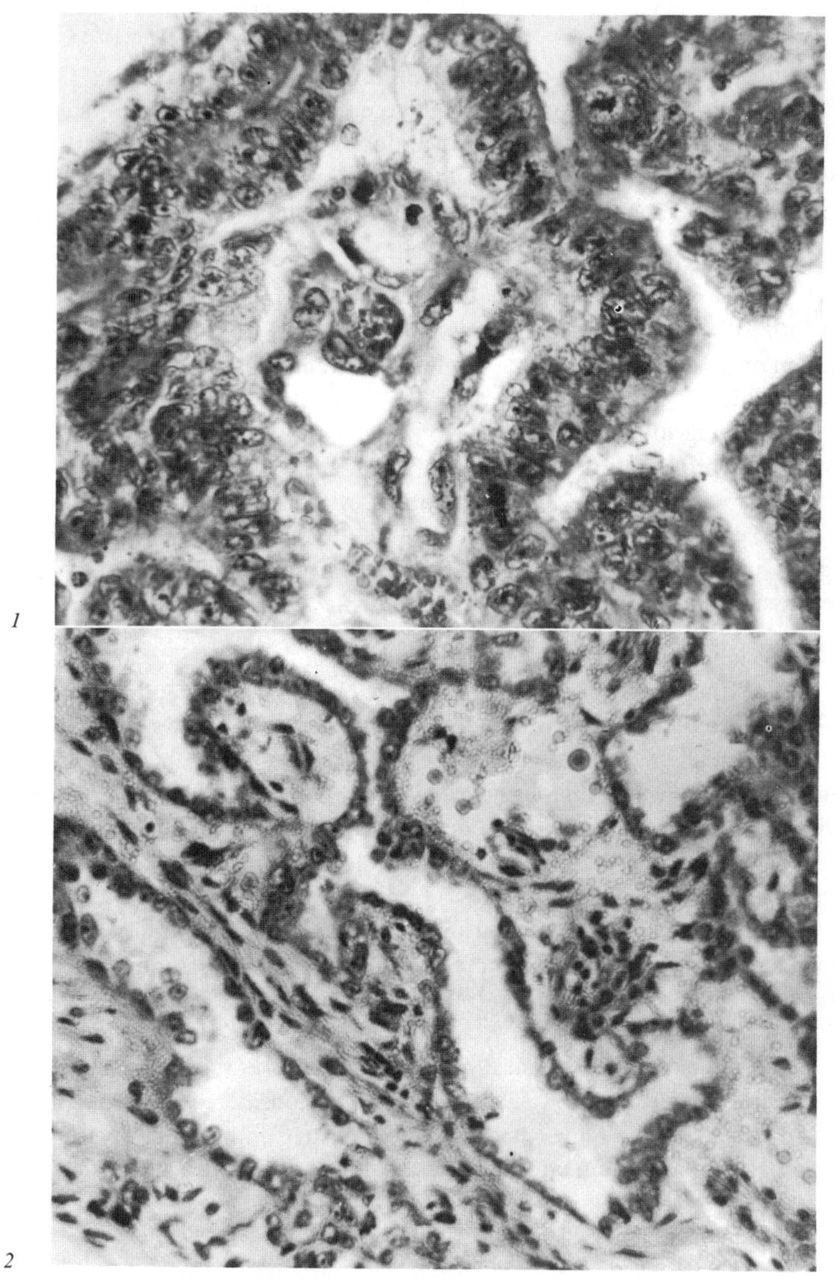

Fig. 1. Exuberant angiogenesis in immature trophoblastic proliferation. (Testicular teratoma 412 098; personal observation) H.-E. × 490.

Fig. 2. Yolk-sac structures. (Ovarian teratoma 12093; PEYRON) H.-E. × 235.

the embryo, than by their morphological differences [see DA COSTA, 1948]. Finally, these proliferations also involve extra-embryonic tissues, certain of which have an ephemeral existence during normal development. They have strange aspects undoubtedly corresponding with formations that pass unnoticed during ontogenesis, becoming conspicuous only in the tumors.

a) The Embryonic Germ Layers

The ectoblastoma has been well-defined by MASSON [1956]. It is characterized by wreaths of basophilic epithelial cells, from which a delamination process can bring forth a single layer of entoblastic cells, lying under the ectoblast.

According to MELICOW and USON [1964] and MARIN-PADILLA [1968], entoblastic proliferàtions would appear frequently, existing as glanduliform structures with more or less differentiated cylindrical cells.

The embryonic mesoblast gives rise to sarcoma-like neoformations, of either tight or loose texture, sometimes resembling undifferentiated mesenchymal blastemas [MARIN-PADILLA, 1968].

b) The Extra-Embryonic Germ Layers (fig. 1, 2 and 3)

The immature trophoblast was identiffed by HARTMANN and PEYRON in 1919. Before that time, its proliferations had been mistaken for carcinomas of Wolffian origin [PILLIET and COSTES, 1895] In reality, their morphology and their angiogenetic (fig. 1) and mesogenetic potentialities are comparable with those of the trophoblast in the lacunar and pre-villous stages, before the end of the third week of placental development [GAILLARD, 1966], i.e., before the appearance of a positive Friedman test.

Various histological patterns can be linked to extra-embryonic entoblast and mesoblast. They appear more frequently in ovarian tumors than in those of the testis, and are rarely found elsewhere [HUNTINGTON and BULLOCK, 1970]. The best known of these patterns is Teilum's 'endodermal sinus tumor', so named because of its resemblance to the rat placenta. In it, striking glomerulus-like structures, previously considered as mesonephric by SCHILLER, mimic Duval bodies. The latter are entodermal diverticula which accompany the fetal blood-vessels into the superficial layers of the

---

Fig. 3. A controverted, immature pattern, interpreted by eminent pathologists as trophoblastic or endodermal. In the author's opinion all primitive germ layers can build similar perivascular structures. Herein, cellular density and tumoral context are arguments for a trophoblastic origin. (Testicular teratoma 12 391; TAYOT) H.-E. × 235.

Fig. 4. An embryoid body. At upper right, note a trophoblastic giant cell and on the right angle yolk-sac, the allantois. (Testicular teratoma 11013; PEYRON) H.-F. × 150.

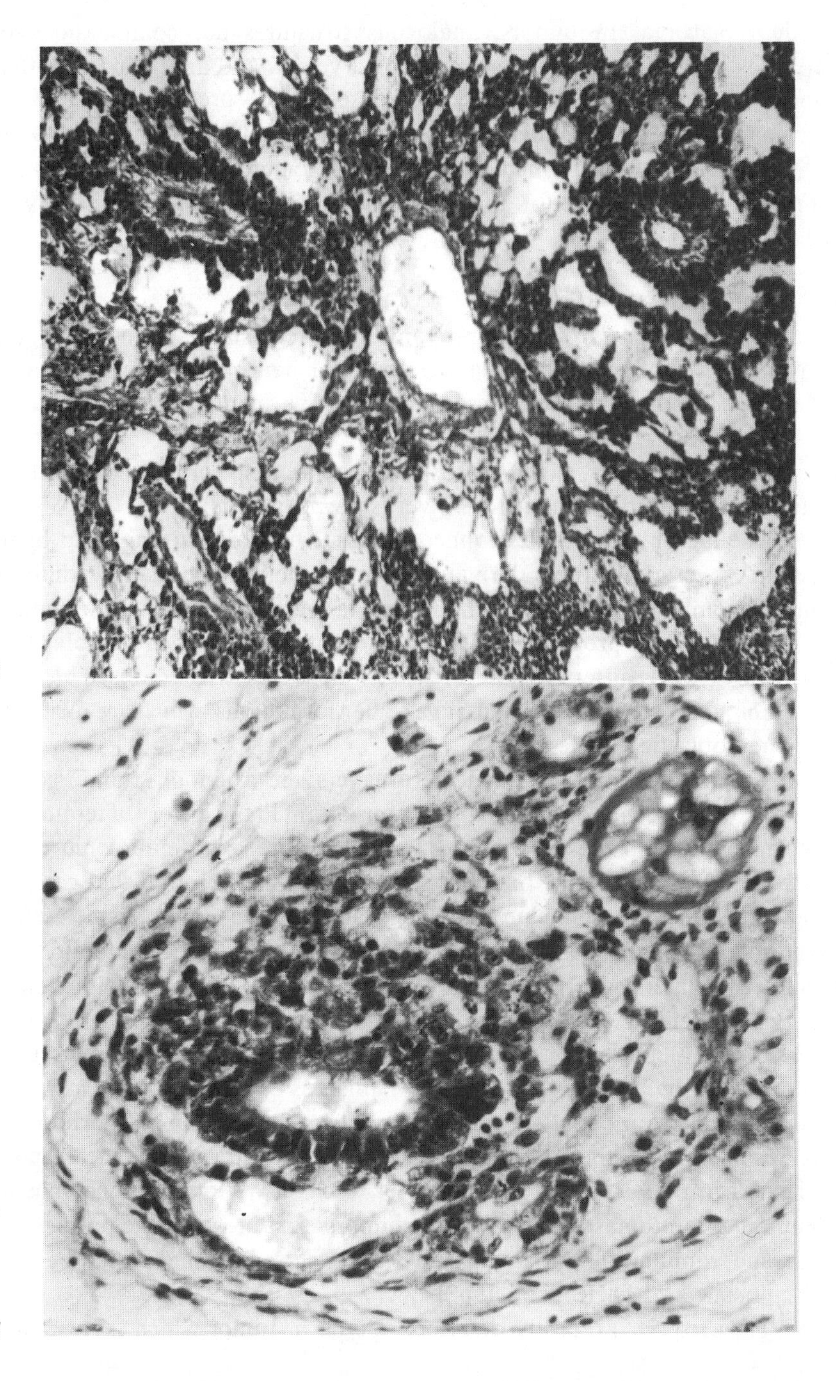

3

4

rodent placenta. In tumors, analogous structures are combinations of neoplastic epithelial elements with stroma vessels and are included within miscellaneous histological contexts. Consequently, preference should be given to the term 'yolk-sac tumors' recently proposed by PIERCE *et al.* [1970].

Another pattern formed by a loosely reticulated tissue resembles the primitive mesenchyme of the chorionic cavity [MARIN-PADILLA, 1968; TEILUM, 1965].

c) Evolution of the Germ Layers

The tissues pass from the embryonic to the fetal stage. Of the embryonic layers, the ectoblast differentiates into either neuroblast or various ectodermal components, often presenting attractive medulloblastic structures [WILLIS, 1936].

The entoblast gives rise mostly to intestinal cysts, or to auxiliary digestive-gland groupings (pancreas, liver, salivary glands).

The mesoblast becomes connective tissue, muscle, cartilage and bone. At times, certain of these derivatives can proliferate abundantly (rhabdomyosarcoma). Strangely enough, no one has yet reported any truly somitic structures. This is probably due to the fact that somites must resemble other structures of a different nature, with which they are confused. In my opinion, it is usually a question of vesicular formations, which are mistaken for neuroblastic vesicles.

Among the extra-embryonic germ-layers, the trophoblast undergoes a characteristic metamorphosis, which is important in the history of teratomas. It made possible the first demonstration of true extra-embryonic tissue existing in these tumors. In 1902, SCHLAGENHAUFER showed that carcinomatous elements observed in certain testicular tumors resembled the uterine chorio-epithelioma. Their morphological and physiological identity was clarified at later date. The structure of these non-gravid chorio-epitheliomas (which undeniably also have extra-genital sites [RITCHIE, 1903; HIRSCH *et al.*, 1946; FRIEDMAN, 1959; COURY and CABANNE, 1959]), is characteristic. It is formed by mosaics of clear cells (Langhans cells) mixed with multinucleated cells (Van Beneden's plasmodiblast-trophoblast). This morphology corresponds with the ultimate evolution of an extra-embryonic trophoblastic germ layer. Thus, it is incorrect to classify chorio-epitheliomas as immature teratomas. However, some tumors exist in which the progressive passage from undifferentiated trophoblast to chorio-carcinoma can be seen [FRIEDMAN and DI RIENZO, 1963].

TEILUM [1965] recently described a type of tumor which corresponds to an evolution of the primitive entoblast. These are polycystic ovarian teratomas formed by a multitude of yolk-sac vesicles. The evolution of the extra-embryonic mesoblast is as little known as are its real potentialities for normal development. One should probably link to certain finely-reticulated and highly-vascularized proliferations, which resemble the mesenchyme surrounding the yolk-sac.

2. Embryogenic Differentiation

In 1935, PEYRON and LIMOUSIN described structures in a testicular teratoma which were analogous to bi-vesicular embryos during gastrulation (fig. 4). Although very meticulous, PEYRON's descriptions have been disputed, and even denied [WILLIS, 1948]. They were confirmed by MELICOW [1940], CUCCIOLI [1950], FRIEDMAN and MOORE [1946], DIXON and MOORE [1952], D'ORLANDO and DELLA PENIA [1954], EVANS [1956], GAILLARD [1956], CABANNE [1957] and MARIN-PADILLA [1965] for testicular teratomas; and by NICOD [1944], CUÉNOD [1944], SIMARD [1950], MASSON [1956] and MARIN-PADILLA [1965] for ovarian teratomas.

Until now, embryoid bodies have been found only in human genital teratomas, (similar to normal human embryo), and in metastases of a spontaneous testicular teratoma in mice (similar to normal mouse embryo) [STEVENS, 1958]. The study of embryoids has threefold interest [GAILLARD, 1968]:

a) To permit little-differentiated proliferations, not easily identified as ectoblast, trophoblast, or entoblast, to be related to the immature germ layers; these proliferations are often confused with each other when they are not a part of embryoid structures [PEYRON, 1940, 1941].

b) To permit a comparative study of them in relation to the precocious development of the normal embryo. In the teratomas concerned, the number of embryoids far exceeds the number of 2- to 3-week-old human embryos observable until now. One can thereby explain certain histogenic processes which are, as yet, not well elucidated. To give one example: embryoid bodies have never been contained in a chorionic cavity. Therefore, it is difficult to concede, as is generally done [HAMILTON *et al.,* 1962; STARK, 1965; see HERTIG, 1968], the trophoblast's role in the formation of extra-embryonic amniotic yolk-sac structures [GAILLARD, 1968].

c) Finally, the embryoid bodies are inconstant; their very existence, however, suggests that they have a particular role and significance in the histogenesis of teratomas [MARIN-PADILLA, 1965].

3. Mature Differentiations and Organogenesis

Almost all the variations of mature tissues have been described in teratomas. However, certain types are more often encountered than others, for example:

a) The skin and its anexes: sebaceous glands, sudoriparous glands, hair.

b) Diverse epithelia: respiratory, urinary, digestive, intestinal (fig. 5) or gastric.

c) Glandular formations: salivary, bronchial, pancreatic, and especially thyroid, frequently observed in ovarian teratomas (ovarian goiters) [NICHOLSON, 1937; BRET *et al.*, 1955; EVANS, 1966].

d) Cerebral or cerebellar tissue, nerves, choroid plexuses, sympathetic ganglia, Pacinian corpuscules and argentaffin tissue, apparently of great interest to pathologists, considering the numerous papers written on the subject [WILLIS, 1949; KELLEY and SCULLY, 1961; see EVANS, 1966; KERMAREK and DUPLAY, 1968].

e) Pigmented neuro-epithelium (fig. 5 and 6) [CORSY, 1927] probably giving rise to an exceptional melanotic retinal anlage tumor [HAMMED and BURSLEM, 1970].

f) Renal tissue: this is always a tissue analogous to that of nephroblastomas (metanephric).

g) Bone tissue: all types of ossification can be seen. There is no skeletal framework, but the medullary tissue can be hematopoietic [CAVAZANNI, 1907].

h) Cartilage [PAGET, 1855; ASTLEY COPPER, 1845].

i) Teeth, whose number can exceed that of a normal dentition (300 in one case observed by PLOUCQUET and quoted by CHEVASSU).

j) Lymphatic ganglia.

k) Adipose tissue. In exceptional cases, this tissue can have the appearance of a hibernoma.

l) Smooth and striated muscle tissue [SENFTLEBEN, 1958] and cardiac muscle [KOSLOWSKI, 1897; KATSURADA, 1901].

Other tissues are less frequent, for example, pulmonary [MONTPELLIER, 1929; HARDING and NAISH, 1935] or hepatic tissue (fig. 5 and 7) [CHEVASSU and PICQUE, 1898; PEYRON *et al.*, 1941; NINARD, 1950; MASSON and THIERAULT, 1956].

Fig. 6. Cystic structure lined by neuroblastic (upper left), retineal (lower left) and chroidal plexus epithelia (right). (Ovarian teratoma 177 553; personal observation.) H.-E. × 87.

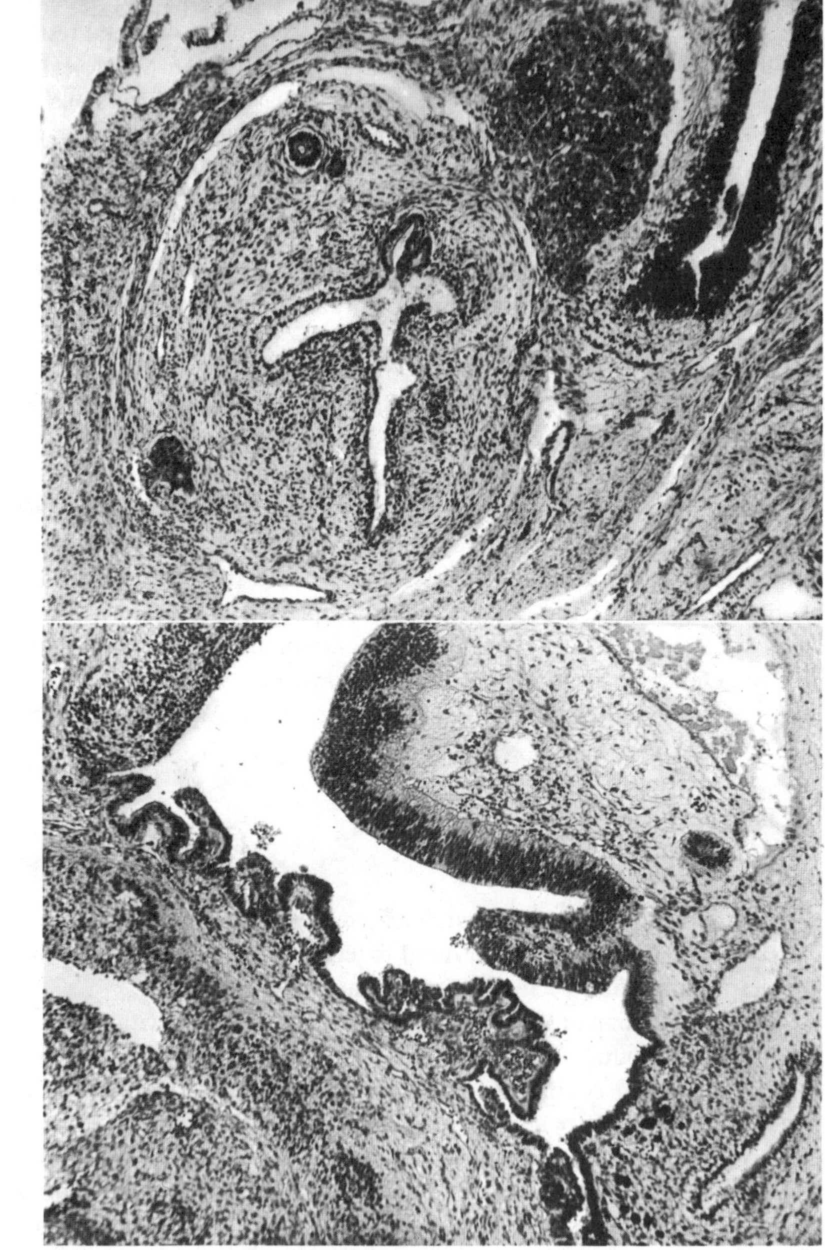

5

6

Fig. 5. From left to right: pigmented retineal anlage, liver, and intestine. Fetal stage. (Sacrococcygeal teratoma 3 468; PEYRON H.-E. × 87.

In very exceptional cases, genital tissue has been described. However, in a sacro-coccygeal teratoma, PEYRON *et al.* [1939] found two fetal testicles (fig. 8), with spermatic ducts and abundant interstitial tissue. MASSON [1956] published photomicrographs demonstrating this case, which poses in a very particular way the problem of organogenesis in teratoma. Also, STEVENS [1967] observed a fetal testis in a unique adult teratoma of strain-129 mice. But in this case the derivation from teratoma may be discussed, since it resembles a dysgenetic adenoma of Pick.

In fact, one must concede that most of the organs reported in these tumors can be traced to an erroneous interpretation by the first observers, of misleading macroscopic forms [SAINT-DONAT, 1696; PROCHAZKA, 1803; ANDRÉ DE PERRONE, 1833; VELPEAU, 1841; ASTLEY COOPER, 1845; VERNEUIL, 1885; see EWING, 1911]. Principally concerned are bone, cerebroid tissue, and rounded cysts, which have been joined too rapidly either to limbs or fragments of limbs, to brain, or to an eye.

These observations have been pertinently criticized [see WILLIS, 1960]. However, a rather remarkable organization of tissues can exist in these tumors, in both the histological differentiation and the perfect correlation of their elements. In this way, respiratory epithelium is often associated with cartilage and serous glands; likewise intestinal epithelium is surrounded by mesenchymal and muscular coats, showing in cross-section a realistic image of intestinal segments; here one can even see Auerbach's plexuses.

B. Synthetic Structure

1. The Framework

All the histological differentiations enumerated above constitute the pieces of a puzzle, which, when fitted together, enable one to reconstruct, topographically, the structure of each teratoma.

The combination of the various parts occurs at random, with no evident metameric distribution around an axis [BUDDE, 1926].

The structural complexity is both qualitative and quantitative. The proportions of immature and mature tissues vary greatly from one tumor to the next. There exist teratomas that are entirely mature; others, entirely immature, and still others containing both mature and immature components. The structure is sometimes reduced to a single layer, that can be blocked in its evolution, and show only one kind of differentiation or structure.

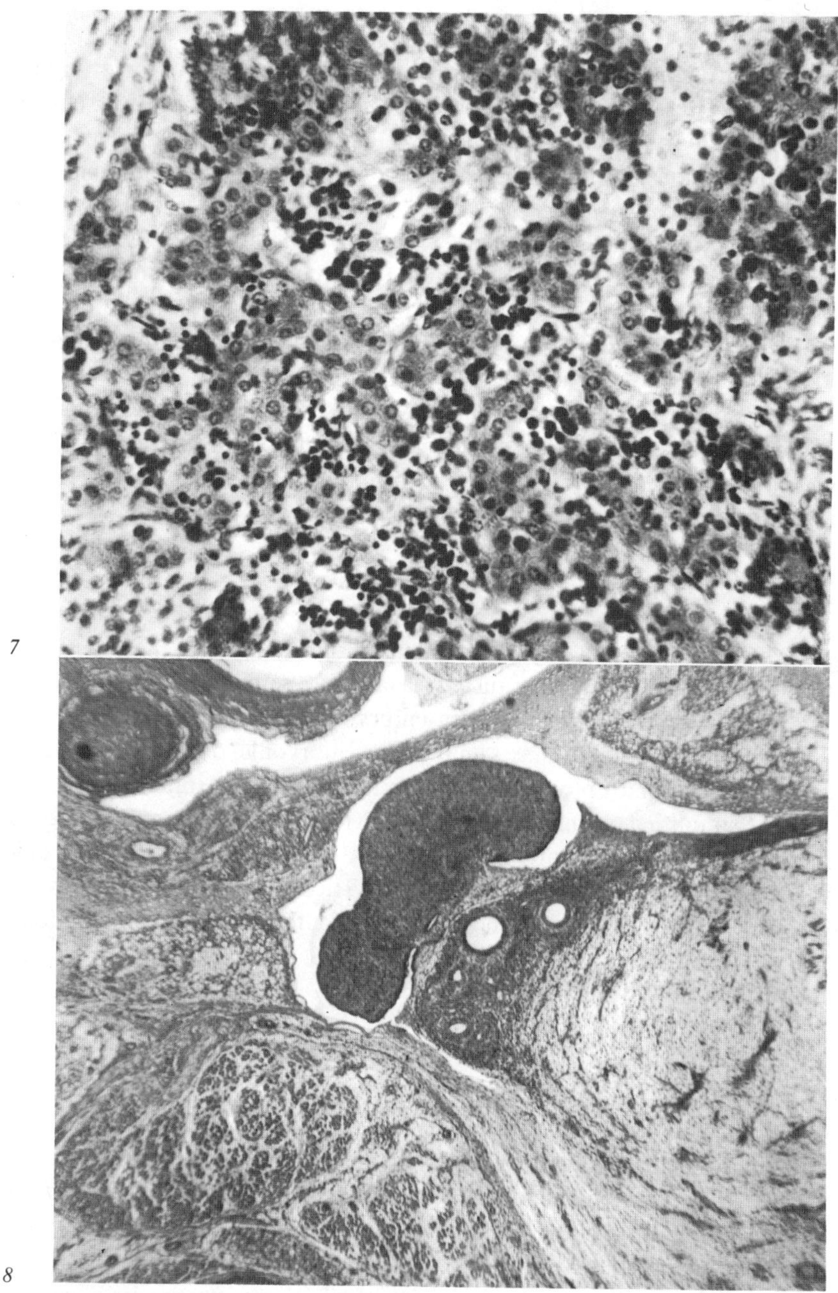

7

8

Fig. 7. Fetal hepatic cords and hematopoiesis. (Sacrococcygeal teratoma 3 468; Peyron.) H.-E. × 290.

Fig. 8. One of the two genital glands in teratoma. (Sacrococcygeal teratoma 3 468; Peyron.) H.-E. × 31.

The different morphological metamorphosis can be found for any given tissue, characterizing its fetal, embryonic and adult stages. The diverse parts build elementary structures composed of either vesicular or cystic glanduliform formations; or, of cellular associations forming more or less compact trabeculae, or colonies.

As a result, the exterior appearance of a teratoma lies within the limits of the two extreme fundamental architectures. One is nothing more than an agglomeration of cysts, varying in volume and form; the other is made up of fleshy masses having different appearances and consistencies [CURLING, 1854; WESTERGREEN, 1872].

If we combine this bivalent macroscopic structure with the findings of microscopic analysis, we can define a certain number of tumorous models which cover practically all the forms of teratomas [CABANNE, 1957; DUGH and SMITH, 1964],

2. The Models

These come under two main headings:

a) Complex, multi-tissued teratomas, either immature, mature (Askanazy's coeatans teratoma) or mixed: semi-solid, semi-cystic, polycystic tri- or didermal, of embryonic or extra-embryonic tissues, fetal or adult.

b) Simplified teratomas: immature, embryonic or extra-embryonic tumors: ectodermal, entodermal or mesodermal growths, solid or cystic, etc.

The simplified, immature teratomas or terato-carcinomas merit special mention. They are made up of one of the 'six' embryonic or extra-embryonic germ layers, or a combination of them. Their simplification is only an illusion. Actually, if one applies the formula for factorial combinations,

$$C_n^r = \frac{n!}{r!\,(n-r)!}$$

one notes that these 6 germ layers can give rise to 57 combinations, in addition to the 6 original forms. However, combinations by two's or three's are the most common. This is in accordance with the order of appearance of these layers during development, as well as with the epigenetic character of this development.

Thus, the study of the structure of teratomas permits a precise inventory of the materials of which they are composed, showing how they are arranged spatially. It is instructive to visualize the way they are joined temporally.

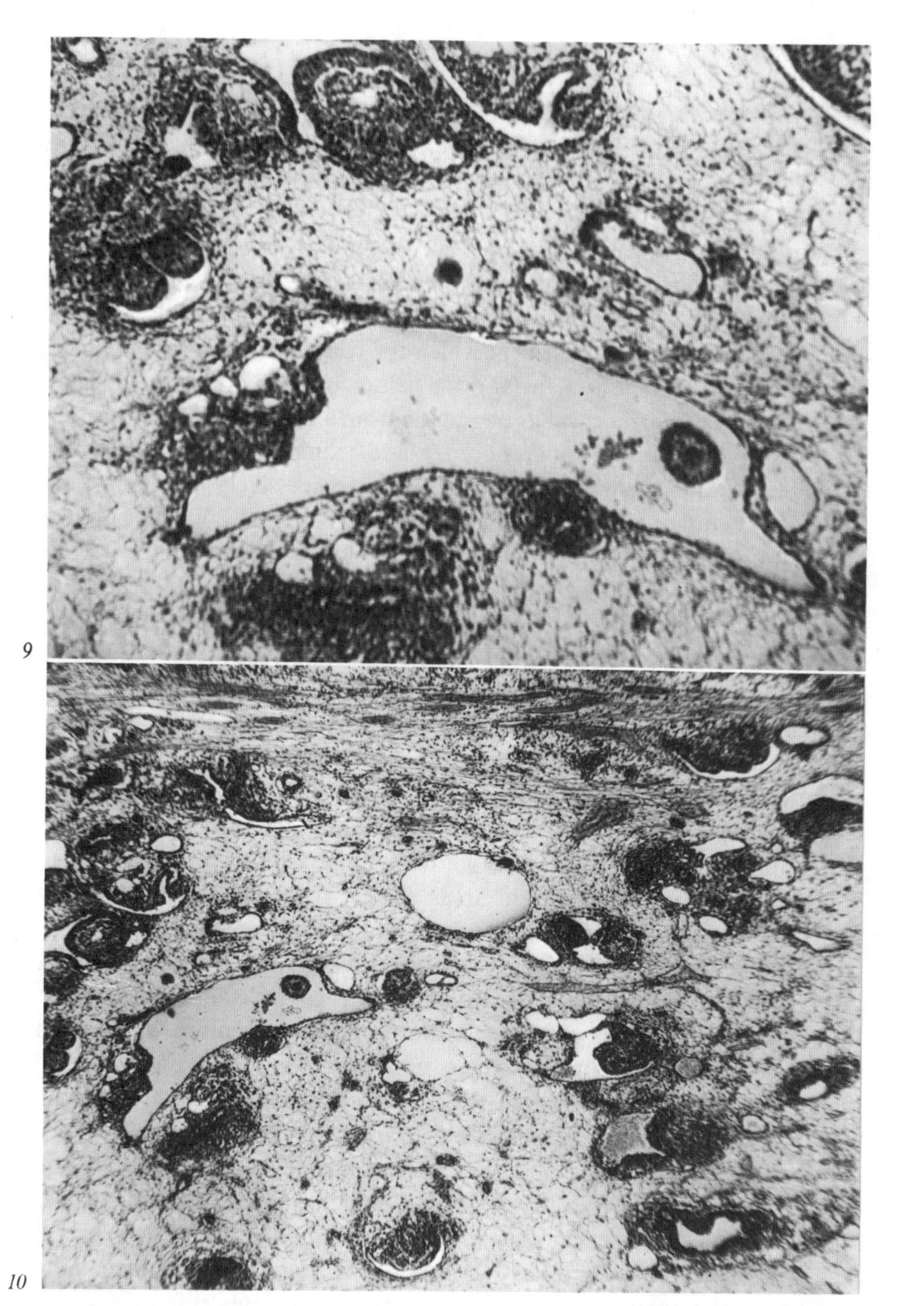

9

10

Fig. 9. Embryogenic cysts of various shapes and polyembryony. (Testicular teratoma 11 013; PEYRON.) H.-E. × 31.

Fig. 10. One of these embryogenic cysts at higher magnification showing their potencies. (Same case.) H.-E. × 87.

III. Kinematics of Teratomas

A. Histogenesis of the Parts

The solid portions of a teratoma form a 'hotch-potch' of tissues [NICHOLSON, 1922], whose appearance and organization are comparable to those of tissues during normal ontogenesis. The same materials are used and assembled to form patterns in which the characteristic differentiations and, the structural motifs of the various tissues are easily recognized. Their histogenesis can be conceived as beginning from a mosaic of morphogenetic fields distributed with no apparent order, and acting according to their own capacities, evolving either simultaneously or successively, depending on whether the tumorous structure is composed of tissues of the same or different ages.

In view of their importance and frequency in the general structure of teratomas, the histogenesis of cysts merits more careful study and reconstruction than has yet been made. Their significance has often been either neglected, or erroneously interpreted. Indeed, the structure of these cysts is often unusual, for within their walls can be observed numerous differentiations (fig. 6). Sometimes ectodermal epithelium, trophoblast, neuroblast, squamous or ciliated cells, and muciparous entodermal cells can be seen in succession, side by side, around the same cavity [FRANK, 1906]. In immature teratomas, certain cysts whose volumes and form are subject to great variations, show thickened areas exactly resembling germinal discs such as those found in the egg during the blastula stage (fig. 9 and 10). One can even see an embryonic or polyembryonic process occurring that takes place sometimes on the exterior of the cavity, sometimes on the interior (fig. 10). This could suggest the proligerous membrane of taenia echinococcus, or the germinating cysts in primitive modes of reproduction. PEYRON [1941], proposed to call these cysts 'blastular embryogenic cysts with fertile membranes'. The modalities of their evolution help explain many paradoxical aspects observed in teratomas [GAILLARD, 1966]. The cystic structure here takes on the very particular significance of an elementary organism, its exchanges with the environment are thereby facilitated, and it has its own primitive interior medium. It is no longer only a question of a spatial arrangement favoring the colonial life of a group of the same, or different cells – but rather of the first stage of a true teratological blastula. The polycystic architecture so frequent in teratomas could thus be explained either by the building out and detaching of an initial cyst, or certainly more often simultaneous or successive formation of different autonomous cysts would represent the fundamental process of their histogenesis.

B. Histogenesis of the Whole

One can imagine that the joining together of the various parts of a teratoma occurs in two ways. Either, these parts together form a coherent, even though chaotic, independent biological system, situated inside the host like a worm in a fruit; or, that each part forms its own system, thus yielding just as many miniature teratomas.

The first of the two processes could apply to adult teratomas wherein all the constituents are of the same age. This suggests at the origin a teratological-ontogenic system, issued from one single totipotential element. That is why the earlv authors looked for embryonic annexes around the periphery of these tumors [CHEVASSU, 1906]. In this way, the periphery of dermoid cysts has at times been considered equivalent to an amniotic membrane. But this viewpoint is not easily applicable to a teratoma whose tissues differ in age. Their histogenesis is better conceived as a successive development of several embryonic complexes whose ontological evolution is different with regard to space, time, and potentialities [GAILLARD, 1956]. One can then explain:

The place and significance of embryoid bodies in the histogenesis of teratomas, because these latter represent the most complete, most spectacular (although inconstant) morphogenetic expression of these potentialities.

The histogenesis of adult teratomas, this time through the simultaneous development of several embryonic complexes having different ontogenic potentialities.

The presence of vestigial structures of the host-organ at the very heart of the teratoma. For example, in testicular localizations, it is not uncommon to find spermatic tubes, more or less well-preserved, mixed in with the tumorous tissues [GAILLARD, 1956].

The general architecture of these teratomas thus involves a double composition: one which results from the potentialities of the tumorous elements; the other, which depends upon the correlations that have been established between these elements and the host-tissues (the latter being sometimes stimulated by the tumorous elements). This is especially clear in the case of primitive trophoblast or yolk-sac, which form characteristic structures when they come into contact with the vessels of the stroma [GAILLARD, 1956]. Connections between nervous elements of teratomas and nerves of the host have been also described [NICHOLSON, 1929; COUTELLE, 1961].

The degree of malignancy of these tumors depends also upon the behavior of the teratoma's own elements, and upon the host's reactions.

C. Evaluation of the Malignancy

The malignancy of a teratoma is generally inversely proportional to the degree of differentiation of its tissues. Thus, theoretically, one could attenuate the malignancy by influencing the differentiation [see PIERCE, 1967]. However, this idea is quite relative, since cases of adult-tissue metastases which start from testicular teratomas, either adult, mixed, or immature, are known [WOGALTER and SCOFIELD, 1962; KARPAS and JAWAHIRY, 1964].

From a dynamic viewpoint, this malignancy is manifested in different ways:

The notion of a pluricentric development of teratomas practically never permits the affirmation of the benign nature of a tumor; some adult teratomas giving rise to metastases of an immature type have been described. However, less dramatic cases are known (for example, mature testis teratomas in infants are almost always benign, and immature forms are not always malignant).

The notion that the more immature the tissue, the more likely it is to be malignant, implies that the mother-cells themselves can give rise to malignant proliferations either totally anaplasic, or involved in primitive embryonic or extra-embryonic differentiations [DIXON and MOORE, 1953]. This is the case for many testicular tumors, incorrectly classed as 'embryonal carcinomas', and usually including ectoblastic or trophoblastic (trophocarcinoma) differentiations [FRIEDMAN and DI RIENZO, 1963].

The notion of a dysontogenetic development leads us to assume that secondary canceration occurs more easily in certain elements, possibly hindering the elaboration of the teratoma. Such proliferations can even remove or prevent the formation of a complex multi-tissued structure by unilateral development of one of its constituents. This hypothesis, mainly upheld by EWING, should be shaded to avoid confusion of secondary malignant proliferations in a teratoma, with those of the mother-cells.

The most convincing examples of secondary canceration of certain elements in teratomas are: a) the epidermoid canceration of a dermoid cyst [BERTRAND and LATAIX, 1952; KELLEY and SCULLY, 1961]; b) the proliferations of choroid plexuses starting from a sacro-coccygeal teratoma [BOLANDE, 1967].

IV. Origin of Teratomas

The evolution of ideas on the origin and nature of teratomas has followed several streams of thought. The greater frequency and the spectacular nature of their genital localizations naturally incline the human mind to draw a parallel between them and the process of reproduction. On the other hand, however, the no-less spectacular character of natural monstrosities, whose most frequent localizations actually coincide with those of certain of these neoformations (for example, the junction of siamese-twins is principally at the cranium and the sacro-coccyx), suggested the simultaneous development of an abnormal twin, or of a disturbance dating back to the embryonic period of the host. Thus, it was generally considered that teratomas had a fetiform structure (foetus in foetu) [HINLY, 1831], and two kinds were finally recognized as having different origins, either parthenogenetic or parasitic-twin inclusion [PEYRON *et al.,* 1922].

Once the cell-theory had established the foundations of biology, an effort was made to identify the cells capable of producing neoformations which thus mimicked the ontogenetic processes. One also sought to know the cause and the mechanisms of their peculiar behavior. Within the period of a century and a half, a series of hypotheses was proposed, which pretended to determine, once and for all, the origin of teratomas, and which found their arguments along with the advances made in histology and embryology [see REPIN, 1891; EWING, 1911; WAELKENS, 1929; NEEDHAM, 1942; WILLIS, 1960].

Overlooking the first hesitant steps toward application of the cell theory to pathological histology (LEBERT's plastic heterotopia, VIRCHOW's cellular indifference, BARD's nodal cell), the following hypotheses have been advanced:

a) A diplogenesis by inclusion of a parasite-twin (GEOFFROY SAINT-HILAIRE, TUMIATI); b) fertilization of a polar corpuscle (MARCHAND); c) parthenogenesis of a sexual cell [BUFFON, MAEKEL, WALDEYER, 1872; LANGHANS, 1887; BLAND-SUTTON, 1889; WILMS' first version 1896; PEYRON *et al.,* 1922, etc.], d) an inclusion of embryonic rests [CONHEIM, 1889] coming from a germ layer or from the primitive blastoderm; e) the exclusion of a blastomere [WILMS' second version, 1898–99; BONNET, 1901]; f) various disturbances of the embryonic mechanism, at the time when germ layers (VERNEUIL's theory of 'enclavement') or the primitive streak [KRAFKA, 1936; BREMER, 1951] are laid down; g) various difficulties in the chemistry of development due either to lack of activity, or to flaws in the organizers, or embryonic morphogenetic substances [ASKANAZY, 1907; BUDDE, 1926; NICHOLSON, 1929; WILLIS, 1935; NEEDHAM, 1942].

All these hypotheses are extrapolated from results in experimental, descriptive and causal embryology: the discovery of polar corpuscles (LARUS and CH. ROBIN); the VON BAER's theory of germinal layers; the conception of separation between soma and germen (WEISMAN); the materialization of the germinal lineage and tracking of its elements in their migration toward the gonads (WITSCHI); the studies on parthenogenesis in mammals (LEO LOEB): the evaluation of blastomere potentialities (ROUX, DRIESCH, SPEMANN); the role of organizers (SPEMAN), and exogastrulation experiments (HOLTFRETER) [see DA COSTA, 1958; HAMILTON *et al.,* 1962, STARK, 1965].

For a long time, every teratoma was considered as being a single ontogenic process, that is, as an embryoma arising either from a sexual cell, or from a segregated embryonic cell.

The first hypothesis was easily applied to ovarian teratomas, which thus represented the neoplastic evolutionary mode of the ovule. But it was less easily applied to testicular tumors, and still less easily to extra-genital teratomas. It was conceivable that an ovule was capable of parthenogenetic development, but this was no longer the case for the male sex cell. Thus, it was supposed that female cells could be found in a testicle, or that there existed an intra-glandular hermaphroditism. In order to explain extra-genital teratomas, some authors suggested an aberrant germinative plasma.

Later, however, the concept of the germinal lineage arose, and one passed from the sexual cell to the primordial germ-cell: the gonocyte, indifferent element which is as yet neither male nor female, and from which comes the expression 'ephebogenesis', proposed by BOSAEUS [1926] to designate the ontogenic process in teratomas.

Next, the demonstration of gonocyte migration toward the gonad provided a plausible explanation for the extra-genital localizations of teratomas [see WITSCHI, 1948; and SIMON, 1960].

The second hypothesis was at first based on a purely hypothetical notion (BARD's nodal cell), then materialized, thanks to the discovery of blastomere properties. After some fluctuations due to experimental results that were sometimes contradictory (depending on whether one was dealing with regulative or mosaic eggs), the properties of blastomeres were applied to teratomas, on the assumption that they develop from an excluded blastomere. This yields a tumor all the more complex, since the blastomere, when segregated, is in a stage closer to the potential egg. In order to explain the sometimes considerable number of constituents in certain teratomas, the advocates of this theory supposed that there could be several blastomeres involved. This is the beginning of the concept of the polyontogenic development in teratomas. It was considerably later that this concept was formulated, chiefly due to PEYRON's discovery of embryoid bodies.

At the present time, everyone agrees with the view that teratomas result from a dysontogenetic development of toti- and pluri-potential cells. But there is disagreement as to the nature of these cells. According to some authors, they are germinal, having the embryonic potentialities of the egg or of blastomeres; for other authors, their potentialities are blastodermic in nature. In particular, the gemellary origin of sacrococcygeal or extragenital teratomas is questioned by many pathologists, since patterns usually characteristic of genital localizations (endodermal sinus pattern, dysgerminoma, etc.) are found more and more often in these tumors [FRIEDMAN, 1951; PACHTER and LATTES, 1964; RAGHUNATHA RAO *et al.*, 1964; NISHIYAMA *et al.*, 1966; TEILMANN *et al.*, 1967; CONKLIN and ABELL, 1967; BESTLE, 1968; KNOWLES and SCULLY, 1969].

We cannot as yet, determine exactly, either topographical placement of the totipotential cells in the host organism, or their place in the time sequence of the host's development. Neither do we know if it is a question of cells set apart during ontogenesis or cells which escape the laws of homeostatic equilibrium in an adult organism, when development has already long been completed [GAILLARD, 1962, 1964]. To solve these problems, it is clear that we must be able to reconstruct teratomas experimentally. This leads us to examine what has been accomplished up to the present in the experimental field.

V. Experimental Teratomas

A. Experimental Teratomas in the Rooster

MICHALOWSKY [1928–29] caused complex teratomas to appear in the rooster, by injecting a zinc chloride solution into the male genital gland. These experiments were reproduced by several authors [BAGG, 1936;

ANISSIMOWA, 1939; FALIN and GROMZEVA, 1936; CARLETON *et al.*, 1953; SMITH POWELL, 1957]. They tend to demonstrate that the chemical induction of teratomas in the rooster is particularly effective during the period of sexual activity after stimulating the sexual function with gonadotropic hormones. Similar successful experiments were performed in the adult mouse [BRESLER, 1959] and in the rat [RIVIERE *et al.*, 1960].

B. Spontaneous Testicular Teratomas in the Mouse [see STEVENS, 1967–1970]

In 1954, STEVENS and LITTLE observed spontaneous congenital testicular teratomas in inbred mice (strain 129). STEVENS undertook a series of well-conducted experiments, whose results can be summarized as follows:

Teratomas in mice are congenital and the neoplastic process is initiated at 12.5 days of gestation. Their structure and composition are comparable to the human counterpart. They are sometimes of an adult type, sometimes of a mixed, adult and immature type. In fetal or infant mice, tumors are composed of undifferentiated embryonal type cells (embryonal carcinoma). At 5 days and older, the tumors contain both differentiated and embryonal elements (teratocarcinoma) and in most adults the tumors are completely differentiated and lack embryonal elements (teratoma). When fragments of these tumors are grafted onto unaffected mice of the same strain-129, one sometimes obtains immature tumors, sometimes adult tumors, and all the intermediaries. Their evolution toward maturity depends upon the evolution time of the neoplastic growth.

Intra-peritoneal grafts give embryoid bodies which have embryonic potentialities. Growth and differentiation of teratoma in mice are under genetic control. Spontaneous tumors depend upon the action of several modifying genes but a single major gene [STEEL] can influence susceptibility for experimental teratocarcinogenesis.

Teratoma in mice originates from primordial germ cells. The starting point of the tumor is inside the fetal spermatic tubes and appears as a carcinomatous proliferation. Genital ridges from strain-129 fetuses develop teratomas when grafted to adult testis of the same or foreign strains. Mice deprived genetically of primordial germ cells do not develop teratomas. However, identical tumors can also develop from disorganized clusters of undifferentiated embryonic cells that originate from primordial germ cells, bi-cellular eggs, or from grafted early embryos. The study of karyotype suggests that the teratomas originate at the expense of diploid germ cells.

Using the same strain-129, and also starting from an ovarian teratoma in the mouse, PIERCE and co-workers [PIERCE, 1961; PIERCE and VERNEY, 1961; KLEINSMITH and PIERCE, 1964] showed that the stem cells of genital teratomas have the embryonic potentialities of the 3 embryonic germ layers. They can express these potentialities by forming embryoid structures. Therefore, they are very probably of a germinal nature.

All of these experiments thus substantiate the conceptual existence of totipotential germ-cell proliferations [KLEINSMITH and PIERCE, 1964]. This was suggested in an early form by FRIEDMAN, DIXON and MOORE, by postulating the interrelationship of the patterns most frequently observed in the testicular tumors, and afterwards applied to other localizations of teratomas[1].

One problem which remains to be solved consists of finding distinctive figures, for example, by using electron microscope [PIERCE, 1966] between the true proliferations of germ-cells (or germinal carcinoma) [FRIEDMAN, 1959; GAILLARD, 1965] and the yet-undifferentiated proliferations corresponding to embryonic cells of the inner cell-mass or embryonic disc, or else to a precursor of the germ-layers [MACGRADY, 1944].

C. Attempt of Production by Graft

Numerous attempts at embryonic homoplastic tissue grafts have been made in birds, amphibia and mammals. Under certain conditions, teratomas were successfully obtained; these, were, however, always benign. In fact, it is a question of *in vivo* culture of embryonic cells which produce neo-

1 The pluripotentiality, i.e. the ability of diverse histological and morphogenetic differentiation, of the stem cells of teratomas has been demonstrated by cloning experiments. KLEINSMITH and PIERCE [1964] established a number of clones using individual teratocarcinoma cells of the substrain 402 AIII. All these clones could proliferate and differentiate into derivative tissues of the embryonic germ layers. FINCH and EPHRUSSI [1967] not only established the clones *in vitro* but also confirmed their ability for diverse differentiation by injecting the cells subcutaneously or intraperitoneally into compatible host mice strains. KAHAN and EPHRUSSI [1970] and ROSENTHAL *et al.* [1970] also obtained considerable differentiation in tumours derived from single cells of spontaneous testicular teratomas. EVANS [1972] has recently reported the isolation of a pluripotent clonal culture strain from a transplantable teratoma. On re-inoculation the cell strain produced teratomas which differentiated into at least ten types of tissue. On being subcloned, two lines could be distinguished, one of which exhibited a spectrum of differentiation comparable to that of the original strain, but the second line was not tumorigenic and did not show multiple differentiation. [Ed.]

formations hardly comparable with tumorous teratomas observed in pathology [see SALAÜN, 1968].

These experiments show to what extent the conceptions of the origin and nature of teratomas can differ according to the viewpoint of the observer or experimenter as an embryologist or a pathologist.

D. Induced Teratomas in Plants

Teratomas have been induced by BRAUN [1965] from normal tobacco cells. In certain experimental conditions, the process is reversible. Therefore, a tumorous state does not seem definitively determined in plants, hence, arguments against the dogma of the irreversibility of the malignant change are reinforced [PIERCE, 1967].

VI. The Sex of Teratomas

Research on the Barr and Bertram corpuscle yielded the following results:

– ovarian teratomas are all female;

– testicular teratomas are either male (2/3), or female (1/3);

– extragenital teratomas give equivalent results, although not as clear-cut (possibility of male tumor in female host: THEISS *et al.,* [1960]);

– mosaic tumors exist, which contain both male and female zones [MYERS, 1959; DAYAN, 1963].

It is generally held that teratomas, at least in their genital localizations, result from a parthenogenetic development of haploid cells, either by fusion or re-duplication [HUNTER and LEMNOX, 1954]. In reality, this question of the sex of teratomas remains obscure, and additional factors must certainly be borne in mind; in particular, the host's age [PIERCE and NAKANE, 1967], the exact nature of the sex chromatin, the possibility of polyploidy [see STEVENS, 1962], and variations of karyotype [LAURENT *et al.,* 1968].

VII. Conclusion

The natural history of teratomas, therefore, occurs on several levels.

A. On the Structural Level

The diversity of teratomal localizations suggests that it is a question of scattered tumors having common characteristics but different origins.

According to many authors [see WILLIS, 1960], they form a single class of tumors; the basic difference between models is in terms of their complexities. In reality, what they have in common is the heterologous nature of their tissues, in relation to those of the organs in which they develop. Variations of structure according to localizations are not always easy to define. They sometimes result from co-relationships formed between the constituents of the teratomas and the host-tissues. In other cases, there exists a quasi-specific regional particularity. For example, dermoid cysts of the ovary have an internal protuberance (mamilla or embryonic eminence of Rokitansky), which distinguish them from other 'dermoids'. Sacro-coccygeal teratomas are generally of a more organoid nature, than are genital teratomas. They share, with the retro-peritoneal localizations, the characteristic of containing some metanephrous structures which are very rare in other sites. Chorio-epitheliomatous differentiations are more frequent in the testicle. Their extra-genital localizations show that some correlations exist between epiphysis, the thymus, and the gonads. The migration of gonocytes does not seem to be sufficient explanation for these localizations, even though in these organs one also encounters homologous tumors with a dysgerminoma pattern, considered to be the archetype of germ-cell proliferation.

In my opinion, the elementary cystic structures represent a characteristic architectural element of teratomas. In this role, efforts to clarify their significance is merited. Their outward appearance of simplicity requires insight and ingenuity to be understood. These hollow structures are something more than crude geometrical constructions. Their composition begins as a blastula, but, in most cases, becomes more simplified, yielding cysts that contain cells of the same, or of different nature.

B. On the Histogenic Level

In teratomas, the tissue metamorphoses are similar to those which take place in the course of normal development. The tumorous ontogenesis thus disposes of the same materials, but achieves a construction which, on the whole, has a poor equilibrium, and does not result in a complete organism. According to DRIESCH, the system is equipotential but disharmonious. That does not mean that the distribution of potentialities is not, in this case, subject to certain laws, since there exist laws for teratogenesis [WOLFF, 1948]. RUYER [1958], compares the chaotic mixture of tissues to a 'Markovian jargon'. It is, thereby, possible to foresee in the study of teratomas, the application of a form of reasoning which would no longer be purely analogical, and which would thus be more rigorous.

The stem-cells of teratomas come from either, a) embryonic cells segregated during embryogenesis, revealing their potentialities only after a period of time, or, b) from adult cells which become totipotential under certain conditions. The first assumption, derived from CONHEIM's hypothesis, presents one major objection: it is difficult to conceive of a static, dormant state, and it seems necessary to postulate that, in reality, it is more a question of a dynamic state, perpetuated by homotypic cell-divisions. On the other hand, the second hypothesis apparently contradicts the restriction of cellular potentialities implied by the adult state and differentiation. In either case, all the cells are genetically identical, and the tissues must be capable of regeneration, as well as ensure the continued existence of the individual through its off-spring. Thus, it must be conceded that in every organism there exist certain cells capable of using their genetic keyboard to the utmost. Perhaps it is necessary to enlarge the concept of germinal tissue, as proposed by WINTREBERT [1963]. This has the additional advantage of establishing a bridge between the process of regeneration and that of reproduction.

The lack of metameric axiation was an argument of historical importance against the germ-cell theory of the origin of teratoma. Its value is diminished in view of recent results of experiments using ova of mice, and frog-eggs grafted in extra-uterine sites. Growths thus obtained are composed of various germ-layer derivatives but show no organization [see STEVENS, 1968, 1970].

On the other hand, the origin of teratoma is probably pluricentric, the first step of which consists of a cancerous, totipotent cellular population. As a consequence, their final structure is a puzzled conglomerate of embryos [MASSON], or of embryonal complexes [GAILLARD]. Therefore, metameric axiation should be sought in every part, but not necessarily in the whole. However, identification of stem-cells could represent but one step in the understanding of teratomas, since that factor alone cannot explain the events occurring on the infra-cellular and molecular levels, i.e., at the true site of the mechanisms of differentiation.

C. On the Causal Level

Whatever may be the nature of the stem-cells, it seems that they possess the ontogenetic plans, but that application of these plans is not coordinated due to an unknown perturbation occurring on the level of the principal plan, or during the many stages through which are transmitted the secondary plans. One can imagine that each part of a teratoma, generated by stimuli of undetermined nature, assumes form according to its own controls, which, in turn is imposed by a specific conditioning. Its organization, however, falls short of complete individuation. NEEDHAM [1950] assumed that the basic process involves a failure of the individuation field early in development, to control the action of evocative substances. Experimentally, a chaotic assemblage of tissues can be induced by implanting heat-coagulated chick-embryo extract into the blastocoele cavity of a newt embryo [HOLTFRETER, 1938].

Information from the primary evocator [KRAFKA, 1936] is certainly not responsible [WILLIS, 1960], but the second and third-level organizers are probably present in most adult tissues. In addition, a perverse competence in the reacting tissue must also be considered. The reacting tissue follows a self-differentiation that includes both cyto-differentiation and histogenesis, but which lacks spatial arrangement. When the morphogenetic process collapses like a house of cards, it signifies an absence of control of physical forces. For example, in human tumors, all embryoid bodies vanish and dislocate after the stage of an embryo 12 or 15 days old. Meanwhile, their components may continue the processes of growth and differentiation, often with a predominant differentiation. In this way yolk-sac carcinoma, trophocarcinoma or myogenic tumors have been experimentally induced by conversion of transplanted teratomas [PIERCE and DIXON, 1959; PIERCE and VERNEY, 1961; STEVENS, 1967]. Outgrowths in which a single-cell type prevails, can be explained either by a selective proliferation among a heterogeneous population, or, by a limitation of embryonic potency of a stem-cells. However, there is no proof that the reacting tissue is embryonic in nature, since teratomas have been induced in adult mice [BRESLER, 1959]. In addition, the parts certainly also possess some inductive effects upon the host's dedifferentiated mesenchymal elements, which can participate to a certain extent in the construction of chimeric structures [GAILLARD, 1966]. This suggests the possible existence of factors stimulating overgrowth, such as those released from Rous sacroma-cells [RUBIN, 1970]. Thus, it is important to gain a better understanding of the tissue interactions between tumor and host. Take, for example, the situation of a dermoid cyst in the ovary. It can be seen to be situated in the same way as various other cysts encountered in this organ. One can thus suppose that a dermoid, far from representing a caricatural and incomplete embryo, results rather from the ɪnodulation of a banal follicular cyst, by evoking substances which come from ovular lysis. The difficulty lies in trying to demonstrate the recovery of varied competences by cells of the cumulus oophorus, or the stromatocytes as are the follicular cells of the cumulus oophorus, or the stromatocytes of the ovarian parenchyma. Perhaps it is a question of a reversed trans-determination process [HADORN, 1966], capable of bringing about allotypic differentiations.

Reactivation of a portion of the cellular genome could also serve to explain the difference between teratomas that are benign from their onset, and those which are from the onset malignant. For the former, genetic contamination would occur through isogenic inductors, whereas for the latter, the inductors would be heterogenic, e.g., viruses.

Thus, the construction of a teratoma can be expressed, as in normal development, in terms of induction and competence. But it is exluded from the rigorous control of the organization directing plan, that establishes for each element its time and place, so as to achieve that specific finality and form which characterizes every living organism.

In my opinion, two main classes of teratomas seem to emerge: tumoral teratomas, and malformation teratomas. The former spring from germinal cells (i.e., cells capable of using extensively their genetic keyboard and which are by no means, necessarily, sexual). The latter are subject to various and unknown embryogenetic defects. Paradoxically, tumoral teratomas may change into a benign condition when cyto-differentiation overcomes cellular multiplication, whereas malformational teratomas may become cancerous as a result of a one-sided anarchic growth of any of their constituents.

Finally, our knowledge of the origin and nature of teratomas remains fragmentary and hypothetical. It deserves to be deepened, because these tumors certainly constitute the most complete expression of tumorous and cancerous processes. What appears certain is that we would have a more exact picture were pathologists and embryologists to be confronted with their respective opinions more often. In the final analysis, however, collaboration in all the biological disciplines is required to discover the secrets of teratomas.

References

Most of the publications before 1930 cited in this paper can be found in the following works:

CALBET, J.: Contribution à l'étude des tumeurs congénitales d'origine parasitaire de la région sacro-coccygienne (Steinheil, Paris 1893).

CHEVASSU, M.: Tumeurs du testicule (Steinheil, Paris 1906).
EWING, J.: Surg. Gynec. Obstet. *12;* 230 (1911).
OBERNDORFER, S.: Hodengeschwülste; in F. HENKE und O. LUBARSCH Handb. spez. Path. Anat. Histol., Bd. VI/3, 768–812 (Springer, Berlin 1931).
WAELKENS, D.: Les embryomes du testicule (Davy, Paris 1929).

More recent studies can be found in the bibliographical index of the following books:

EVANS, R. W.: Histological appearances of tumours (Livingstone, Edinburgh/London 1966).
WILLIS, R. A.: Pathology of tumours; 3rd ed. (Butterworth, London/Washington 1960).
WILLIS, R.A.: The borderland of embryology and pathology; 2nd ed. (Butterworth, London/Washington 1962).

Basic concepts of embryology and references to studies in embryology are given in the following books:

DA COSTA, A.C.: Eléments d'embryologie (Masson, Paris 1948).
HAMILTON, W. J.; BOYD, J. D. and MOSSMAN, H. W.: Human embryology (Hefter, Cambridge 1962).
HERTIG, A. T.: Human trophoblast (Thomas, Springfield 1968).
NEEDMAN, J.: Biochemistry and morphogenesis (Cambridge University Press, Cambridge 1942).
STARK, D.: Embryologie; 2. neubearb. Aufl. (Thiemie, Stuttgart 1965).

Most of PEYRON's papers are cited in:

DUFAU, R.: Les tumeurs du testicule et les syndromes de masculinisation (Le François, Paris 1941).

Fundamentals of biology and experimental researches on teratoma can be found in:

STEVENS, L.C.: The biology of teratomas; in M. ABERCOMBIE and J. BRACHET Adv. in morphogenesis, pp. 1–31 (Academic Press, New York/London 1967).

The following is a complementary but not exhaustive list:

ABELL, M. R.; FAYOS, J. V. and LAMPE, I.: Cancer *18:* 273 (1965).
BERTRAND. I. and LATAIX, P.: Sem. Hôp. Paris *54:* 102 (1952).
BESTLE, J.: Acta path. microbiol. scand. *74:* 214 (1968).
BRAUN, A. C.: In Canad. Cancer Conf., vol. 4, p. 89 (Academic Press, New York 1961).
BREMER, J. L.: quot. GROSS, *et al.* [1951].
BRET, A. J.; DUPERRAT, B. and GRENIER, A. J.: Sem. Hôp. Paris *57:* 2963 (1955).
CABANNE, F.: Arch. Anat. path. *33:* A 165 (1957).
CAMPBELL, J. G.: Tumours of the fowl (Heinemann, London 1969).
COLLINS, D.H. and PUGH, R.C.B.: The pathology of testicular tumours (Livingstone, London 1965).
CONKLIN, J. and ABELL, M. R.: Cancer *20:* 2105 (1967).
CORSY, E.: Bull. Cancer *16:* 90 (1927).
COURY, CH.: Les kystes dermoïdes intra-thoraciques (Doin, Paris 1945).
COURY, CH. and CABANNE, F.: J. franç. Méd. Chir. thor. *13:* 675 (1959).
EVANS, M.J.: J. Embryol. exp. Morph. *28:* 164–176 (1972).
FINCH, B.W. and EPHRUSSI, B.: Proc. Nat. Acad. Sci. U.S.A. *57:* 615–621 (1967).

FRANK, R. T.: J. Amer. med. Ass. *4:* 248 (1906a).
FRANK, R. T.: J. Amer. med. Ass. *5:* 348 (1906b).
FRIEDMAN, N.B.: Cancer *4:* 265 (1951).
FRIEDMAN, N.B.: Ann. N.Y. Acad. Sci. *80:* 161 (1959).
FRIEDMAN, M. and DI RIENZO, A.J.: Cancer *16:* 868 (1963).
GAILLARD, J.A.: Ann. Anat. path., Paris *10:* 283 (1965).
GAILLARD, J.A.: Arch. Anat. micr. exp. *1:* (1966a).
GAILLARD, J.A.: Ann. Anat. path., Paris *11:* 285 (1966b).
GAILLARD, J.A. Ann. Anat. path., Paris *13:* 97 (1968).
GROSS, R. E.; CLATWORTHY, H. W. and MEEKER, I. A.: Surg. Gynec. Obstet. *92:* 341 (1951).
HADORN, E.: In Major problems in developmental biology, pp. 85–104 (Academic Press, New York 1966).
HAMEED, K. and BURSLEM, MR. G.: Cancer *25:* 564 (1970).
HARTMANN, H. and PEYRON, A.: Bull. Acad. nat. *81:* 733 (1919).
HUNTINGTON, R. W. and BULLOCK, W. K.: Cancer *25:* 1356; 1368 (1970).
KAHAN, B.W. and EPHRUSSI, B.: J. natl. Cancer Inst. *44:* 1015–1036 (1970).
KARPAS, C. M. and JAWAHIRY, K. I.: J. Urol. Baltimore *91:* 387 (1964).
KERMAREC, J. and DUPLAY, H.: Arch. Anat. path., Paris *16:* A 56 (1968).
KNOWLES, J. and SCULLY, R.E.: New Engl. J. Med. *281:* 434 (1969).
KLEINSMITH, L.J. and PIERCE, G.B.: Cancer Res. *24:* 1544–1567 (1964).
KRAFKA, J.: Arch. Path. *21:* 757 (1936).
LAURENT, M.I.; ROUSSEAU, M.F. and NEZELOF, C.: Ann. Anat. path., Paris *13:* 413 (1968).
MAIER, R.: Virchows Arch. path. Anat. *20:* 536 (1861).
MARIN-PADILLA, M. Virchows Arch. path. Anat. *340:* 105 (1965).
MARIN PADILLA, M.: Arch. Path. *85:* 614 (1968).
MASSON, P.: Tumeurs humaines (Maloine, Paris 1956).
MASSON, P. and THIERAULT, J.P.: Ann. Anat. path., Paris *1:* 7 (1956).
MICHALOWSKY, I.: Cbl. allg. Path. path. Anat. *38:* 585 (1926).
MICHALOWSKY, I.: Virchows Arch. path. Anat. *267:* 27 (1928).
MICHALOWSKY, I.: Virchows Arch. path. Anat. *274:* 319 (1929).
MONTPELLIER, J.: Bull. Cancer *18:* 643 (1929).
NINARD, B.: Tumeurs du foie (Le François, Paris 1950).
NISHIYAMA, R. H.; BATSAKIS, J. G.; WEAVER, D. K. and SIMRALL, J. H.: Arch. Surg. *93:* 342 (1966).
PACHTER, M.R. and LATTES, R.: Dis. Chest. *45:* 301 (1964).
PEYRON, A.: C. R. Soc. Biol., Paris *135:* 860 (1941).
PIERCE, G.B.: In Canad. Cancer Conf., vol. *4,* pp. 119–137 (Academic Press, New York 1961).
PIERCE, G.B.: In A. MONROY and A. MOSCONA Current topics in developmental biology pp. 225–246 (Academic Press, New York/London 1967).
PIERCE, G.B. and NAKANE , P.K.: Nature *214:* 820 (1967).
PIERCE, G.B. and VERNEY, E.L.: Cancer *14:* 1017–1029 (1961).
PIERCE, G.B.: BULLOCK, W.K. and HUNTINGTON, R.W.: Cancer *25* 644 (1970).
PORTUGAL, S.E.: Contribution à l'étude du tératome rétro-péritonéal; thèse Bordeaux (1919).
RAGHUNATHA, R. N.; VELIATH, G. D. and SRINIVASAN, M.: Cancer *17* 1604 (1964).
REPIN: Origine parthénogénétique des kystes dermoïdes de l'ovaire (Steinheil, Paris 1891).
RIVIERE, M.R.; CHOUROULINKOV, I. and GUERIN, M.: Bull. Cancer *47:* 55 (1960).

ROSENTHAL, M.D., WISHNOW, R.M. and SATO, G.H.: J. natl. Cancer Inst. *44:* 1001–1014 (1970).
RUBIN, H.: Science *167:* 1271 (1970).
RUYER, R.: La genèse des formes vivantes (Flammarion, Paris 1958).
SALÜN, J.: Arch. Anat. micr. Morph. exp. *57:* 11 (1968).
SMITH, A.G. and POWELL, L.: Amer. J. Path. *33:* 653 (1957).
STEVENS, L.C.: J. Embryol. exp. Morph. *20:* 329 (1968).
STEVENS, L.C.: Develop. Biol. *21:* 364 (1970).
TEILMANN, I.; KASSIS, H. and PIETRA, G.: Acta path. micro. biol. scand. *70:* 267 (1967).
TEILUM, G.: Cancer *12:* 1092 (1959).
TEILUM, G.: Acta path. microbiol. scand. *64:* 407 (1965).
WINTREBERT, P.: Le développement du vivant par lui-même (Masson, Paris 1963).
WOGALTER, H. and SCOFIELD, G. F.: J. Urol., Baltimore *87:* 573 (1962).
WOLFF, E.: La science des monstres (Gallimard, Paris 1948).

Author's address: Dr. J.A. GAILLARD, Department of Pathology, Hospital Center, *Evreux* (France)

Neoplasia and Cell Differentiation, pp. 350–379
(Karger, Basel 1974)

The Association of Tumor and Embryonic Cells *in vitro*

M-F. SIGOT-LUIZARD
Institut d'Embryologie et de Teratologie expérimentales du C.N.R.S., Nogent sur Marne

Contents

Introduction

One of the most characteristic properties of cancer cells is their ability to invade and destroy normal cells. The mechanism of interaction between normal and tumorous cells has been abundantly studied *in vivo* and *in vitro.* Here we shall limit ourselves to work using culture methods and especially the association of cancerous and normal embryonic tissues. Since CARREL in 1926, who achieved, experimentally, the survival of Rous sarcoma on embryonic tissues, many authors have observed that cancer cells are more easily maintained *in vitro* when associated with embryonic cells [SANTESSON, 1935; GEY and GEY, 1936; SCHLEICH, 1955; DE BRUYN, 1958]. Moreover, one is often tempted to compare certain properties of tumors and embryonic cells. This method of association has proved useful, enabling us to reveal some of the characteristics of the growth, invasive capacity and degree of malignancy of the tumor cells.

I. Culture Techniques

The invasion of normal cells by cancerous cells occurs in three dimensions *in vivo;* some authors have pointed out, however, that approaching the problem by means of culturing tissue on a flat surface, does not closely resemble the *in vivo* conditions. LEIGHTON in 1951, perfected a sponge matrix tissue culture in which the invasion of normal tissue by cancerous cells could take place in three dimensions. He notes that not only the malignant cells, but also the embryonic cells multiply in these conditions and form organized cellular aggregates. At the same time, WOLFF and HAFFEN [1952], perfected an organotypic method which allows the growth both of the isolated cells and of whole embryonic organs. The success of the xenoplastic embryonic organ chimera, has incited WOLFF and his co-workers to associate *in vitro* some mammalian tumors and embryonic bird organs. We shall describe in detail these two techniques; the other techniques which we shall mention will be described in the course of the other sections.

A. Sponge Matrix Method

1. Simple Cellulose Sponge

Fragments of commercial cellulose sponge are cut into pieces approximately 1 mm thick and 1 cm square and thoroughly washed and sterilized.

Tissue fragments 1 to 3 mm square are placed on the exposed surface of the sponge and impregnated with a plasma clotting system. When the explants are firmly adherent to the cellulose support, the sponge fragments are placed in tubes and 6 drops of nutrient liquid are added. The tubes are then incubated at 37.5° C in a roller drum at 15 rph. Three times weekly, the nutrient liquid is removed and replaced by fresh fluid (fig. 1).

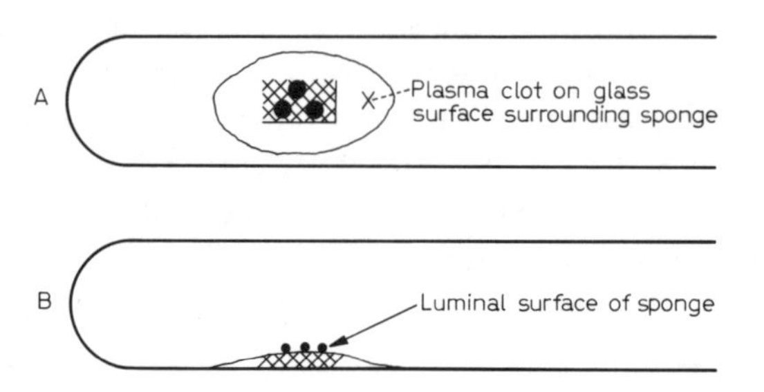

Fig. I. A. Culture tube showing the luminal surface of a sponge from above, through the opposite side of the tube. B. Culture tube showing the lateral surface of the sponge [LEIGHTON, 1951].

The studies of interactions between two tissues by means of this technique involve the establishment of the host culture of normal tissue in the sponge and then, inoculation of the sponge by malignant cells. After various periods, the sponge is fixed, embedded flat and stained with hematoxylin-eosin.

2. Collagen-Coated Cellulose Sponge

A number of tissues, and especially many carcinomas, induce a plasma clot liquefaction. The authors modified their technique by using cellulose sponge in which the trabeculae are coated with collagen. The collagen fibers are dissolved in a 50% methyl alcohol solution. Pieces of cellulose sponge are impregnated in this solution and as the solvent dries, the collagen fibers adhere to the cellulose trabeculae. The matrix offers two essential qualities: first, it shows a great physical stability due to the cellulose skeleton of the sponge; secondly, the collagen membrane coating the cellulose trabeculae allows cell growth without the use of the plasma clot.

3. The Culture Medium

The basic medium used by LEIGHTON is as follows:
1 vol 50% chick embryo extract;

2 vol horse serum;

2 vol balanced salt solution (Eagle's).

A number of variations of this medium are used, particularly one for solid tumors, which consists of calf serum and Hank's balanced salt solution to which is added 30% of Eagle's minimal essential medium. 50 u/ml of penicillin and streptomycin mixture are added, as well as 100 u/ml of mycostatine.

B. Culture on Agar-Agar Medium

1. Standard Medium [WOLFF and HAFFEN, 1952]

The constitution of the medium is as follows:

7 vol gel made up of 1% agar solution in Gey's solution;

3 vol Tyrode's solution with 2 drops of penicillin added;

3 vol 9-day chick embryo extract, diluted 50% in Tyrode's solution.

1 cm^3 of hot sterile medium is poured into a culture vessel, where it cools and solidifies. Fragments of the tumor and the embryonic organs about 1 mm square are placed side by side on the medium (fig. 2). The explants are dried with small pieces of filter paper. After 7 days culture, the tumor fragments are replaced on a new medium and reassociated with fresh embryonic organs. The most frequently used stain is hematoxylin-eosin.

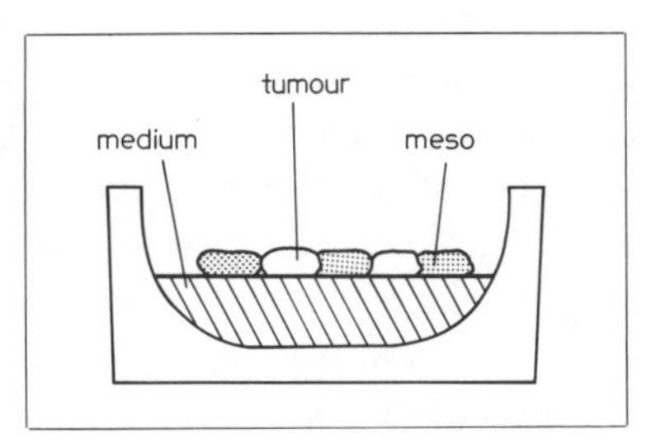

Fig. 2. Standard culture technique: mosaic association of the different explants.

2. Technique Using the Vitelline Membrane [WOLFF, 1961]

This modification of the technique has been employed by WOLFF and WOLFF [1961 b], chiefly for human cancer cultures. The 3 drops of Tyrode's solution are replaced by 3 drops of horse serum. The explants are wrapped in a piece of hen-egg vitelline membrane in one of two different ways (fig. 3a and b).

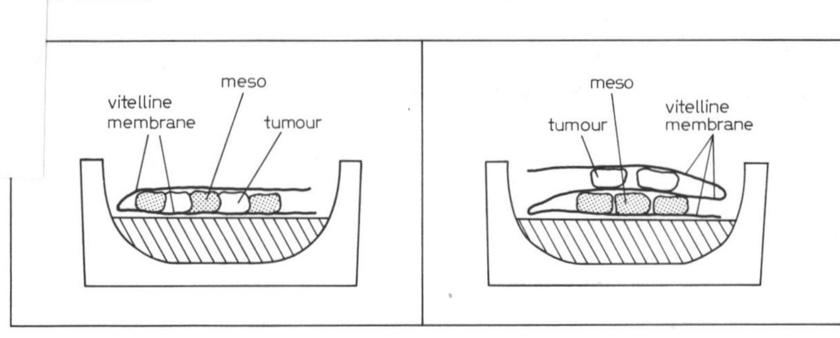

a,b

Fig. 3. a Mosaic association of the different explants wrapped in a piece of hen-egg vitelline membrane. *b* The tumoral and embryonic explants are separated by a double membrane.

II. Growth and Metabolism of Tumor Cells in vitro

A. Growth

1. Favorable Factors

This problem has been mainly studied by WOLFF and his co-workers. Indeed, the culture medium used shows the following characteristics: cancerous cells do not multiply when cultured alone; they degenerate rapidly. However, they do grow and can be maintained in culture during several months, even several years in the most favorable cases, when associated with chick embryonic organs. The aim of this work was to point out which organ is the most favorable to the growth of the different tumor cells, and to see if any of these tumors possess a selective affinity for a definite organ.

A number of animal and human tumors have been associated with chick embryonic organs such as 8- to 9-day mesonephros, 7-day liver, 8- to 9-day gonads ♂ and ♀, 15-day lungs, 8-day small intestine, and 7-day skin.

Rodent tumors. The following 6 tumors have been used: the mouse sarcoma S180 [SCHNEIDER, 1958]; a mouse mammary carcinoma T2633; the uterine and mammary rat epithelioma T8 and G6; a rat myelosarcoma T58 [WOLFF and SIGOT, 1961 a and b], and the Zajdela rat ascites hepatoma HZ [WOLFF, ZAJDELA and SIGOT-LUIZARD, 1964]. The authors found that there is no incompatibility between the two associated tissues. Moreover, they observed a different behavior of the tumors toward the different associated organs. For all of these tumors, the mesonephros appears to be the most favorable organ for cancerous growth (fig. 4). It also permits the maintenance of the adenoid structure of both T2633 and HZ tumors (fig. 5, 6, 7). This is also the case for embryonic liver, especially when it is associated with the rat ascites heptoma HZ. However, the gonads, both ♂ and ♀, as well as the lungs are less favorable for the multiplication of these different tumors and finally, the intestine, and especially the skin, allow the survival of the tumor cells only with difficulty (fig. 8). A good illustration of these results comes from the study of the number of mitoses per unit surface of the different associations (table I and diagram).

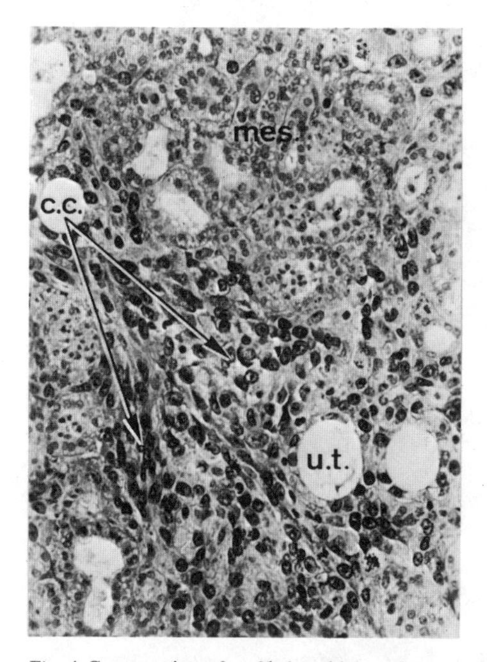

Fig. 4. Cross section of an $8\frac{1}{2}$-day chick embryo explant associated with cells of the mouse sarcoma S180. c.c. = Cancerous cell; mes. = mesonephros; u.t. = urinary tubule. × 588. [WOLFF and SCHNEIDER, 1957].

Fig. 5. Biopsy of the mouse mammary carcinoma T2633. The histological section shows the adenoid structure of the tumor. × 450.

Fig. 6. Association of explants of the mouse mammary carcinoma with fragments of the chick embryonic mesonephros, after 7 days of culture. The tumoral tissue shows the same structure as the biopsy. c.c. = Cancerous cell; mes. = mesonephros. × 600.

Fig. 7. The same culture after 135 days of culture. The glandular aspect of the tumor is still visible. c.c. = Cancerous cell: × 600. [SIGOT, 1969].

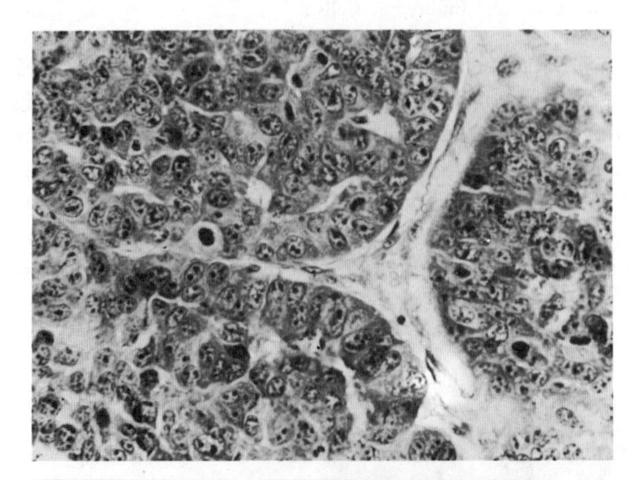
5

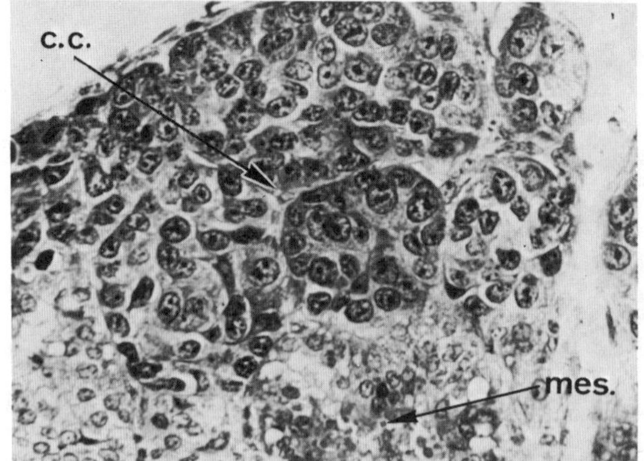

6

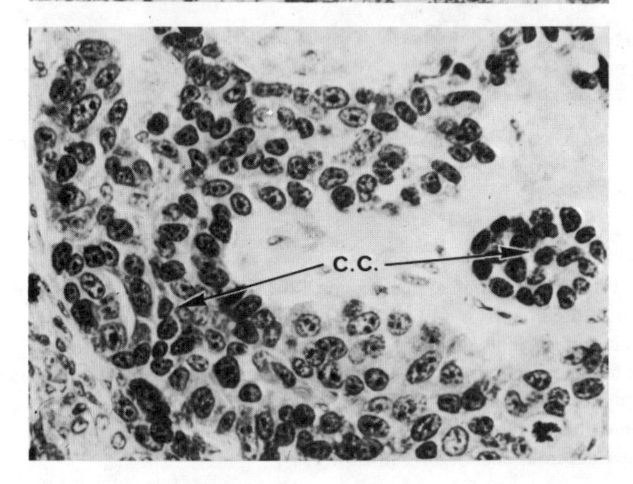

7

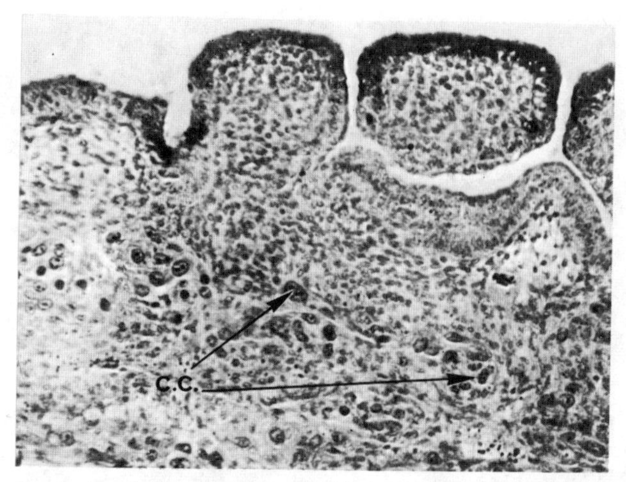

Fig. 8. Association of embryonic skin explants with cell of the sarcoma S180. c.c. = Cancerous cell. × 315. [WOLFF and SCHNEIDER, 1956.]

Table I

Embryonic organs Tumors	Meso-nephros		Gonads		Liver		Intestine		Skin	
	mit.	tr.	mit.	tr.	mit.	tr.	mit.	tr.	mit.	tr.
T2633	12	25	9	6	9	5	8	4	5	0
HZ	11	22	5	2	11	6	6	3	1	0
T8	10	10	8	4	10	5	4	2	3	0
G6	10	6	7	3	8	3	6	3	4	0
T58	7	3	0	0	0	0	0	0	0	0

mit. = Number of mitoses/mm^2; tr. = number of transfers [SIGOT, 1969].

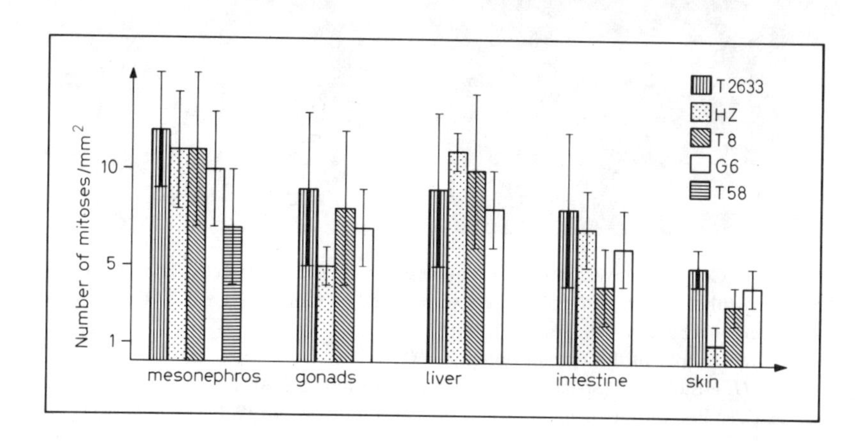

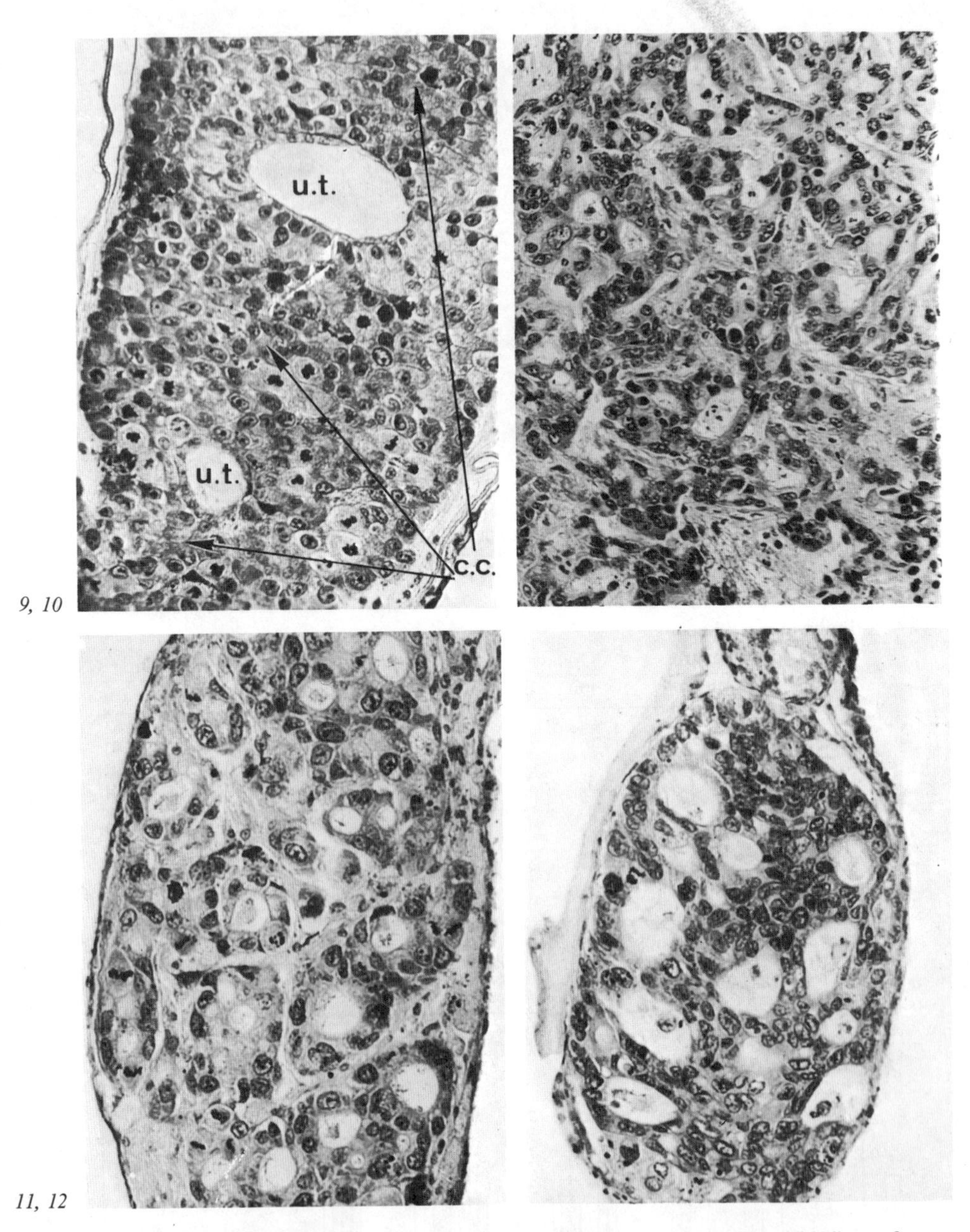

Fig. 9. Complete invasion of the mesonephros by the HeLa cells. The epithelium of the urinary tubule is reduced and entirely surrounded by tumor cells. c.c. = Cancerous cell; u.t. = urinary tubule. × 320. [WOLFF, 1961b]

Fig. 10. Biopsy of the human tumor Z 200. × 240.

Fig. 11. Histological aspect of the explants of the tumor Z 200 associated with fragments of the chick embryonic mesonephros after 7 days of culture. × 340. [WOLFF and WOLFF, 1961a]

Human tumors. Human malignant cell lines [WOLFF and WOLFF, 1959] and fragments of human tumors taken directly from the patient [WOLFF and WOLFF, 1961 a and b] have also been cultured in association with the mesonephros (fig. 9, 10, 11, 12). A study of one of these fragments of a human tumor associated with different organs has been undertaken by DUREL, WOLFF and WOLFF [1965]. This human tumor, Z200, is a hepatic metastasis of a tumor of gastric origin. Cultured since 1962 by WOLFF and WOLFF, it is composed of epithelial cords surrounding mucous-filled cavities. The cells are more or less regularly orientated at the edges of these cavities. Here again, the mesonephros and the liver, supply all the factors necessary for the survival and the proliferation of the tumor. They also preserve its epithelioid glandular structure of adeno-carcinoma with alveoli. The other organs and especially the gonads, allow growth and survival of the tumor, but the histological structure of the explants is often transformed. The alveoli tend to disappear and the tumor forms compact cords without characteristic organization.

Conclusions. Some organs seem more favorable than others for the growth of tumor cells *in vitro.* Still others are clearly unfavorable, even if they remain perfectly healthy. Mesonephros and liver represent the best nutritive support tested up to now. We acknowledge the fact that they contain substances stimulating growth and nutrition. These two organs represent the center of important metabolic activity, synthesis and degradation. Certain factors common to both tissues, would be essential to the growth of the cancerous nodules. The contact between both the normal and the cancerous tissues, is not necessary, since the tumor cells multiply actively despite the interposition of the vitelline membrane (fig. 13, 14). Likewise, we obtained good proliferation of malignant cells when they are explanted on 'conditioned media'. In this case, prior to tumor cell explantation, fragments of mesonephros are placed on the medium for 2 to 4 days. This indicates the presence of one or several substances necessary for the growth of malignant cells. The analysis of these factors has been undertaken by WOLFF and his co-workers (1967–1971). Some active fractions have been obtained from dialyzates of a yeast extract and chicken liver (less than 6

Fig. 12. The same association after 35 days of culture. We can recognise the structure of the tumor with alveoli surrounded by epithelial cords. × 300. [WOLFF and WOLFF, 1961a]

months old) after gradual fractionation through a chromatographic column (fig. 15). Many amino acids are present in these fractions. Some are essential, others are only favorable for the human malignant cell growth *in vitro*. But they must be used at a higher concentration than those present in embryonic chicken extract. Some other essential heat-labile substances,

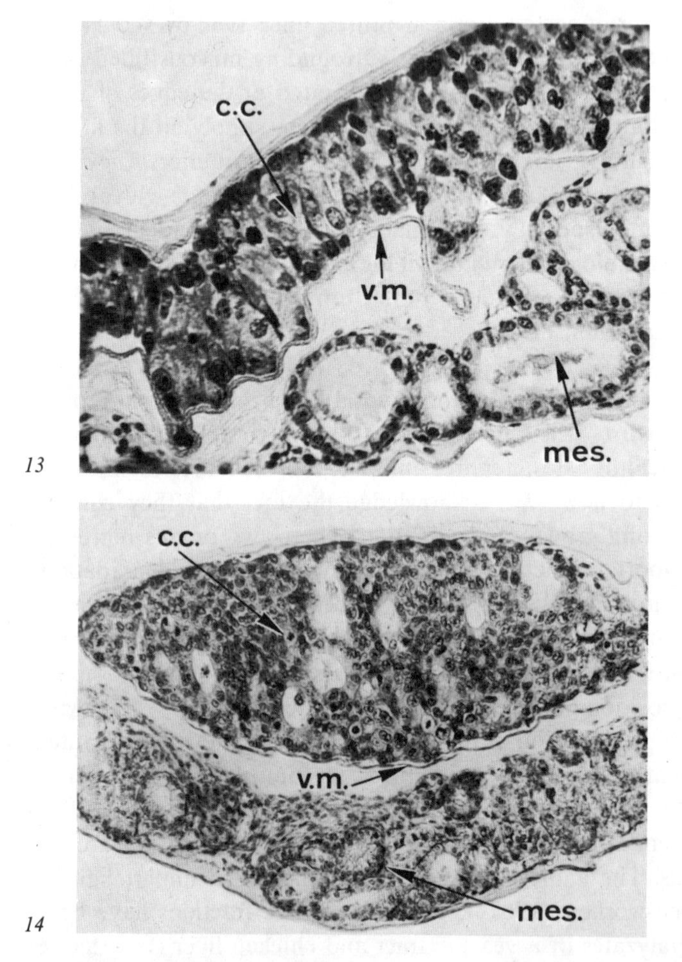

Fig. 13. Colonies of Kb cells associated with chick embryonic mesonephros by means of the double membrane technique. c.c. = Cancerous cell; v.m. = vitelline membrane; mes. = mesonephros. × 340. [WOLFF and WOLFF, 1961 b]

Fig. 14. Association of the tumor Z 200 with chick embryonic mesonephros after 514 days of culture. The explants are separated by the vitelline membrane. c.c. = Cancerous cell; v.m. = vitelline membrane; mes. = mesonephros. × 192. [WOLFF, 1964.]

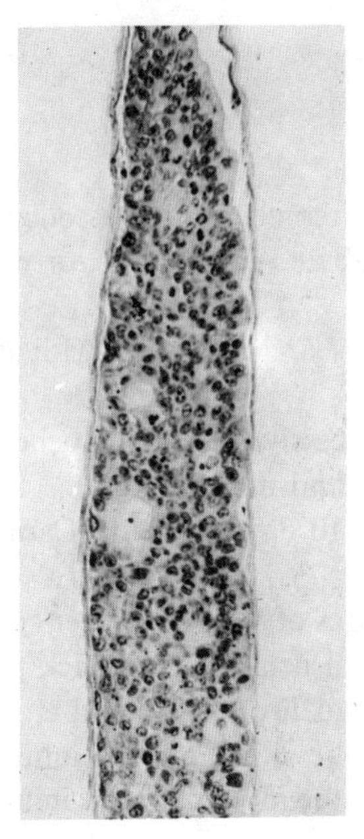

Fig. 15. Cross section of a fragment of the tumor Z200 cultured on a medium added with yeast extract, after several months of culture. × 180. [WOLFF and WOLFF, 1965.]

the nature of which is still unknown, are present in the chicken liver dialyzate and in the embryonic extract. Horse serum also seems essential to the proliferation of the malignant cells *in vitro.*

2. Inhibitory Factors

Here we find two different experimental approaches. The inhibition of tumor growth can be obtained either by inducing the differentiation of the cancer cells, or by the use of anti-mitotic drugs.

Inhibition by differentiation [LUSTIG, LUSTIG and JAUREGUI, 1968]. During cell differentiation, the rate of growth and division is reduced. The tumoral dedifferentiation process would arise from a partial or total loss of some specific properties. These authors think that, if the malignancy of the tumor is not a consequence of a genetically changed material, a tumoral population would again be able to undergo an identical differentiation to the one from which it was derived, when exposed to the influence of a controlling system such as an organizer center. They associate in organ culture, mouse and human tumors with mouse and chicken embryonic tissues. The composition of the medium is as follows:

2 vol nutrient agar;
1 vol medium 199;
1 vol human serum.

These constitute three groups of associations:

a) mouse mammary carcinoma or mouse sarcoma S180 is associated with the primitive streak of a chick embryo incubated for 19–20 h, or with 7- to 9-day embryonic mesonephros.

b) Human or mouse giant cell bone tumors are combined with the notochord of a 3½-day chick embryo.

c) Mixed parotid tumors are associated successively with notochord, primitive streak or embryonic mouse salivary gland mesenchyma.

All of these cultures are fixed between 3 to 10 days in Bouin-Carnoy and stained in hematoxylin-eosin or PAS.

We have observed that as in the organ cultures of WOLFF, cancer cells, when cultured alone, degenerate and die. But, in contact with chick embryonic mesonephros, the tumor cells divide and invade the embryonic tissue. The presence of the primitive streak of the chick embryo inhibits the invasive capacity of the sarcoma S180, and of the mammary carcinoma, inducing the malignant cells to organize some tubular structures with basal membranes and secretion of PAS-positive substances in the lumen of the ducts. The most effective embryonic tissues to induce differentiation, and hence, the expression of latent properties present in malignant cells are the primary organizer of 19–20 h embryos; the notochord of 3½-day embryos, when the tumor has an osseous origin (fig. 16, 17), and the mouse salivary gland mesenchyme when it is associated with parotid glands. All these observations show that cancer cells that have lost their capacity to undergo a specific differentiation *in vivo*, are able, when cultured in the presence of embryonic inducers, to regain a certain level of morphogenetic and functional activity, which reflects their origin. In certain tumors, the malignant properties are apparently reversible and the tumor is able to halt its invasive growth. Other experiments of similar design, undertaken by ELLISON, AMBROSE and EASTY [1969], although for different reasons, show that the differentiation characteristics of an anaplasic tumor are not irreversible (see also pp. 391–393 of this volume).

Inhibition by the action of drugs. LETTRE [1968] has used mixed cultures of normal rat fibroblast associated with the cells of Yoshida tumor in order to show the selective action of 6-purinyl histamine. Indeed, the cells of the rat Yoshida tumor do not survive alone in culture, but when associated with normal fibroblast, they proliferate. The addition of 6-purinyl histamine to the culture medium of these associations, shows the disintegration of the malignant cells while normal fibroblasts form a pure culture in the absence of tumor cells. The mechanism is as follows: the normal cell possesses an enzyme that transforms this compound by oxidation, into an inactive metabolite. The malignant cells studied by LETTRE do not possess

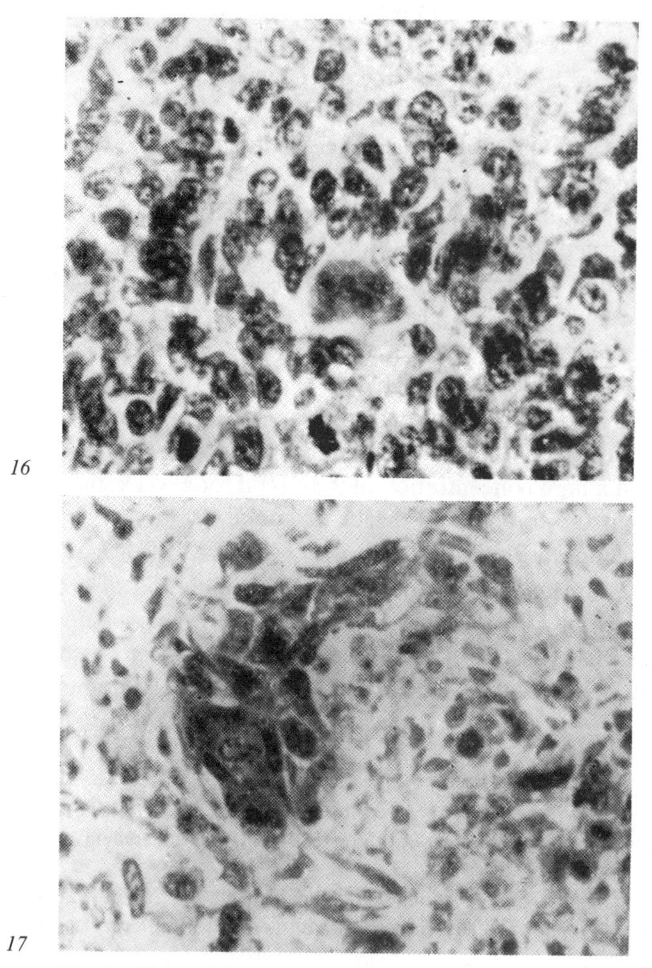

Fig. 16. Giant cell bone tumor cultured on agar.

Fig. 17. The same tumor associated with notochord of a $3\frac{1}{2}$-day chick embryo. Giant cell bone now reduced. Some masses of tumoral cells embedded in osteoid substance are dispersed as focal cellular areas. [LUSTIG, LUSTIG and JAUREGUI, 1968.]

this enzyme. The 6-purinyl histamine adheres to the cell surface, reducing the permeability of the cell membrane.

LEIGHTON [1968] has also observed the effect of acetylpodophyllotoxine-w-pyridinium chloride (NCI 3022) on HeLa cells associated with chick embryonic heart or strain D (fig. 18, 19, 20). Following a single application, and after repeated applications during a 4-week period, with a 6 h exposure (1.0 μg/ml), he found a complete metaphase arrest in both normal and tumorous cells. When the drug is removed, the effect disappears in 18 h. But when a dose of 1.0 μg/ml is given to the associated tissues on alternate days for 4 weeks, the HeLa cells are affected more severely than the normal cells.

Another group of authors also studied the selective action of an alkylating agent, melphalan, used in the chemotherapy of cancer. It has an inhibitory effect on a great number of animal tumors. STUBBLEFIELD and HSU [1957], having tested the substance on tissues cultured *in vitro*, found that normal tissues are more resistant than malignant tissues. SIMPSON [1969] studied the comparative effects of melphalan on chick embryonic gonads and on a human cancer in organ culture. The gonads are of great interest because they include somatic tissue and germ cells which are generally very sensitive to the destroying agents. SIMPSON noted a great sensitivity of the cancer cells and the embryonic gonads to the melphalan. Their treatment with a high concentration of 0.67 g/ml for a week, induced important necrosis in the tumors and a lack of gonocytes in the gonads. The ♀ germ cells seem more sensitive than the ♂ germ cells: the somatic tissues of the gonads are more resistant, showing an inhibition of growth and a lack of mitoses, but no observable necrosis as in cancer tissue.

The selective action of such anti-mitotic drugs and the higher resistance of normal tissues to their effects is an essential point for the chemotherapy

---

Fig. 18. Microscopic section of a thick sponge culture in which HeLa and cells of chick embryo heart were combined. In this control culture, HeLa cells are undergoing normal divisions. × 1170.

Fig. 19. Microscopic section of a thick sponge culture treated with 1.0 g/ml of NCI-3022 for 6 h. Metaphase arrest is seen in many HeLa cells with a splattering of chromosomal substance in the cytoplasm. × 1170.

Fig. 20. Microscopic section from a culture of chick heart and HeLa fixed 18 h after a single 6 h exposure to the drug. Recovery from the drug effect is indicated by the presence of many normal dividing figures in both the HeLa cells and the embryonic cells. One normal division in a HeLa cell is seen below the center of the field. × 670. [LEIGHTON *et al.*, 1957.]

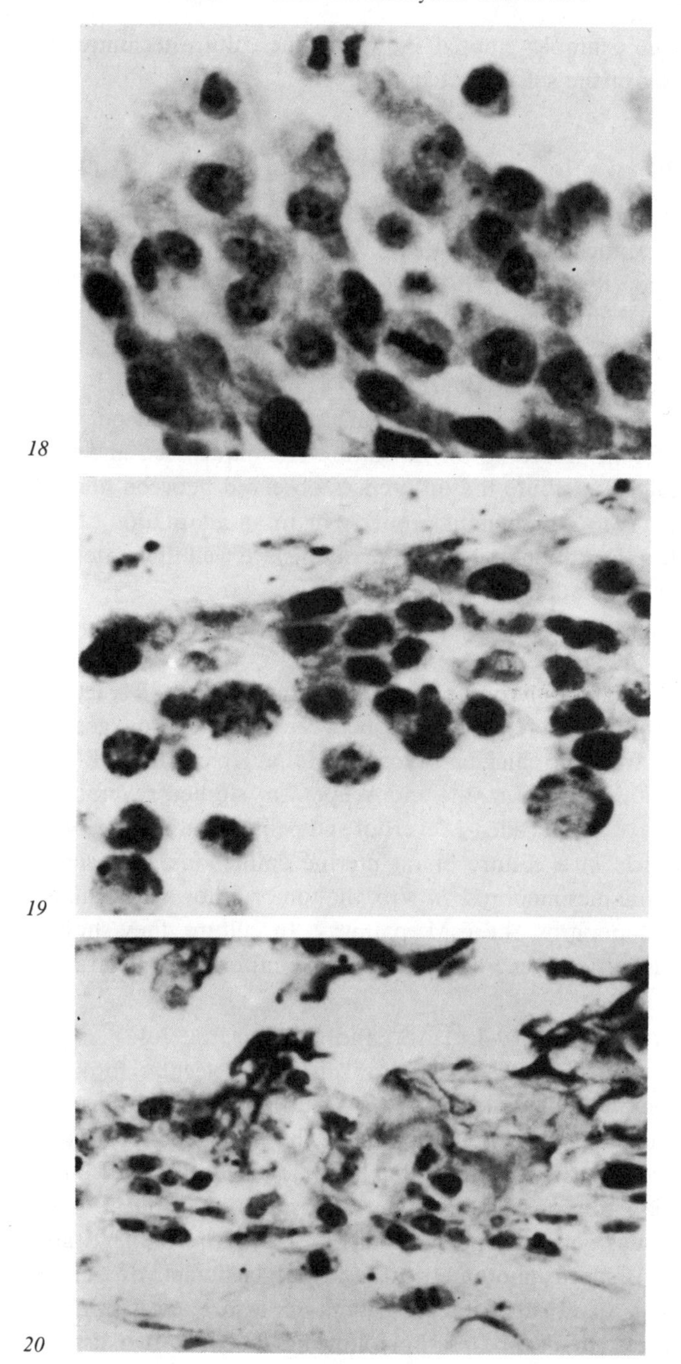

18

19

20

of cancer. The three examples quoted show that the culture technique is useful for the study of the selective action.

B. Metabolism

Since the first studies of WARBURG in 1930, on the glycolysis increase in malignant cells, much work has dealt with the respiration and the glycolytic activity of these cells. POTTER and his co-workers [1964] in particular, made an important contribution to our knowledge in this domain. Work done *in vitro* is also very abundant [HORI, TAKAOKA and KATSUTA, 1958; ZIMMERMAN *et al.*, 1960; MORGAN and PASIEKA, 1960; EASTY *et al.*, 1964], and some key points arise from it. We must point out that the results are different according to the culture technique used and that it is always difficult to attribute the differences observed between normal and cancerous tissues to a malignant property or to an adaptation of the tissues to the culture conditions. These observations are well illustrated by experiments using associations in organ culture.

1. Histochemical Studies

The PAS, methylgreen-pyronine, Feulgen or bromophenol reactions revealed no discernible difference *in vitro* between mesonephros, gonads or thyroid explants from the chick embryo, and HeLa, Kb cell lines [BEAUPAIN, 1963]. In 1970, SIGOT-LUIZARD and WEGMANN studied several enzymes characteristic of the Embden-Meyerhof and pentose pathways as well as of the Krebs cycle in a culture of rat uterine epithelioma, associated with chick embryonic mesonephros. *In vivo*, the tumor follows the pentose shunt, and the mesonephros, the E.M. pathway. In culture, they show a decrease of these activities after 5 days and become uniform in appearance.

2. Spectrophotometry [SIGOT-LUIZARD and CROISILLE, 1970]

Lactic-dehydrogenase (LDH) exists in 5 distinct molecular forms in most vertebrate organs [MARKERT and MOLLER, 1959]. Two forms are generally known: one, which is predominant in the heart, the form H or LDH1; the other, predominant in the skeletal muscle, the form M or LDH5. *In vivo,* the chick embryonic mesonephros possesses the LDH1 from whereas cancer tissues generally possess the LDH5 from [GOLDMAN and KAPLAN, 1963]. The spectrophotometrical study of lactic-dehydrogenase activity of fragments of rat uterine epithelioma T8 associated with chick embryo mesonephros, demonstrates the following fact: the two tissues,

Table II [SIGOT-LUIZARD and CROISILLE, 1970]

	in vivo				*in vitro*			
	LDH1 (Heart)	LDH5 (Muscle)	T8	mes.	T8 cs	mes. cs	T8 ca	mes. ca
R	1.7	1.0	1.0	1.7	1.0	1.0	1.0	1.0

T8 = uterine rat epithelioma; mes. = mesonephros; cs = cultured alone; ca = cultured in association; R = the ratio of activity at 5.10M to activity at 100.10M. This is characteristic for each form and each organ extract [WILSON *et al.*, 1963].

which show a difference in their LDH activity *in vivo,* have identical behavior when cultured, whether alone or associated (table II).

This change then is connected with the setting of the culture and not with an interaction of one tissue on the other. The conditions of culture influence the respiratory metabolism and especially the LDH which changes its form.

3. Autoradiography

MCCULLY [1968] measured by means of incorporation of *L* leucine ^{14}C, the protein synthesis of cultured tissues. He associated two human neoplastic tissues and normal thyroid with chick embryonic mesonephros in organ culture. After a week's culture, these tissues had retained their morphological integrity and their metabolic and synthetic activities. After freezing and thawing both tissues, he found that contact between normal and cancer tissues is necessary to the recovery of the protein synthesizing activity. It seems that interposing the vitelline membrane between normal and malignant explants prevents the restoration of this activity. MCCULLY suggests that high molecular weight polymers of an undeterminated nature are released by the mesonephros which allow the protein synthesizing recovery of the frozen and thawed tumor cells.

All of the observations described here, on the one hand, and numerous studies which demonstrate the spontaneous cancerization *in vitro*, on the other hand, seem to point to the fact that the metabolism adopted by normal cells *in vitro* is similar to the metabolism of cancer cells *in vivo.*

III. Interactions between Normal and Tumor Cells in vitro

A. Intercellular Exchanges

Many observations have been made on the interactions between cells *in vitro,* involving either cellular organelle transfer [ROSE, 1960] or the fusion between entire cells [LEWIS, 19274 WEISS, 1944; HSU, 1960; BENDICH, 1967; HILL, 1968]. BARSKI and his co-workers [1961] observed in cultures of different cell lines associated *in vitro,* the appearance of new cellular types with 'hybrid' characteristics. They investigated the cytological mechanism involved in this phenomenon by associating normal mouse embryonic explants in a plasma clot culture, with a highly malignant mouse cell line NI [BARSKI and BELEHRADECK, 1963]. The areas of contact between the two tissues are observed with the aid of phase contrast time-lapse microcinematography. They noted two cases of transfer of the nucleus from one cell to another. In the first case, a momentary constriction of the nucleus and a displacement of the nucleolus through its constriction is visible, and in the second, some advanced and rotative movements of the nucleus. All these observations demonstrate the participation of the nucleus in the intercellular exchanges between cells of different histological origins such as normal and malignant cells.

B. Invasive Property of the Tumor Cells

This property, so characteristic of malignant cells, is closely connected with the problems involved in the previous sections, i.e., the growth and nutrition of cancer cells. However, many other factors interfere in the process: mechanical factors such as the structure of the associated organs, or physiological and biochemical factors, such as, secretion of toxic substances, surface properties and immunological reactions. Many authors have used the method of association of normal and malignant tissues *in vitro,* in order to investigate the mechanism; we shall consider successively the various aspects of this phenomenon.

1. Mechanical Aspect

WILLIS in 1952 and YOUNG in 1959, postulated a purely mechanical theory of invasion. The pressure which develops in the cancer tissue by cellular multiplication, would induce the invasion of normal tissue by the malignant cells, due to the lower resistance of the former. But many experiments in this field show that, if the organ structure and the pressure due to multiplication play an important part, on the other hand, an intrinsic mobility of the cancer cells during the invasion is also evident. This is well illustrated when a tumor is cultured under less favorable conditions of nutrition, i.e., with fragments of embryonic intestine. WOLFF and his co-workers note the presence of isolated cancerous cells which are sufficiently far away from the initial nodule to demonstrate locomotor mobility of these cells. But when the nutritive conditions are favorable, as in the case of an association with the mesonephros (fig. 21), the multiplication of the mali-

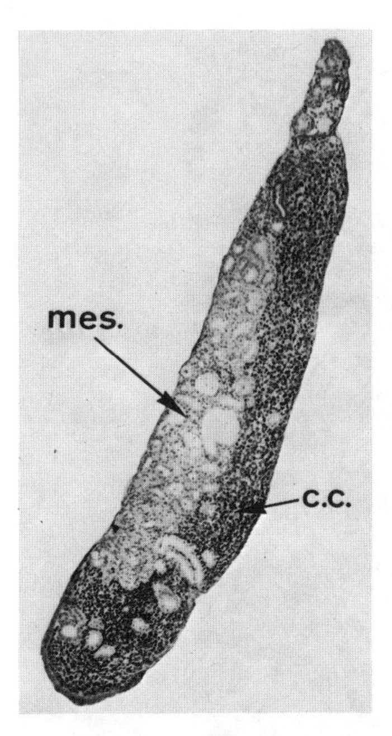

Fig. 21. Association of the mouse mammary carcinoma T2633 with the chick embryonic mesonephros. The tumor cells severely invade the embryonic tissues. c.c. = Cancerous cell: mes. = mesonephros. × 75. [SIGOT, 1969.]

gnant cells is very intensive, and the invasion takes place by means of massive cords. The cancerous nodules progressively invade the surrounding tissues and advance partly by pressure in the embryonic tissues. These authors also observe that the epithelial tissues offer a resistance to the penetration of the cancer cells which travel round them and go through tissues of looser structure. This is well illustrated by the associations of embryonic intestine (fig. 22) or skin. The tumor cells travel along the muscle fibers and invade the dermis, but never penetrate, through the intestinal epithelium and the epidermis. On the other hand, the mesonephrotic mesenchyme has a loose structure which facilitates the invasion (fig. 23).

Some identical results were obtained by LEIGHTON and his co-workers [1956], by associating HeLa cells and rapidly growing embryonic cells such as heart, frontal and tibia bone, brain, intestine and lung. A liquefaction of the plasma clot makes the histological interpretation difficult, but the authors observe different modalities of migration according to the associated

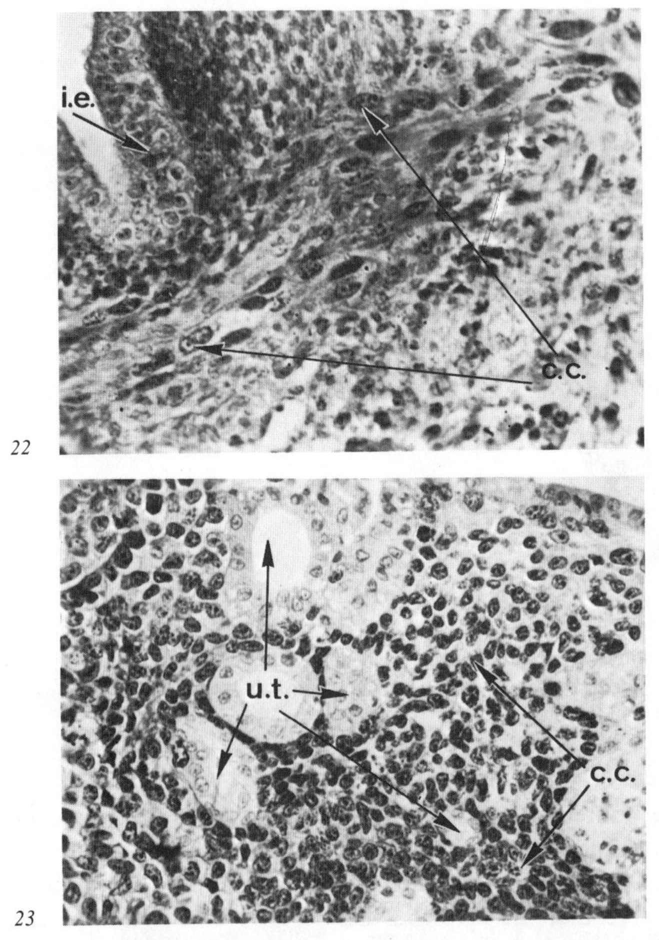

Fig. 22. Invasion of the embryonic intestinal tissue by the cells of the sarcoma S180. The tumor cells travel along the muscle fiber and do not penetrate through the intestinal epithelium. c.c. = Cancerous cell; i.e. = intestinal epithelium. × 400. [WOLFF and SCHNEIDER, 1957.]

Fig. 23. The cells of the mammary carcinoma T2633 infiltrate through the mesonephrotic mesenchyme and surround the urinary tubule of the mesonephros. c.c. = Cancerous cell; u.t. = urinary tubule. × 460. [SIGOT, 1969.]

organs. The embryonic mesenchyme, the heart and bone connective tissues are severely invaded by the HeLa cells, whereas the embryonic liver and brain offer more resistance. Lastly, they note that the intestinal and lung epithelia are not invaded by the malignant cells (fig. 24 and 25).

SCHLEICH [1970], using the reaggregation technique *in vitro,* showed that HeLa cells associated with fetal muscle or human endometrium, invade

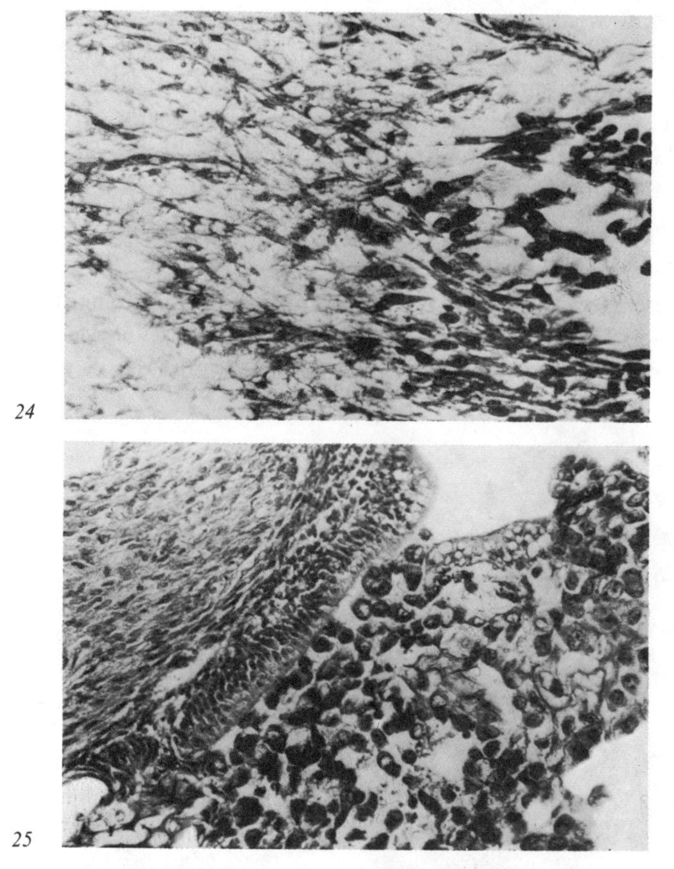

24

25

Fig. 24. Histologic section of a culture of a cardiac tissue of the chick embryo combined with HeLa cells. Tumor cells in the upper half of the field are invading connective tissue outgrowth from the heart. × 350.

Fig. 25. Explants of HeLa in gelfoam and embryonic intestinal epithelium of the chick in close contact. The tumor cells have not disturbed the integrity of the epithelial surface-lining of the 'intestinal organism'. In the upper part of the figure, a layer of embryonic columnar epithelium has spread out to cover part of the surface of the gelfoam explant. × 300. [LEIGHTON *et al.,* 1956.]

the normal tissues through the areas of least resistance, whereas the coherent epithelial structures stop the invasion.

BARSKI and BELEHRADEK [1965], made a cinematographic study of such associations. They followed the area of contact between the normal and tumor cells using a highly malignant cell line NI, and either fibroblasts or endothelium as normal tissue. They found that the endothelial structure offers a definite resistance to invasion by tumor cells, whereas the normal fibroblasts represent weak points of considerable invasion. The highly malignant cells invade more easily the areas of loose fibroblastic growth, but are inhibited, or even entirely stopped in their progression when they meet endothelial growth with coherent and well ordered structures.

This part played by the organ structure is, therefore, a general and typical phenomenon which supports the 'contact guidance' theory of WEISS [1958].

2. Physiological and Biochemical Aspects

However, these mechanisms are not sufficient to explain the invasive phenomenon. When 2 normal tissues are associated, the 2 cell types gather according to certain affinities, and form chimerae (fig. 26). In no case, does

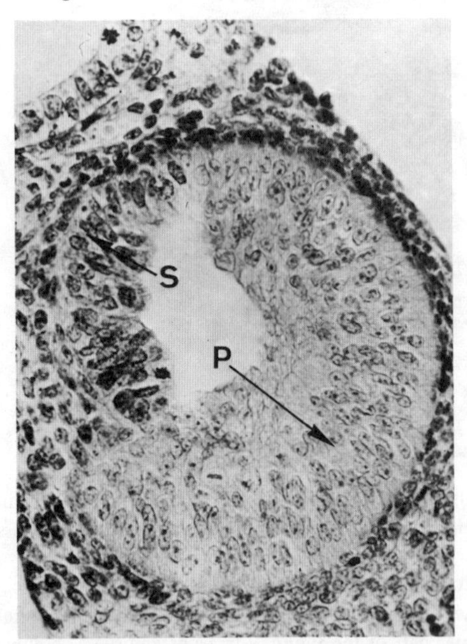

Fig. 26. Xenoplastic association of mouse lung fragments with chick lung fragments. s = Mouse; p = chick. × 450. [WOLFF, 1954.]

one tissue predominate at the expense of the other. But in the described associations, no chimeric organization takes place, and the normal cells degenerate and disappear, in most cases leaving no visible trace. A number of hypotheses have been proposed to explain these observations. The most important is certainly the 'contact inhibition' theory of ABERCROMBIE [1966]. The normal cells are able to stop their migratory movements when they come into contact with one another. The malignant cells would have lost this property. This phenomenon brings us to the question of surface properties. We shall simply mention this hypothesis, since this problem is the subject of another chapter of this book. Moreover, the disappearance of normal cells suggests either a metabolic competition for the medium, or a toxic effect due to the cancer cells, because lysis of the normal cells allows the progression of the tumor cells.

LEIGHTON [1968] suggests a destruction of the normal cells by suffocation. He associated three animal tumors with chick embryonic organs: a mammary and a squamous cell carcinoma of the mouse, and a human transplantable carcinoma that had its origin in the colon. The results are identical to those he obtained with human lines. He notes a considerable growth of the cancer cells which spread out over the surface of the degenerating embryonic cells. But, in areas where the normal cells are not covered by malignant cells, the embryonic tissues are healthy and in contact with the medium (fig. 27). The sheet of cancer cells covering the normal cells prevents the metabolic exchanges of the latter with the medium. The loss of adhesion of the malignant cells allows them to migrate and attain a favorable position for their nutrition and multiplication. Proliferation leads to a 'neoplastic blockade', that is to say, starvation and death of the underlying normal cells. This mechanism may also occur *in vivo*. LEIGHTON showed by intravenous injection of a cell suspension of the Yoshida ascites hepatoma into 11-day chick embryo, that, after one week, the hepatoma cells invade the brain of the embryo and form a metastasis. He saw that the capillaries near the brain are surrounded by a layer of carcinomatous cells, which block the metabolic exchange between the circulation and the nervous tissue.

The mechanism of 'neoplastic blockade' could account in part for the destruction of normal cells by the malignant cells, but a release of toxic substances by the tumor cells probably also interferes in this process. The cytoplasmic exchanges seen by microcinematography demonstrate this fact. SYLVEN [1961] extracted some peptides of low molecular weight from a number of tumors and found that they are toxic for normal tissues. Also,

KATSUTA *et al.* [1968] showed that cancer cells attack normal cells by direct contact (fig. 28). The latter degenerate and are eventually destroyed by phagocytosis. He extracted a substance from the hepatoma which is toxic

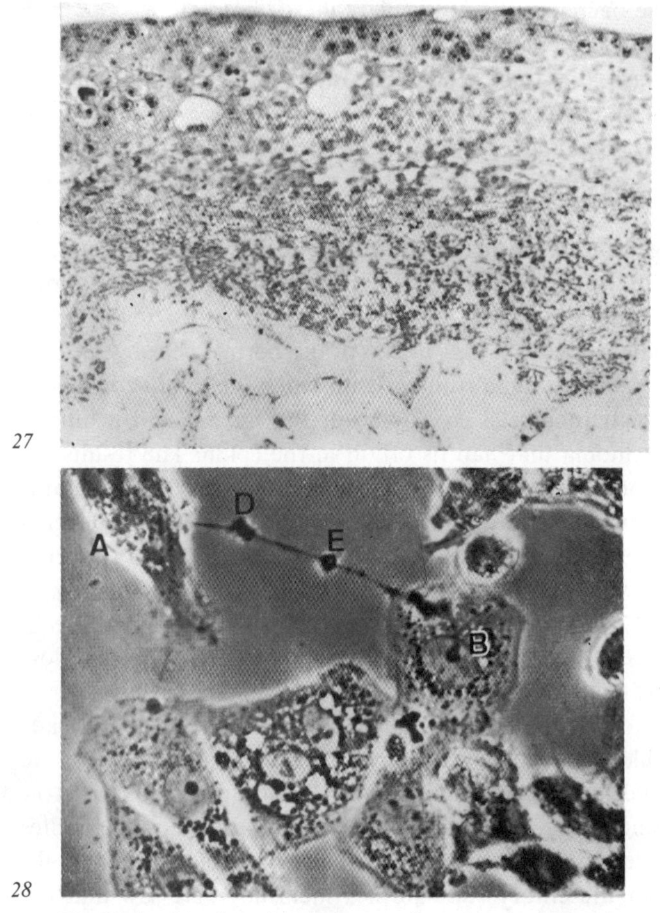

Fig. 27. Combination of a squamous cell carcinoma of the mouse and chick embryonic heart. Tumor has spread from the left side of the field, where it is many cell layers thick. The underlying heart tissue is completely necrotic. In the right half of the figure, carcinoma cells are only 2 to 4 cell layers thick, and the heart tissue immediately below is viable. × 210. [LEIGHTON, 1968.]

Fig. 28. Phase contrast photomicrograph of a mixed culture of normal liver cells of strain RLC-1 and ascite hepatoma AH-130 cells of rats. Hepatoma cells are adhering to liver cells at A and B. Small swelling (D and E) in the cytoplasmic projections of the hepatoma cells are due to material absorbed by pinocytosis or phagocytosis from liver cells and are being transported toward the cytoplasm of hepatoma cells. [KATSUTA, TAKAOKA and NAGAI, 1968.]

for normal liver. It is heat-labile and of low molecular weight. This substance does not seem to be identical to that described by NAKAHARA and FUKUOKA [1954], that is, the toxohormone which suppresses the catalase activity of normal liver cells.

3. Conclusions

From the analysis of all these studies, the complexity of the problem becomes apparent. The chief characteristics of the invasive phenomenon, on the one hand, e.g., the inherent properties of the cancerous cells such as disordered proliferation, cell mobility, destruction of the cell barrier; on the other hand, properties inherent to the host-tissue, such as nutrition and structure, have been studied in culture. But part of the mechanism depends largely on the defensive reactions of the host-tissue, and this aspect mainly escapes study *in vitro*. Some authors have demonstrated the role of polymorphonuclear leukocytes in the invasive process [LEIGHTON, 1970]. Others, i.e., HALPERN and his co-workers [1967], have shown the importance of the reticulo-endothelial system. Finally, the discovery of the embryonic α feto-protein in certain tumors by ABELEV *et al.* [1963], opens up great possibilities in immunology.

IV. Degree of Malignancy of a Tissue in vitro

Many authors have observed in long-term cultures that the cells are very different in many respects from those present in newly explanted tissues. Particularly, they acquire invasive properties as shown by their capacity to produce malignant tumors after inoculation into an isologous animal. Is it possible to predict the degree of 'cancerization' of cells, according to their behavior outside the organism, in suitable *in vitro* conditions? Two types of study have contributed to this problem. The first using a cellular line system; the second, hepatic explants of the rat after having received a diet containing para-dimethylaminoazobenzene (DBA).

A. Evaluation of the Malignancy of Different Cell Lines *in vitro* [BARSKI and WOLFF, 1965]

In several experiments, BARSKI and his co-workers have developed many cellular lines and clones characterized by different degrees of malignancy, from pulmonary tissue of a normal C57 black adult mouse [1963].

For example, we shall mention the clonal lines C191 and C192 which give no tumor after 18 to 21 months, when inoculated into a C57 black adult mouse, and the PTT12 line and its derived clone 61 which produce in the same conditions (3.10^4 inoculated cells) invasive tumors characterized by rapid and fatal growth. When the C91 clonal cells are associated with the chick embryonic mesonephros explants, they are unable to invade these cultures, whereas the PTT12 cell line extensively infiltrates the mesonephros and can be maintained during at least 90 days in these conditions. This technique permits the evaluation of the malignancy of transformed cells outside the organism.

B. Demonstration of a Step in the Cancerization Process

SIGOT-LUIZARD [1970] used liver fragments of rats which had received a diet supplemented with DAB. In the organ culture method of ET. WOLFF, the normal adult tissues and benign teratomas did not survive in culture [SALAÜN, 1964]. Only the malignant cells were able to survive in long-term culture. The culture of such liver fragments, when removed at different stages of tumoral evolution, may enable us to point out the precise stage at which the cells acquire a malignant characteristic, that is to say, the possibility of maintaining themselves in a long-term culture.

We define three stages:

Stage I: The liver explants are removed after a diet of 30 days; the macroscopic and microscopic aspects of the explants are generally normal.

Stage II: The liver fragments are removed after a diet of 60 to 90 days, i.e., between the 2nd to 4th stages of LACASSAGNE [1968].[1] The liver shows a generally granular aspect with lenticular and punctiform nodules.

Stage III: The liver explants, removed after a diet of 100 days, i.e., between the 4th–5th stages of LACASSAGNE, show very cancerous nodules.

These different liver explants are associated with chick embryonic mesonephros fragments in organ culture. The explants of stage I cannot be cultured whereas 24% of the stage II and III cultures remain *in vitro* after 6 transfers, and 10% of the stage III cultures survived to about 77 days of culture (fig. 29, 30).

These experiments of associations *in vitro* reveal a so-called 'pre-cancerous' stage, where adult tissues which have apparently not yet developed cancerous aberrations, can survive in culture.

1 A step has been determined for each stage of the intoxication. These steps have been observed respectively at 25 (1), 41 (2), 82 (3), 105 (4) and 145 (5) days.

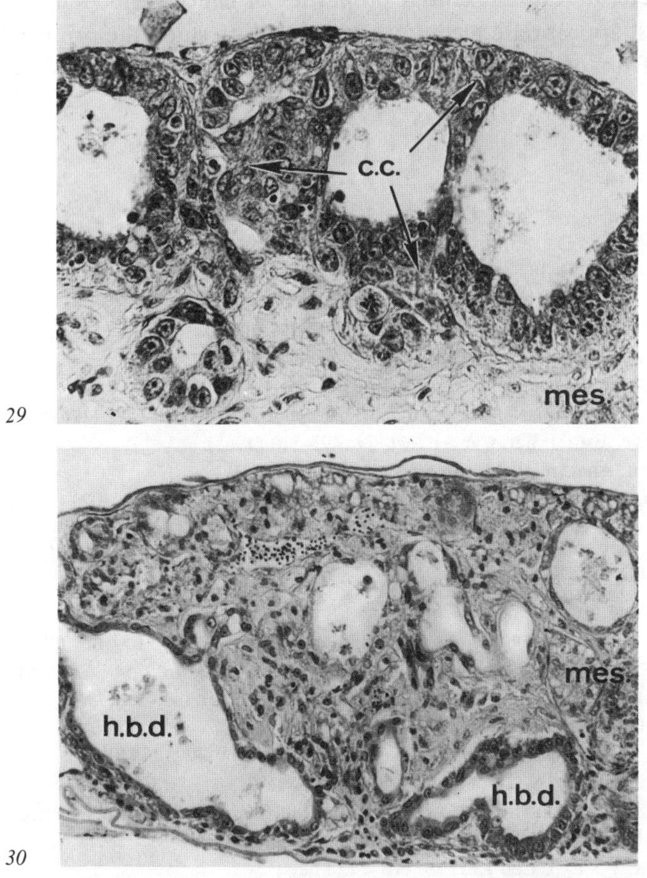

Fig. 29. Association of chick embryonic mesonephros with fragments of rat liver removed at stage III, after 77 days of culture. c.c. = Cancerous cell; mes. = mesonephros. × 450. [SIGOT, 1969.]

Fig. 30. Association of chick embryonic mesonephros with fragments of rat liver removed at stage II, after 14 days of culture. Note the presence of hypertrophic bile ducts. mes. = Mesonephros; h.b.d. = hypertrophic bile duct. × 240. [SIGOT-LUIZARD, 1970.]

Conclusions

It is difficult to compare with certainty the behavior of a cell *in vivo* and *in vitro.* We are still unable to affirm whether an *in vitro* result is really an equivalent of the *in vivo* process. However, the experiments we have described, using the association technique in organ culture have enabled us to determine the essential properties of cancer cells and it is certainly the most appropriate technique with which to study the principal characteristics of tumor invasion.

References

ABELEV, G. I.; PEROVA, S. D.; KHRAMKOVA, N. I.; POSTNIKOVA, Z. A. and IRLIN, I. S.: Transplantation *1,2:* 174–183 (1963).

ABERCROMBIE, M.: Nat. Cancer Inst. Monogr. *26:* 246–277 (1966).

BARSKI, G. and BELEHRADECK, J.: Exp. Cell Res. *29:* 102–111 (1963).

BARSKI, G. and BELEHRADECK, J.: Exp. Cell Res. *37:* 464–480 (1965).

BARSKI, G. and CASSINGENA, R.: J. nat. Cancer Inst. *30:* 865–883 (1963).

BARSKI, G.; SORIEUL, S. and CORNEFERT, F.: J. nat. Cancer Inst. *26:* 1269–1291 (1961).

BARSKI, G. and WOLFF, EM.: J. nat. Cancer Inst. *34:* 495–510 (1965).

BEAUPAIN, R. R.: Arch. Anat. micr. Morph. exp. *52:* 75–82 (1963).

BENDICH, A.; VIZOSO, A. D. and HARRIS, R. G.: Proc. Nat. Acad. Sci., U.S. *57:* 1029 (1967).

BILLARDON, CL.; JULLIEN, P. M. and CARSWELL, E.: Int. J. Cancer *1:* 541–556 (1966).

BRUYN, W. M. DE: Jaarboek van kankeronderzoek en kankerbestrijding in Nederland, p. 94 (1958).

CARREL, A.: J. exp. Med. *43:* 647–668 (1926).

DUREL, J.; WOLFF, EM. and WOLFF, ET: Arch. Anat. micr. Morph. exp. *54:* 659–669 (1965).

EASTY, G. C.; YARNELL, M. N. and ANDREWS, R. D.: Brit. J. Cancer *18:* 354–367 (1964).

ELLISON, M. L.; AMBROSE, E. J. and EASTY, G. C.: Exp. Cell Res. *55:* 198–204 (1969).

GEY, G. and GEY, M.: Amer. J. Cancer *27:* 45–76 (1936).

GODLESKI, J. J.; LEE, R. E. and LEIGHTON, J.: Cancer Res. *30:* 1986–1993 (1970).

GOLDMAN, R. D. and KAPLAN, N. O.: Biochim. biophys. Acta *77:* 515–518 (1963).

HALPERN, B. N.; GIOZZI, G. and STIFFEL, C.: In P. DENOIX Mechanisms of invasion in cancer, UICC Monogr. Ser., vol. 6, pp. 149–162 (Springer, Berlin 1967).

HILL, M. and SPURNA, V.: Exp. Cell Res. *50:* 208–222 (1968).

HORI, M.; TAKAOKA, T. and KATSUTA, H.: Jap. J. Exp. Med. *28:* 259–288 (1958).

HSU T. C.: Tex. Rep. Biol. Med. *18:* 31–33 (1960).

KATSUTA, H.; TAKAOKA, T. and NAGAI, Y.: In H. KATSUTA Cancer cells in culture. Proc. Int. Conf. Tissue Culture in Cancer Res., pp. 157–168 (University of Tokyo Press, Tokyo 1968).

LACASSAGNE, A.; JAYLE, M. F. and HURST, L.: C. R. Acad. Sci. *267:* 137–140 (1968).

LEIGHTON, J.: J. nat. Cancer Inst. *12:* 545–561 (1951).

LEIGHTON, J.: In H. KATSUTA Cancer cells in culture. Proc. Int. Conf. Tissue Culture in Cancer Res., pp. 143–156 (University of Tokyo Press, Tokyo 1968).

LEIGHTON, J.; KLINE, I.; TETENBAUM, Z.: BEIKIN, M.: J. nat. Cancer Inst. *16:* 1353–1373 (1959).

LEIGHTON, J.; KLINE, I.; BEIKIN, M. and ORR, H. C.: Cancer Res. *17:* 336–344 (1957).

LEIGHTON, J.; MARK, R. and JUSTH, G.: Cancer Res. *28:* 286–296 (1968).

LETTRE, R.: In H. KATSUTA Cancer cells in culture. Proc. Int. Conf. Tissue Culture in Cancer Res., pp. 99–104 (University of Tokyo Press, Tokyo 1968).

LEWIS, W. H.: Amer. Rev. Tuberc. *15:* 616 (1927).

LUSTIG, E. S.; LUSTIG, L. and JAUREGUI, H.: In H. KATSUTA Cancer cells in culture. Proc. Int. Conf. Tissue Culture in Cancer Res., pp. 135–142 (University of Tokyo Press, Tokyo 1968).
MCCULLY, K. S.: Int. J. Cancer *3:* 142–149 (1968).
MARKERT, C. L. and MOLLER, F.: Proc. Nat. Acad. Sci. U.S. *45:* 753–763 (1959).
MORGAN, F. J. and PASIEKA, A. E.: Canadian J. Biochem. Physiol. *38:* 399–408 (1960).
NAKAHARA, W. and FUKUOKA, F.: GANN *45:* 77–84 (1954).
OKADA, T. S.: J. Embyol. exp. Morph. *13:* 299–307 (1965).
POTTER, V. R.: Cancer Res. *24:* 1085–1098 (1964).
ROSE, G. G.: Tex. Rep. Biol. Med. *18:* 103–115 (1960).
SALAÜN, J.: Arch. Anat. micr. Morph. exp. *53:* 387–396 (1964).
SANTESSON, L.: Acta path. microbiol. scand. *24:* 1–237 (1935).
SCHLEICH. A.: Naturwissenschaften *42:* 50 (1955).
SCHLEICH, A.: UICC, 8ème Réu., avril 1970.
SIGOT, M. F.: thesis (1969). Thèse d'Etat. Paris.
SIGOT-LUIZARD, M. F.: C. R. Acad. Sci. *270:* 3170–3172 (1970).
SIGOT-LUIZARD, M. F. and WEGMANN, R.: Ann. Histochem. *15:* 87–92 (1970).
SIGOT-LUIZARD, M. F. and CROISILLE, Y.: Ann. Histochem. *15:* 93–96 (1970).
SIMPSON, P.: Europ. J. Cancer *5:* 331–337 (1969).
SMITH, J.; WOLFF, EM. and WOLFF, ET.: C. R. Acad. Sci. *272:* 1465–1468 (1971).
SPURNA, V. and HILL, M.: Exp. Cell Res. *50:* 223–232 (1968).
STUBBLEFIELD, E. and HSU, T. C.: Antibiot. Chemother.; N. Y. *7:* 493 (1957).
SYLVÉN B.: Biochim. Biol. Sper. *1:* 8–20 (1961).
WARBURG, O.: The metabolism of tumor (Constable, London 1930).
WEISS, P.: Anat. Rec. *88:* 205 (1944).
WEISS, P.: Int. Rev. Cytol. *7:* 391–423 (1958).
WILLIS, R. A.: The spread of tumors in the human body (Butterworth, London 1952).
WILSON, A. C.; CAHN, R. D. and KAPLAN, N. O.: Nature *197:* 331–334 (1963).
WOLFF, EM.; CROISILLE, Y.; MASON, J. and WOLFF, ET.: C. R. Acad. Sci. *265:* 2157–2160 (1967).
WOLFF, ET.: Bull. Soc. Zool. France *79:* 357–368 (1954).
WOLFF, ET.: Develop. Biol. *3:* 767–786 (1961).
WOLFF, ET. and HAFFEN, K.: Tex. Rep. Biol. Med. *10:* 463–472 (1952).
WOLFF, ET. and SCHNEIDER, N.: Arch. Anat. micr. Morph. exp. *46:* 173–197 (1957).
WOLFF, ET. and SIGOT, M. F.: C. R. Soc. Biol. *155:* 265–267 (1961a).
WOLFF, ET. and SIGOT, M. F.: C. R. Soc. Biol. *155:* 960–962 (1961b).
WOLFF, ET. and WOLFF, EM.: C. R. Soc. Biol. *153:* 1898–1900 (1959).
WOLFF, ET. and WOLFF, EM.: C. R. Acad. Sci. *252:* 1873–1875 (1961a).
WOLFF, ET. and WOLFF, EM: J. Embryol. exp. Morph. *9:* 678–690 (1961b).
WOLFF, ET. and WOLFF, EM.: C. R. Acad. Sci. *256:* 1173–1174 (1963).
WOLFF, ET. et WOLFF EM.: La presse médicale no. *20:* 1137–1162 (1965).
WOLFF, ET.: ZAJDELA, F. and SIGOT, M. F.: C. R. Acad. Sci. *258:* 4633–4634 (1964).
YOUNG, J. S.: J. Path. Bact. *77:* 321–339 (1959).
ZIMMERMANN, M.; DEVLIN, T. M. and PRUSS, M. P.: Nature *185:* 315–316 (1960).

Author's address: Dr. M-F. SIGOT-LUIZARD, Institut d'Embryologie et de Teratologie expérimentales du C.N.R.S., *F-94130 Nogent sur Marne* (France)

Neoplasia and Cell Differentiation, pp. 380–399
(Karger, Basel 1974)

Embryonic and Tumour Cell Interactions

M. S. LAKSHMI and G. V. SHERBET

Department of Biochemical Pathology, University College Hospital Medical School, London

Contents

I. Introduction

Cellular interactions that occur when embryonic and tumour cells are brought into contact with each other in *in vitro* conditions or by grafting of tumour cells into embryos or of embryonic cells into tumours, may prove to be a rewarding field of investigation for a variety of reasons. In the course of development embryonic cells undergo characteristic changes in their surface, which is closely associated with their ability to organise into well-defined structures. One reason, therefore, is the possibility that one might glean useful information as regards the surface properties of tumour cells by analysing their interactions with embryonic cells. For instance, studies on the association of tumour cells and embryonic cells *in vitro* [SIGOT in this volume] have produced a good deal of information about the compatibility of their cell surfaces, not to say similarities between their surfaces. This data has to be viewed in the light of what we know about the sorting out of tissues in a reaggregating mass of embryonic cells.

The cell surface also plays a considerable role in the invasiveness of neoplastic cells. Information on the mobility of tumour cells in an embryonic environment and the specificity of localisation in embryonic tissues might be relevant to the understanding of invasiveness and metastasis of tumours [see pp. 189–233 in this volume].

Tumours show a varying degree of histotypic differentiation, more often than not they show a loss of differentiation. There are also, however, neoplastic tissues that retain, to a large extent, histotypic differentiation and even some of their original functional characteristics. It would be interesting if the state of differentiation of tumours were found to be related to the degree of their progression or the degree of their deviation from the normal state. The loss of differentiation in tumours has often been erroneously described as dedifferentiation. For an embryologist the term dedifferentiation means much more than a lack of the morphological and structural make-up. Most significantly it would imply a reversion to a state from which the tissue could redifferentiate into a definable new, structurally and functionally different tissue. The interacting systems encountered in embryonic inductions or regeneration phenomenon could be profitably employed to study the question of the state of differentiation of tumours, or the question of acquisition of new functional characteristics, such as certain cancers of the kidney being able to synthesise hormones normally produced by the parathyroid gland, or the cancers of the bronchus being able to secrete hormones normally produced by the anterior pituitary gland, etc. These latter findings support the view that the neoplastic cell contains an intact genome and that it has undergone an abnormal genetic activation. The important point, of course, is: will it be possible to confer normal differentiated function to these cells by interacting them with embryonic inductors which have the ability to induce differentiation of specific cell types?

II. Tissue Implantation into Embryos

In the recent history of embryology the use of implantation of tissues into embryos as an experimental tool may be said to have begun with the classical experiments of SPEMANN [summarised in SPEMANN, 1936] in which he demonstrated the ability of the dorsal lip of blastopore of early amphibian gastrulae, to induce the formation of a secondary embryonic axis on implantation into the blastular cavity of a late segmentation stage of an amphibian embryo. Beginning with these experiments performed in the 1920s, several kinds of tissues have been implanted into amphibian or chicken embryos. This experimental system afforded an opportunity to analyse, what later came to be known as interacting induction systems, the processes involved in the induction of differentiation and organisation in embryonic systems.

A. Induction of Tissue Differentiation by Neoplastic Cells in Amphibian Embryos

Following the observation that several adult tissues possessed the ability to induce the differentiation of a variety of cell types, i.e., act as inductors when implanted into amphibian early gastrula stage embryos, TOIVONEN and SAXÉN [1957] made the first implantation of tumour tissue into embryos. These experiments were undertaken to study the inductive capacity of tumour tissue. The object was to see if differences existed between the inductive capacities of neoplastic cells and the corresponding normal tissues.

TOIVONEN and SAXÉN [1957] implanted bone marrow cells from 4 rats with typical myeloid leukemia induced by gastric administration of methylcholanthrene, and bone marrow tissue of 4 normal rats of the same age. Their experiments showed that normal bone marrow induced differentiation of mesodermal-type cells in 98% of the implants. As is normal with these grafts, no cephalic structures or sensory organs were induced. In contrast to this high degree of inductive activity, leukemic bone marrow produced positive inductions in only 6% of the implants. On the basis of these results they have suggested that the leukemic bone marrow used in their experiments appeared to possess a substance which inhibited the differentiation of mesodermal tissues, especially since implantation of normal and leukemic bone marrows simultaneously produced the normal induction spectrum.

Of 9 cases of positive induction using leukemic bone-marrow at least one case involved induction of cephalic structures. This is significant when compared with the absence of cephalic induction in 98 cases of positive induction encountered with normal bone marrow implants. Whether this is due to the appearance of a second new factor responsible for the cephalic inductions or if one and the same factor might be responsible for this, as well as for the suppression of the mesodermal inductions is not clear. BECKER *et al.* [1959] also noticed a similar reduction in the inductive capacity of tissues from leukemic rats. However, the lack of inductive capacity in tumour tissue is not a general phenomenon. In fact SAXÉN and TOIVONEN [1962] have given a list of tumour tissues which do possess inductive capacity. Among those mentioned are Walker rat carcinoma 256; Rous sarcoma; rat-ascites hepatoma AH 7794 and AH 130; Ehrlich's ascites and solid tumour, a malignant thymoma from mouse, and human hypernephroma. No significant differences were apparent between Ehrlich's ascites and solid tumours. This might not be meaningful in itself since in these experiments no proper control could be maintained. Neither were any differences encountered in the inductive activity of normal kidney tissue of Man and a human kidney tumour [BECKER *et al.*, 1959].

These experiments, which show that differences existed in the abilities of normal and neoplastic tissue to induce cellular differentiation and organisation in competent totipotent ectoderm of the amphibian embryonic system, are subjected to criticism on the grounds that the amphibian tissues are highly sensitive to environmental stimuli, and that a wide variety of unrelated tissues often produced a similar inductive spectrum [SHERBET, 1970]. This is not to detract from the merits of the differences demonstrated by TOIVONEN and SAXÉN. It is difficult indeed to have to sift the available evidence, part of which supports and part of which rejects the observed differences, in order to pin down observations that might be significant. For this reason the authors favour the chick embryonic system as being more refractory and therefore more discriminatory in its responses to inductive stimuli.

We studied several types of neoplastic cell as regards their inductive effects on implantation into 16–18 h old chick embryo. With the exception of three cases out of 22 of Yoshida ascites sarcoma, and one case out of 17 untreated and 11 γ-globulin-treated HeLa cells, no inductive effects were observed. In one case of the exceptions mentioned (with HeLa cells) ectodermal tissue proliferation was noticed. The proliferated areas resembled neural 'palisade' formations produced by implantation of weak inductors between the embryonic germ layers. The reaction, however, differed a good deal as regards the cellular arrangement. In a normal palisade formation the cells are arranged in regular alternating layers, but in the palisades induced by HeLa and Yoshida ascites sarcoma cells, this distinct cellular arrangement was completely absent. We are therefore not inclined to attach much significance to these palisade formations [SHERBET and LAKSHMI, 1970a].

MAREEL *et al.* [1968] implanted HeLa cells into chicken blastoderms and found that the movement of the graft inside the blastoderm left behind a 'streak' phenomenon in the ectoderm. This streak-like formation was similar to the one described by us as the palisade formation [SHERBET and LAKSHMI, 1970a]. MAREEL *et al.* [1968] have indicated the possibility that this might be induction of a primitive streak ('blastopore'). This is difficult to ascertain unless a true invagination of cells resulting in the formation of the mesoderm is shown to occur at this point.

SAXÉN and TOIVONEN [1958] and TOIVONEN [1959] also reported on the inducing capacity of HeLa cells which had been cultivated on culture medium containing human sera. These cells appeared to have considerable capacity to induce the differentiation of hindbrain with associated structures, and caudal structures but not archencephalic structures. When HeLa cells that had been cultured on media not containing human serum or, on media containing inactivated serum were implanted they showed no inductive ability. Apparently, HeLa cells themselves have no inductive-capacity but some factor present in the human serum which was used in the culture medium conferred this ability on the cells.

TOIVONEN and SAXÉN [1959] have admitted that it is not at all clear how the information on the inductive abilities of tumour tissues could be related to the neoplastic changes that have occurred in them. An interesting point which has come to the surface when we were analysing these various results, is perhaps worth noting. VAHS [1957] examined the inductive abilities of kidney and liver tissue from adult and embryonic rats. Embryonic tissue appeared to possess more archencephalic inductive capacity than the corresponding adult tissues. The data of TOIVONEN and SAXÉN [1955] showed that leukemic bone marrow cells produced archencephalic inductions in at least one out of 9 instances, while none was encountered in experiments with normal bone marrow. One wonders if there is any significance in the fact that in the spectrum of inductions produced by normal embryonic kidney and liver tissues and leukemic bone marrow, an archencephalic ability was noticed which was absent in normal adult kidney, liver or bone marrow tissues. Whether this archencephalic ability is lost in the adult tissue and is reacquired during or as a result of neoplastic changes is worth investigating. It is unfortunate that there are no proper controls in any of these experiments, which is not due to an error of omission on the part of the investigators but is due to the difficulty of obtaining tissue materials which can strictly serve as control implants.

B. Implantation of Neoplastic Cells into Chick Embryos

1. Embryo Culture and Implantation Methods

Much of the recent work on the interactions between embryonic and tumour cells has been done using the chick embryonic system. Apart from the scientific merits of using this system, two other factors have encouraged the use of this system: (a) this embryonic material is available all through the year unlike amphibian material whose availability is either restricted to certain seasons of the year or which requires procedures involving induction of ovulation through injection of gonadotropic hormones, etc.; (b) two excellent methods are available for culturing very young chick embryos (10–18 h of age) *in vitro.* These have been described by NEW [1955] and a modification of NEW's technique by GALLERA and CASTRO-CORREIA [1960]. Both these techniques are simple in the set up and the embryos can be kept in culture for at least 3 to 5 days.

An added advantage of these techniques is the accessibility of the embryo for operative procedures for implantation of the neoplastic and normal heteroplastic grafts. Implantation of tissues to be studied is made by separating the epiblast and hypoblast layers of 16–18 h old chick embryo near the margin of the area pellucida, slightly above the level of Hensen's node, using fine tungsten needles. The tissue is placed in this locally separated area. The incision made in the endoderm in order to separate the germ layers rapidly heals up. This method of implantation was described by WADDINGTON [1932].

2. Embryonic Responses to Tumour Grafts

a) The Host Mesodermal Response

In 1965 at the suggestion of Professor (now Sir) ALEXANDER HADDOW, one of us (SHERBET) began experiments on the interaction between embryonic and tumour cells. The initial experiments were performed with a benzpyrene-induced sarcoma which is maintained in the Chester-Beatty Research Institute in London. The fact which most strikingly came to our notice during these experiments was that the tumour cells showed considerable mobility inside the embryos. What was also obvious was the sharp contrast to the response which the embryo makes when a normal heterograft is implanted in this way. Our experience with several kinds of normal heterografts had shown that they are invariably isolated by the embryo by covering them up with abundant masses of mesoderm. Following this observation a few more tumour cell types were implanted. The mesodermal response (HMR) which they evoked varied from 0–25%. The nature of this host response is not yet understood. The embryonic cells are immunologically noncompetent at this stage of development and obviously, therefore, the HMR is not a consequence of immune reaction. There are indications, however, that the net negative charge present on the surface of the implanted cells may be partly or wholly involved in producing the response from the host embryo. HeLa cells, for instance, produced HMR only in 3% of the implants. When they were treated with chicken γ-globulin before implantation into the embryo, the host response reached 81%. Measurement of the surface charge of the cells after such treatment showed an increase in the net negative charge. This is presumably due to the adsorption of the negatively charged molecules of γ-globulin on the cell surface. The exact mode of participation of the surface charge in producing HMR, however, remains to be elucidated.

Recent experiments have shown that the loss of the ability to elicit HMR might be associated with fundamental changes that have occurred in the cells which, presumably, are also closely linked with the process of neoplastic change. These experiments were performed with minimum deviation hepatomas, showing a spectrum of growth rate and karyotype deviation. It has been observed that the incidence of HMR decreases with increase in the degree of deviation of the tumour karyotype from normal [SHERBET *et al.*, 1970].

The mesodermal response such as that described by us [SHERBET and LAKSHMI, 1970; SHERBET *et al.*, 1970] has not been encountered by MAREEL *et al.* [1968] who implanted HeLa cells and mouse testis tissue into chick blastoderm. They claim, on the contrary, that mesoderm cells usually crowded up against both types of graft, and that a greater proportion of the embryonic mesoderm was present on the side of the blastoderm where the grafts were situated. Moreover, they claim that the endoderm fulfils the role that we ascribe to the mesoderm cells. They have described cases, for example, where the endoderm covered up and isolated the normal heterografts of mouse testis, but it did not thus isolate HeLa cells. However, there is supporting evidence from the work of PALAYOOR and BATRA [1971] for our observation that the mesodermal response is not produced by neoplastic cells.

b) The Endodermal Response

Contrary to the observations made by MAREEL *et al.* [1968] we were unable to find any effect on embryonic endoderm as a consequence of the implantation of HeLa, Yoshida ascites sarcoma, Landshütz ascites cells, and benzpyrene-induced sarcoma cells. We did, however, observe distinct proliferation and characteristic organisation of the cells in the proliferated regions, as a response to the presence of implants of Morris hepatoma cells. Frequently, we also noticed the overgrowth of these endodermal cells to cover up the graft either as a whole or individual cells of the graft. What seems more significant is the fact that the frequency of incidence of this embryonic endodermal proliferation was also related to the degree of deviation of the karyotype of the hepatoma, but unlike the HMR, the endodermal proliferation increased with increase of karyotype deviation.

The mesodermal and the endodermal response seem to complement each other. It appears from this that the loss of HMR suffered by hepatomas tissue, presumably as a result of the neoplastic changes, is compensated by an acquisition of the ability to induce endodermal proliferation which also, apparently, is a consequence of the neoplastic changes. The correlation of both these responses with chromosomal deviation therefore, seems to be far from fortuitous. It may be that the neoplastic conversion of the liver cells has involved some genetic change on which both the mesodermal and the endodermal responses are dependent.

Further evidence for the close association between the neoplastic state and the endodermal response evoked by tumours, comes from the observation that this response is histogenetic in nature. In other words, only those tumours that are histogenetically related to the embryonic endoderm are able to evoke this response. The response was not produced by sarcomas which are mesodermal in origin. Recent work [SHERBET, unpublished] shows that gliomas, which are tumours of neuroectodermal origin, are also not capable of inducing endodermal response from the host chick embryo.

c) Chromosomal Deviation in Neoplasia

Whether the relationship between the parameters of mesodermal and endodermal responses to the degree of chromosomal deviation is relevant to an assessment of the deviation of the tumour itself from normal, would, of course, depend upon the relevance of chromosomal changes to the basic neoplastic alterations in a cell (see pp. 9–14, in this volume). Even if the chromosomal changes were acquired subsequent to the initiation of neoplasia but during its development and progression, we might yet be able to use this relationship as an indication of the state of progression of a particular tumour.

Neoplastic change is a heritable change, i.e., the daughter cells derived from a neoplastic cell also carry the neoplastic characters with them. It has therefore been suggested that neoplasia might be a result of mutations either in the chromosomal or extra-chromosomal genetic factors [HADDOW, 1944]. Tumour cells frequently showed abnormal nuclei, enlarged nucleoli, abnormal mitoses, and abnormalities in the number and structure of chromosomes. Even if the chromosomal mutation hypothesis of neoplasia were considered as acceptable, one can hardly assume that mutations occurring in the molecular structure of the genetic material will necessarily show up as gross abnormalities in the structure of the chromosomes.

Abnormal mitosis is associated with several pathological conditions [SIMSON, 1963]. WILLIS [1967] has argued that chromosomal abnormalities have no relevance to the cause or effect of neoplastic alterations in a cell. That abnormal mitosis can be produced by subjecting growing or regenerating tissues to heat and chemical agents, seems very much beside the point. More valid is the argument that abnormal mitoses are seen in regenerative hyperplasia. However, the work of SIMSON [1963], on which WILLIS has based his arguments, appears to suggest that the giant polyploid cells and syncitia are produced as a result of secondary fusion of cells rather than by mitotic irregularities. Besides, multinucleate giant cells have not been encountered in experimental regenerating rat liver, where, in fact, a reduction in binucleate cells is noticed during the period of rapid regeneration [ASHWORTH and REID, 1947].

There is considerable evidence which associates cytogenetic abnormalities with leukemias, such as, for instance, the occurrence of the Philadelphia chromosome (Ph^1) in chronic granulocytic leukemia [NOWELL and HUNGERFORD, 1960; TJIO *et al.,* 1966]. A higher incidence of leukemia has been noticed in patients with syndromes such as Down's syndrome, the Bloom syndrome, congenital anaemias, trisomy D-syndrome, etc., which are closely associated with congenital cytogenetic abnormalities [KRAVIT and GOOD, 1957; GARRIGA and CROSSBY, 1959; SCHROEDER *et al.,* 1964; BLOOM *et al.,* 1965]. The chromosomal abnormalities in leukemias do not, of course, show a consistent pattern and there are even cases reported where no detectable cytogenetic abnormalities were found. This does not mean that chromosomal abnormalities are not relevant to neoplasia. The abnormalities may not be among the causal factors of neoplasia but might be a result of neoplastic changes or they may confer a 'proneness' to neoplasia in the defective cells. (A critical discussion of this aspect of the neoplastic disease may be found in the Chapter by SIMNETT.)

ATKIN and BAKER [1966] have studied several human malignant tumours taking advantage of the presence of marker chromosomes. A high proportion of the metaphase plates of these tumours revealed the presence of the marker chromosomes. They have suggested, therefore, that frequently the tumours arise from a 'single cell in which chromosome changes have occurred'. Though the tumour studied by ATKIN and BAKER did not show much variation in karyotype, changes do occur in karyotypes of tumours which are serially transplanted [NOWELL *et al.*, 1967; NOWELL and MORRIS, 1969]. KEMP *et al.* [1964] have reported that in one case of myeloid leukemia Ph^1 + cells were already present in the bone marrow before leukemia was manifest in the peripheral blood. Initially, the modal number of chromosomes was 46, but subsequently, the proportion of hyperdiploid cells increased, evidently with the progressive development of the disease. WHANG-PENG *et al.* [1969] have reported that aneuploid cells are encountered more frequently in the terminal stages than in the early stages of myeloid leukemia, and finally, according to SPIERS and BAIKE [1970], chromosomes of group 17–18 are especially affected in some tumours of the reticuloendothelial system. Anomalies occurred in this group more frequently than in others, and these were detected in every cell of a tumour. The chromosome alterations included structural changes such as short arm deletions and long arm deletions. The authors believe that such distinct association between the tumours and the occurrence of anomalies in a particular chromosome group might have aetiological significance, though the alterations may sometimes be secondary to the inception of neoplasia.

Without prejudging the question of whether neoplastic change is a result of primary chromosomal alterations or such alterations are primary manifestations after the initiation of incipient neoplasia, it may be said that the alterations in the structure or number of chromosomes, represented as total karyotype deviation, might reflect the state of development of the neoplasia or the degree of progression of the tumour. Very little data are available indicating whether or not changes in the cytogenetic picture accompany tumour progression.

In preneoplastic or incipient neoplasia the chromosomal constitution appears to be essentially normal [BANERJEE and DEOME, 1963; DI PAOLO, 1965]. BANERJEE and DEOME occasionally observed aneuploid cells in tumours arising out of transplantation of preneoplastic hyperplastic nodular mammary tissue. DI PAOLO observed that spontaneous lung adenomas showed progressive chromosomal alterations on being serially transplanted.

STANLEY and KIRKLAND [1969] found no changes in the cytogenetic profiles in the progression of preinvasive carcinoma to the invasive stage. However, CELLIER *et al.* [1970] have reported such differences. *In situ* carcinomas consisted mainly of diploid–triploid–quintuploid cells. In the invasive phase however, cells changed into either (a) triploid dominance in diploid–triploid–tetraploid class, or (b) diploid dominance in diploid–triploid–tetraploid class, or (c) triploid dominance in triploid–tetraploid class. The authors have suggested that a shift in the chromosome constitution occurs during progression from the preinvasive to the invasive phase.

As mentioned already, the chromosomal abnormalities not only do not present a consistent pattern but there are also instances where none were encountered in leukemic cells [WHANG-PENG *et al.*, 1969] or in certain animal tumours induced by viruses [BAYREUTHER, 1960; HELLSTRÖM *et al.*,1963]. Chromosomal alterations might yet be occurring in them but cannot be revealed by the techniques available at the present time. It may be noted, however, that human adenovirus type 12 induces a high incidence of chromosomal and chromatid breaks [STITCH *et al.*, 1964; ZURU HANSEN, 1967]. Such chromosomal aberrations are characteristic of cells subjected to high doses of ionising radiation.

The relationship between the karyotype deviation of Morris hepatoma and the embryonic responses which they elicit, therefore, appears to afford a tool to assess the degree of progression of a given tumour. In the development of neoplasia different characters of the tissue progress independently of one another [FOULDS, 1969]. The parameter of karyotype deviation is one of the characters used in our system. In a later section of this chapter it will be shown how, in addition to this parameter, the parameters of growth and differentiation will also be available as tools in the assessment of tumour progression.

d) Growth Rate of Tumours, the Displacement of Tumour Implants, and Surface Charge

Chick embryos at the primitive streak stage consist of two germ layers, the epiblast and the hypoblast. At this stage of development cellular movements occur in the epiblast, in the direction of the primitive streak where the cells invaginate and form the mesoderm which spreads between the epiblast and hypoblast layers. When heterografts are placed between the germ layers at an early primitive streak stage, the grafts shift their position to correspond with the direction of cellular migrations occurring in the epiblast [SHERBET *et al.*, 1970. Similar graft movements were also described earlier by MAREEL *et al.* [1968].

The frequency with which such graft displacement occurred in the case of Morris hepatomas correlates closely with the growth rate recorded for these tumours in their original hosts. Experiments, which were devised to elucidate the nature and significance of graft displacement, have shown that the surface charge of the grafted cells might be inolved in this phenomenon. Increased growth rates encountered in neoplastic or regenerating systems are accompanied by an increase in the net surface electric charge borne by the growing tissues. Experimental alteration of the net surface charge of cells before implanting them into the chick blastoderms resulted in marked changes in the frequency of displacement of these grafts. For instance, Hensen's node homografts which show 100% displacement on implantation, showed a reduction in the displacement after the net negative charge on the grafted cells was lowered by treatment with cationic substances. On the other hand, treatment of HeLa cells with γ-globulin produced a marked increase in the frequency of their displacement. Yoshida ascites cells showed such displacement more frequently than did even the γ-globulin-treated cells [SHERBET *et al.*, 1970]. An examination of the electrophoretic mobility of these cells has shown that the net surface charge of the cells also increased in that order. The electrophoretic mobility values (at 25 °C and 0.145 M ionic strength) for these cells were -0.98, -1.07, and -1.08 $\mu m\,sec^{-1}V^{-1}$ respectively [SHERBET and LAKSHMI, unpublished results]. It appears, therefore, that the correlation between the rate of graft displacement, the rate of growth of the implanted tissue, and the electrophoretic mobilities of the corresponding cells might afford us a reliable indicator in the assessment of growth characteristics and the associated surface charge phenomena of tumour tissue.

3. Tumour Cell Responses

The behaviour of tumour cells placed in an embryonic environment is quite different from their behaviour in tissue or cell culture, in their original host, or in a compatible host animal. Alterations in the behaviour appear to be dictated by the surface specificities or 'compatibilities' between the embryonic and tumour cells.

a) Pseudopodial Activity

HeLa and Landshütz cells show a rounded appearance in tissue culture. When these cells were implanted into 16–18 h old chick embryos, some of them appeared to assume a spindle shape. It was most interesting to see that the pseudopodial activity was directed only towards the mesodermal cells. Such activity was rarely directed toward either epiblast (ectoderm) cells or the endoderm cells. A severe reduction in the pseudopodial activity was noticed after the cells had been treated with γ-globulin before implantation [SHERBET and LAKSHMI, 1970]. Obviously, the surface charge of the cells is involved in this activity. The role played by the net negative charge residing on a cell surface in cellular mobility has been discussed in detail by ABERCROMBIE and AMBROSE [1962].

Theoretical considerations suggest that a parity of net surface charge might be a key factor that enables a cell, or a sheet of cells, to move over a substratum. It is possible that such a parity of surface charge dictates the specific pseudopodial activity of HeLa and Landshütz cells towards the embryonic mesoderm. Some of the benzpyrene-induced sarcoma grafts that we made into embryos were found to be lodged in the heart region. In these instances it was noticed that the sarcoma cells were migrating along the endothelial lining of the heart. The endothelial lining, it might be recalled here, has a net number of negatively charged groups on its inner surface. Further investigations are needed to see if this specificity of activity towards embryonic mesoderm cells might be relevant to the dissemination and metastasis of tumours, particulary in view of the fact that transplantation of cells or groups of cells liberated from tumours occurs mainly by means of the blood and lymph vessels.

b) Organisation in Implanted Tumour Grafts

Recently, there have been a few reports of acquisition of some kind of organisation by tumour cells implanted or injected into embryos. WHISSON injected Yoshida ascites cells into rat embryos at 10 days gestation. When the tumour cells were examined later, they showed a tendency to become arranged into layers of cuboidal cells in the vicinity of host mesenchyme [WHISSON, 1967]. When implanted into chick embryo (16–18 h old), however, Yoshida ascites cells showed no organisation, but were, in fact, very widely distributed in the entire blastoderm. In the chick embryonic system HeLa cells appear to show some acquired organisation. MAREEL *et al.* [1970] explanted embryos by NEW's [1955] method described earlier in this article, placed a suspension of HeLa cells on the endodermal surface, and then aspirated the culture medium leaving, in this process, a few HeLa cells adhering to the endoderm of the embryo. As the embryos grew, some cells were incorporated into the pharyngeal cavity, and others apparently infiltrated through the embryonic endoderm. Though the cells had been applied as a single cell suspension, they subsequently showed distinct nodular organisation inside the embryo. LUSTIG [1968] reported a sheet-like arrangement of HeLa cells when in contact with primitive streak stage chick embryos. The present authors have made a similar observation of HeLa cells [SHERBET and LAKSHMI, 1970]. While cells in the neighbourhood of the embryonic mesoderm appeared to detach themselves from the main body of the graft and show pseudopodial activity towards the mesoderm, those cells which were in the vicinity of the embryonic ectoderm, showed epithelial arrangement.

It is not clear whether the nodular organisation described by MAREEL *et al.* [1967, 1970] is a result of embryonic influences acting on the cells. HeLa cells, we have noticed, show considerable clumping, and if suspended in a medium containing calcium and magnesium and centrifuged lightly, they pellet very easily, and these pellets are palpably stiff. On the other hand, a similar treatment of Yoshida ascites cells produces very unstable pellets, i.e., the cells remain in a pellet form only for a short while. In fact, Yoshida ascites cells have a higher net negative surface charge than do HeLa cells [SHERBET *et al.*, 1970]. The high electric charge produces repulsion between the cells. In our opinion the 'compatibility', or the lack of it, between the interacting tissues plays a considerable part in the organisation of the tumour cells in an embryonic environment.

4. Induction of Differentiation in Tumours

The degree of differentiation attained by tumours or the loss in differentiation sustained by them is not only required in the assessment of the extent of progression that a tumour has undergone since the initiation of incipient neoplasia, but also in understanding the relationships between morphological differentiation of tumours and the retention or non-retention of their normal functional characteristics. Its importance in restoring normal functional and morphological characteristics cannot be emphasized too much.

The loss of differentiation or 'undifferentiation' as KNOX *et al.* [1970] prefer to call it, has often been considered as possible 'dedifferentiation'. Dedifferentiation as an embryological term means that a cell has reverted to its non-differentiated state with pathways of differentiation available, one of which it could choose depending upon the tissue interactions it undergoes (see chapters by Burgess on p. 106 and by Carlson on p. 60 of this volume for a more informative discussion on dedifferentiation in regenerating systems). If anaplastic tumours were in fact in a 'totipotent' nondifferentiated state, it might be possible to bring about some kind of redifferentiation in them by combining the tumours with appropriate homologous (i.e., belonging to the same embryonic induction system) embryonic inductors.

De Lustig and De Matrajt [1961] and De Lustig and Lustig [1964] combined primitive streak pieces of chick blastoderm with sarcoma 180 or mammary carcinoma of mice. They found that, as a result of this co-cultivation, the tumour tissues showed differentiation of secretory tubules with basal membranes. A loss of invasive ability of the tumour cells was also noticed. These authors also examined the effects of combining 3-day-old chick embryo chorda and neural tube with an osteogenic sarcoma. In this combination also the embryonic tissues appeared to induce the differentiation of osseous trabeculae in the sarcoma. Seilern-Aspang and colleagues have done a good deal of work in which they have studied the effects of older chick embryos on tumour differentiation. Seilern-Aspang and Weissberg [1963] induced spindle cell sarcomas in chicken by intramuscular administration of dibenzanthracene. Extracts prepared from 2- to 4-day-old chick embryos were then injected into the sarcomas. The authors claim that this produced redifferentiation of the tumours in several instances. The ability to produce such redifferentiation was found to diminish progressively when older embryos were used. Instead of using whole embryo extracts, Seilern-Aspang and Kratochwil [1963] then injected different regions or particular organs of the embryos into the sarcoma. When injected with chordal tissue derived from 4-day-old chick embryos, the sarcoma cells seemed to be differentiating into cartilage. This was confirmed in a later piece of work by Seilern-Aspang *et al.* [1963a, b]. In the latter work the tumours were cultured *in vitro* in association with chorda mesoderm obtained from 5-day-old chick embryos.

In amphibians the redifferentiation of tumours occurred during regression. *In situ,* the tumours differentiated into pigment cell layers, cornified epithelium, mucous glands, and epithelium of the integument. The invasive tumours, however, differentiated according to the surrounding tissue. Metastases differentiated like *in situ* tumours. The authors also noticed normal epidermal differentiation of epithelial tumours placed in the tail region of *Triturus.* Powerful morphogenetic forces are acting in this region during regeneration processes. The induction of differentiation has been observed in tumours which were subjected to the influence of these forces.

Another way of looking at this problem of the 'differentiating potentiality' of tumours will be to inquire whether they have acquired abilities that characterize embryonic tissue. It is, of course, difficult to assay these abilities, but one might be able to provide indications as to acquisition of such 'embryonic' characters, by analysing effects of tumour cells on embryonic cells. (See also discussion on the acquisition of embryonic antigens by tumours in the chapter by EASTY in this volume.)

In the work on induction using tumour cells implanted into amphibian embryos, no specificity of action could be identified. However, in implantation of hepatomas the induction of endodermal proliferation might provide an interesting parameter. As mentioned earlier, this response is a histogenetically specific response. Several tumours histogenetically not related to endoderm did not have this ability. Furthermore, Morris hepatomas could induce differentiation of blood islands at the sites where the grafts were located. The induction of blood cell differentiation is a function performed by the embryonic endoderm [see SHERBET, 1970]. One is tempted to suggest that this hepatoma has acquired, during its development, certain properties characteristic of the embryonic endoderm.

The acquisition of this embryonic character is also closely associated with specific changes in the chromosomal constitution of these hepatomas. The ability to support blood cell differentiation was largely associated with an extra chromosome in group 4–10 of the hepatoma genome. It appears possible that this phenomenon might be a result of the linkage of genes that are responsible for the neoplastic behaviour of cells and those responsible for the synthesis of a factor, or factors, that induce the differentiation of blood cells, on a chromosome of group 4–10 [SHERBET and LAKSHMI, 1971].

5. The Independent Progression of Characters in Neoplastic Development

FOULDS [1969] has proposed 6 general principles for tumour progression. Though they were based on the study of mammary carcinoma, they probably have general applicability. One of the rules proposed by FOULDS deals with the independent progression of characters in neoplastic development. One might most appropriately quote FOULDS [1954] himself: 'The structure and behaviour of tumours are determined by numerous characters that, within wide limits, are independently variable, capable of highly varied combinations and assortments and liable to independent progression.'

The tumour characters that we are most concerned with in this discussion are: (a) the tumour karyotype; (b) state of differentiation; (c) the biochemical make-up, and (d) growth rate of the tumour—the latter is an important feature of progression of the disease although tumour progression is said to be independent of the growth rate. There is an almost complete lack of correlation between any of these characters in tumours, which is apparently due to the independent progression of the characters. For instance, there is no apparent correlation between chromosomal anomalies and the functional activity or morphological differentiation of tumour. SORRENSCHEIN *et al.* [1970] have maintained 6 clone strains of rat pituitary tumour in culture for over 4 years; the tumours have retained their ability to secrete growth hormone but vary considerably in morphological differentiation. The Morris hepatomas show a spectrum of karyotype deviation, growth rate, morphological differentiation, and enzymic make-up, without any one character being related to the other [MORRIS and WAGNER, 1965; NOWELL *et al.*, 1967; POTTER and WATANABE, 1968]. The biochemical correlation of the adenyl cyclase activity and growth rate of some Morris hepatomas [BROUN *et al.*, 1970] is an exception. It should be mentioned, however, that KNOX *et al.* [1970] have described a series of spontaneous, and chemically induced mammary tumours which can be graded according to their degree of differentiation. The gradation is related to their growth rate as well as their glutaminase content.

6. Assay of Tumour Progression

Information on the degree of progression of a tumour, whether benign or malignant, is invaluable in the prognosis of the disease, and its treatment. Prognosis, at the present moment, is based mainly on the clinical manifestations of the diesease, the histological picture of the tumours, and the behaviour of the tumour on implantation into a compatible host animal. It is doubtful, however, whether a reliable prognosis could be made using a parameter such as the histological differentiation which, on account of its independent progression, is not a reliable parameter in itself. BLOOM and RICHARDSON [1959] have argued that a very accurate grading and prognosis can be made in breast cancer, using histological criteria. The state of progression of one and the same tumour should be assessed using more than one, and if possible several tumour characters, for greater reliability of prognosis.

BARKER and SANFORD [1970] have attempted to establish cytological criteria to follow neoplastic development of cells cultured *in vitro*. The neoplastic state was separately assessed by implanting the tissue of the same tumours into compatible hosts, and examining their growth, invasiveness, and serial transplantability. These authors observed a progressive increase in cytoplasmic bosophilia, increase in the number and size of nucleoli, increase in nuclear: cytoplasmic ratio, retraction of the cytoplasm, and formation of clusters or cords of cells. These changes were associated with the ability of these cells to produce tumours on implantation into suitable hosts. ISRAELI and BARZILAI [1970] studied the growth of tumour tissues from breast, lung, and urinary system. They obtained highly variable results even within one histological group, and have, therefore, tried to define cultural conditions which allow opitmal growth of tumours. These conditions, of course, vary from tumour to tumour, and they are hopeful these could be used as a pointer to the biological behaviour of the tumours.

ATKIN [1962] and ATKIN *et al.* [1966] have employed Feulgen microphotometric estimation of DNA to assess karyotype deviation for prognosis in carcinoma of the cervix. TAVARES *et al.* [1966] found a clear correlation between ploidy of carcinoma of bladder and the prostate and the average length of survival of the patients post-operative. The average life expectancy was greater in patients with diploid-tetraploid tumours than those with triploid-hexaploid tumours. It was also noticed that the patients who had diploid-tetraploid tumours responded to œstrogen therapy while patients with higher-ploidy tumours were found to be resistant.

The chick embryonic system which is used by us over the past few years, might provide a suitable experimental system to assess the progression of heterologous tumours. It is gratifying to note that the embryo at this stage of development makes specific cellular responses. These, and the specific patterns of interaction between the embryonic and tumour cells, are related to important tumour characters such as karyotype deviation, differentiation, growth rate and surface charge of tumour cell, etc. It is possible, therefore, by studying the behaviour of solid tumours or tumour cells separated from ascitic fluid, on implantation into this embryonic system, to draw a composite picture of the state of progression of a tumour.

Acknowledgements

The authors' work was supported by grants from the Damon Runyon Memorial Fund for Cancer Research, Tenovus, and the Central Research Fund of the London University.

References

Abercrombie, M. and Ambrose, E. J.: Cancer Res. *22:* 525–548 (1962).
Ashworth, C. T. and Reid, H. C.: Amer. J. Path. *23:* 269–282 (1947).
Atkin, N. B.: Cytogenetics *1:* 113–122 (1962).
Atkin, N. B. and Baker, M. C.: J. nat. Cancer Inst. *36:* 539–558 (1966).
Atkin, N. B.; Mattinson, G. and Baker, M. C.: Brit. J. Cancer *20:* 87–101 (1966).
Banerjee, M. R. and DeOme, K. B.; Cancer Res. *23:* 546–550 (1963).
Barker, B. E. and Sanford, K. K.: J. nat. Cancer Inst. *44:* 39–63 (1970).
Bayreuther, K.: Nature *186:* 8–9 (1960).
Becker, V.; Dostal, V.; Tiedemann, H. and Tiedemann, H.: Z. Naturforsch. *146:* 260–264 (1959).
Bloom, G. E.; Warner, S.; Gerald, P. S. and Diamond, L. K.: New Engl. J. Med. *274:* 8–14 (1965).
Bloom, H. J. G. and Richardson, W. W.: Brit. J. Cancer *11:* 359–377 (1959).
Broun, H. D.; Chattopadhyaya, S. K.; Morris, H. P. and Pennington, S. N.: Cancer Res. *30:* 123–126 (1970).
Cellier, K. M.; Kirkland, J. A. and Stanley, M. A.: J. nat. Cancer Inst. *44:* 1221–1230 (1970).
De Lustig, E. S.: In H. Katsuta Cancer cells in culture. Proc. Int. Conf. Tissue Culture in Cancer Res., pp. 135–142 (University of Tokyo Press, Tokyo 1968).
De Lustig, E. S. and De Matrajt, H. A.: Rev. Soc. argent. Biol. *37:* 180–186 (1961).
De Lustig, E. S. and Lustig, L.: Rev. Soc. argent. Biol. *40:* 207–216 (1964).
Di Paolo, J. A.: J. nat. Cancer Inst. *34:* 337–343 (1965).
Foulds, L.: Cancer Res. *14:* 327–339 (1954).
Foulds, L.: Neoplastic development (Academic Press, London 1969).
Gallera, J. and Castro-Correia, J.: C. R. Soc. Biol. *154:* 1278–1282 (1960).
Garriga, S. and Crossby, W.H.: Blood *14:* 1008 (1959).
Haddow, A.: Nature *154:* 194–199 (1944).
Hellström, K. E.; Hellström, I. and Sjögren, H. O.: J. nat. Cancer Inst. *31:* 1239–1253 (1963).
Israeli, E. and Barzilai, D.: Cancer Res. *25:* 824–834 (1970).
Kemp, N. H.; Stafford, J. L. and Tanner, R.: Brit. med. J. *i:* 1010–1014 (1964).
Knox, E. W.; Linder, M. and Friedell, G. H.: Cancer Res. *30:* 283–287 (1970).
Kravit, W. and Good, R. A.: Amer. J. Dis. Child, *94:* 289 (1957).
Mareel, M.; Vakaet, L. and De Ridder, L.: Europ. J. Cancer *4:* 249–253 (1968).
Mareel, M.; Vakaet, L. and De Ridder, L.: Virchows Arch. Zellpath. B. *5:* 277–287 (1970).
Morris, H.P. and Wagner, B.P.: Methods in Cancer Res. *4:* 125–152 (1965).
Nowell, P. C. and Hungerford, D. A.: Science, 132: 1497 (1960).
Nowell, P. C. and Morris, H. P.: Cancer Res. *29:* 969–970 (1969).
Nowell, P. C.; Morris, H. P. and Potter, V. R.: Cancer Res. *27:* 1565–1579 (1967).
New, D. A. T.: J. Embyol. exp. Morph. *3:* 326–331 (1955).
Palayoor, S.T. and Batra, B.K.: Ind. J. exp. Biol. *9:* 303–306 (1971).
Potter, V. R. and Watanabe, M.: In Ch. Zarafonetis Proc. Leukemia-Lymphoma Conf., pp. 33–46 (Lea & Febiger, Philadelphia 1968).
Saxén, L. and Toivonen, S.: J. Embryol. exp. Morph. *6:* 616–633 (1958).
Saxén, L. and Toivonen, S.: Primary embryonic induction (Academic Press, London 1962).

SCHROEDER, T. M.; AUSCHUTZ, F. and KNOPP, A.: Humangenetik *1:* 194–196 (1964).
SEILERN-ASPANG, F. and KRATOCHWIL, K.: J. Embryol exp. Morph. *10:* 337–356 (1962).
SEILERN-ASPANG, F. and KRATOCHWIL, K.: Acta biol. med. germ. *10:* 443–446 (1963).
SEILERN-ASPANG, F. and WEISSBERG, M.: Acta biol. med. germ. *10:* 439–442 (1963).
SEILERN-ASPANG, F.; HONUS, E. and KRATOCHWIL, K.: Acta biol. med. germ. *10:* 447–452 (1963a).
SEILERN-ASPANG, F.; HONUS, E. and KRATOCHWIL, K.: Acta biol. med. germ. *11:* 281–285 (1963b).
SHERBET, G. V.: Adv. Cancer Res. *13:* 97–168 (1970).
SHERBET, G. V. and LAKSHMI, M. S.: Oncology *24:* 58–67 (1970).
SHERBET, G. V.; LAKSHMI, M. S. and MORRIS, H. P.: J. nat. Cancer Inst. *45:* 419–428 (1970).
SHERBET, G. V. and LAKSHMI, M. S.: Oncology *25:* 558–563 (1971).
SIMON, I. W.: J. Path. Bact. *85:* 35–39 (1963).
SPIERS, A. S. D. and BAIKIE, A. G.: Brit. J. Cancer *24:* 77–91 (1970).
SORRENSCHEIN, C.; RICHARDSON, U. I. and TASHJIAN, A. H.: Exp. Cell Res. *61:* 121–128 (1970).
SPEMANN, H.: Experimentelle Beiträge zu einer Theorie der Entwicklung (Springer, Berlin, 1936).
STANLEY, M. A. and KIRKLAND, J. A.: Acta cytol., Balt. *13:* 76–80 (1969).
STITCH, H. F.; VAN HOOSIER, G. L. and TRENTIN, J. J.: Exp. Cell Res. *34:* 400–403 (1964).
TAVARES, A. S.; COSTA, J.; DE CARVALHO, A. and REIS, M.: Brit. J. Cancer *20:* 438–441 (1966).
TJIO, J. H.; CARBONE, P. P.; WHANG, J. and FREI, E., III.: J. nat. Cancer Inst. *36:* 567–584 (1966).
TOIVONEN, S.: In Biological organisation: cellular and subcellular, p. 208 (Pergamon Press, London 1959).
TOIVONEN, S. and SAXÉN, L.: J. nat. Cancer Inst. *19:* 1095–1106 (1957).
VAHS, W.: Roux' Arch. Entw. Mech. *149:* 339–364 (1957).
WADDINGTON, C. H.: Phil. Trans. Roy. Soc. B. *221:* 179–230 (1932).
WHANG-PENG, J.; FREIREICH, E.J.; OPPENHEIM, J. J. and FREI, E., III.: J. nat. Cancer Inst. *42:* 881–897 (1969).
WHISSON, M. E.: In A. V. S. DE REUCK and J. KNIGHT Cell differentiation. Ciba Found. Symp., pp. 219–232 (Churchill, London 1967).
WILLIS, R. A.: Pathology of tumours (Butterworth, London 1967).
ZURU HANSEN, H.: J. Virol. *1:* 1174 (1967).

Authors' address: Dr. M. S. LAKSHMI and Dr. G. V. SHERBET, Department of Biochemical Pathology, University College Hospital Medical School, *London WC1E 6JJ* (England)

Subject Index